Engineering Dynamics and Vibrations

Recent Developments

Editors

Junbo Jia
Aker Solutions
Bergen, Norway

Jeom Kee Paik
University College London
UK and Pusan National University
Republic of Korea

CRC Press
Taylor & Francis Group
Boca Raton London New York

CRC Press is an imprint of the
Taylor & Francis Group, an **informa** business
A SCIENCE PUBLISHERS BOOK

CRC Press
Taylor & Francis Group
6000 Broken Sound Parkway NW, Suite 300
Boca Raton, FL 33487-2742

First issued in paperback 2020

© 2019 by Taylor & Francis Group, LLC
CRC Press is an imprint of Taylor & Francis Group, an Informa business

No claim to original U.S. Government works

ISBN-13: 978-1-4987-1926-1 (hbk)
ISBN-13: 978-0-367-78044-9 (pbk)

Library of Congress Cataloging-in-Publication Data

Names: Jia, Junbo, editor.
Title: Engineering dynamics and vibrations : recent developments / editors Junbo Jia, Aker Solutions Bergen, Bergen, Norway, Jeom Kee Paik, University College London, London, UK and Pusan National University, Busan, Republic of Korea.
Description: Boca Raton, FL : Taylor & Francis Group, [2018] | "A Science Publishers Book." | Includes bibliographical references and index.
Identifiers: LCCN 2018041504 | ISBN 9781498719261 (acid-free paper)
Subjects: LCSH: Dynamics. | Vibration.
Classification: LCC TA352 .E54 2018 | DDC 620.1/04--dc23
LC record available at https://lccn.loc.gov/2018041504

Visit the Taylor & Francis Web site at
http://www.taylorandfrancis.com

and the CRC Press Web site at
http://www.crcpress.com

To our families

Foreword

This book has introduced and developed some recent research results which provide important insight and increased understanding in the fields of engineering dynamics and vibrations. The general topic has featured prominently in recent years because of the economic and environmental pressure to produce more efficient structural designs which consume less material and, therefore, lead to lighter products. As a consequence, it is evident that vibrations and dynamic effects will feature more prominently in the optimum behaviour of structural designs than they have in the past. In addition, dynamic loadings (which includes wind, hydrodynamic, explosive, impact and seismic loadings) are significant factors in the hazard assessments of engineering plant, and the safety of various structural systems such as the crashworthy protection of transportation systems. The development of numerical systems has allowed these often complex effects to be incorporated reliably, in many cases, into the design stages, in addition to the consideration of statistical effects.

The dynamic behaviour of structures is a challenging area, not only because of the obvious time-dependent aspects, but also because of the difficulties in specifying the exact details of the external dynamic loadings, the connection characteristics (e.g., welding effects) between various components and the dynamic properties of the associated materials. Thus, it is important to recognise and utilise fully the contributions and understanding emerging from theoretical, numerical and experimental studies on structures and components, as well as investigations into the material properties under dynamic loadings and various environmental conditions.

This book covers all these aspects in varying degrees in the individual chapters and associated case studies. The chapters are written by international experts from China, Greece, Italy, South Korea, Norway, UK and USA. It covers a large and important engineering area and I trust that readers will be inspired to incorporate some of these ideas into their own structural design studies in order to achieve more efficient and environmentally friendly designs without jeopardising safety.

Norman Jones

Emeritus Professor, PhD, DSc, FREng, FNAE, FIMechE, FASME, FRINA
Impact Research Centre, University of Liverpool

Preface

The contents of this book is a result of a team effort by international experts in the field of engineering dynamics, vibrations, and impacts. It contains both essential basics and new developments in this area.

The subject of dynamics and vibrations originated from Sir Isaac Newton's monograph *Philosophiæ Naturalis Principia Mathematica*. Lord Rayleigh paved the way for its further development with his *Theory of Sound*. These provided the basis for the unique position of the field of dynamics in mechanics. Since then, many scientists and engineers have applied and furthered this knowledge in various fields of applied science and technology.

With the enormous investment made in civil, mechanical and aerospace engineering during the 20th century, designs were pushed to the limits of their performance capacity, with the trend being toward high-speed operations, adverse environment capability, light weight, etc. With functionality becoming an essential requirement in an unpredictable, highly uncertain environment, practicing engineers encountered more problems with regard to dynamics. Although dynamics as a scientific topic is by no means fully understood (and perhaps never will be), the great amount of activity in this field during the last century has made it possible to form a practical subject in a fairly systematic, coherent, and quantitative manner. All these factors have pushed applied dynamics to a greater complexity than before, and also made the subject essential. Thanks to the rapid development of computer technology, more portable and accurate testing equipment and techniques, as well as breakthroughs in computation algorithms, during the last 50 years applied dynamics has developed significantly. This raised a vast amount of challenges in implementing designs in reality, while also putting greater demands on engineers, requiring a thorough understanding of the subject. In spite of advances in engineering, the practical problems regarding dynamics and vibrations are often unsuccessful. Moreover, even if engineers can perform sophisticated computer-based dynamic analysis tasks, many of them lack an understanding of the essential principles of dynamics, and hence of the links between theory and application. This leads to an insurmountable barrier when they are requested to validate/verify and provide insightful explanations of analysis results, or to further improve the engineering designs with regard to dynamics, vibrations, and impacts. This poses a significant safety hazard and can also result in significant economic loss.

With the objective of providing principles and up-to-date knowledge of engineering dynamics, vibrations, and impacts for solving practical and challenging engineering problems, this edited volume covers topics on the concepts, principles and solutions in this area. The book provides wide-ranging treatment of the subject. It contains advances from essentials on dynamics, vibrations, and impacts to more advanced topics on nonlinear dynamics, stochastic dynamics, from basic principles to its applications associated with various dynamic actions due to wind, ocean wave, earthquake, and explosion loadings. A special chapter is dedicated to elaborate the principles of noise control.

The book is intended to serve as an introduction to the subject and also as a reference book with advanced topics. A balance between the theoretical and practical aspects is sought. All the chapters are addressed to practicing engineers who are looking for answers to their daily engineering problems, and to students and researchers who are looking for links between theoretical and practical aspects. It should also be of use to other science and engineering professionals and students with an interest in general dynamics, vibrations, and acoustics. Ultimately, the book should also be useful for designing tolerant structural systems against extreme and accidental conditions.

While the book does not seek to promote any specific "school of thoughts," it inevitably reflects chapter authors' "best practice" and "working habit." This is particularly apparent in the topics selected and level of detail devoted to each of them, their sequences, and the choices of many mathematical treatments and symbol notations etc. The editors hope that this does not deter the readers from seeking to find their own "best practice".

We are indebted to many individuals and organizations for assistance of various kinds, such as participation in book reviews and copyright clearance. Moreover, this book has an extensive list of references reflecting both the historical and recent developments of the subject. We would like to thank all the authors in the book references for their contribution to the area.

Junbo Jia
Jeom Kee Paik

Contents

1

Introduction to Dynamics, Vibrations and Impacts

Junbo Jia

1.1 General

Dynamics, vibrations, and impacts are due to time-varying forces and/or motions. If they are experienced by a dynamic system comprising stiffness, inertia and damping, it can exhibit responses different from those of a system subject to only static loading. Inertia and time-varying excitations are major characteristics that distinguish a dynamic problem from its static counterparts. Such time-varying forces and/or motions are called dynamic loading. Examples of dynamic loading include loads caused by winds (Jia, 2011), earthquakes (Jia, 2012), ocean waves (Jia, 2008), ice impacting (Jia et al., 2009), machinery or human-induced vibrations, blast or impact effects, or even the sudden change in stiffness of a structural system that is subject to a static load (Jia, 2014), among others. All such problems are called dynamic problems, even though some of them can be simplified as an equivalent static problem. Design and analysis associated with dynamics, vibration and impact is, in many ways, more of an art than a science. Even though they must be thoroughly checked in a rigorous scientific manner, intuition, imagination as well as a synthesis of experience and knowledge play important roles in the process of the relevant design and analysis. It is essential in approaching dynamic analysis and design that one develops an "intuition" to solve the relevant problems at hand.

Aker Solutions, Bergen, Norway. Email: junbojia2001@yahoo.com

1.2 Experiencing Dynamics

Every day, we are surrounded by environments full of dynamics. The ringing of the alarm clock in the morning, the voice of your loved ones, the sound from radio and TV, noise due to traffic, the swaying of masts and trees in the wind, and even our heart beating (Jia, 2014).

A fundamental example of dynamics is a playground swing. In order to increase the amplitude of a swing (as shown in Fig. 1.1), either the rider or the external excitor must excite the swing in phase with the movement of the swing, i.e., the period of excitation is close or identical to the natural period of the swing, leading to a resonance condition. For the swing shown in Fig. 1.1, if its amplitude is small, the natural period T_n, defined as the time that the swing spends in its arc back and forth once, is constant:

$$T_n \approx 2\pi\sqrt{\frac{L}{g}} \tag{1.1}$$

where L is the length of a single hanging rope in meters and g is the acceleration of Earth's gravity.

It may be noted that, in the calculation above, the rope of the swing is assumed to be weightless and the child on the swing is idealized as a point mass. For a swing with a rope 1.5 m in length, its natural period is 2.5 s. This equation is often referred to as the Law of Pendulum, which was originally discovered in 1583 by Galileo Galilei at the age of 19, when, while sitting in Pisa Cathedral, he noticed that a lamp was swinging overhead with a constant period (Fig. 1.2).

Figure 1.1. A child on a playground swing.

Figure 1.2. The lamp hanging in Pisa Cathedral (the photo was taken in 2013, it may not represent the exact status of the lamp in 1583).

Even though we utilize them all the time, we do not often think about the many dynamic phenomena that occur in our daily life, perhaps precisely because they are so common. However, we may find that our life can be safer, cheaper, more enjoyable, convenient, and environmentally-friendly if we were to devote just a little more consideration to these matters.

In the engineering world, the design or maintenance of many engineering structures require dedicated considerations of their dynamic responses: high-rise buildings, bridges, ships, offshore structures, aircrafts, land-based and space vehicles, mechanical equipment, and even tiny electronic components.

A sense of the importance of dynamics can be conveyed through the example of a few representative accidents, all of which were due to improper accounting for dynamics.

In 1985, an 8.1 magnitude earthquake occurred in Mexico City. As a result, 412 buildings collapsed and another 324 were seriously damaged. A large percentage of the damaged buildings in the downtown area were between 8 and 18 storeys high. Not surprisingly, those buildings had a resonance vibration period of around 2.0 s, indicating general resonance with the soft soils under the city's ground, through which the seismic waves were transmitted to the ground with a dominant period of around 2.0 s (Elnashai and Sarno, 2008). The 1994 Northridge earthquake also caused similar structural damage due to the resonance of the upper structure with the seismic ground excitations (Broderick et al., 1994).

Resonance is frequently responsible for the failure of engineering structures. For example, many leaks in pipes are caused by cracks due to vibrations or

Figure 1.3. An eight-storey building was broken into two during the 8.1 magnitude earthquake that occurred in Mexico City in 1985.

acoustical resonance with excessive excitation forces. Even if the resonance does not cause an immediate structural failure, it can be responsible for significant deflection or acceleration on a structure, leading to objects falling, failure or instability of mechanical and electronic equipment installed on the structure, or human discomfort, injury or casualty. Figure 1.4 shows the chaotic situation caused by a large storm onboard an offshore platform through the platform's excessive movements due to the resonance vibrations. The excess of rolling (rotation around the longitudinal axis of the ship) can be attributed to the resonance of ship roll motions with the sea waves. It has been reported that half of all serious accidents on board ships are caused by vibrations, either directly through structural failures or indirectly through symptoms of fatigue among the crew (ISO, 1997; Berg and Bråfel, 1991). The human body, especially the abdomen, head and neck, are also sensitive to vibrations. When subject to vibrations with a frequency range of 1–30 Hz, people experience difficulty in maintaining correct posture and balance (ISO, 1997). Even if it may be difficult to relate some vibration effects on the human body to specific frequencies, the resonance of human organs is an important

Figure 1.4. The chaotic situation in the office (left) and archive room (right) of an offshore platform after a significant storm (courtesy of Equinor and Aker Solutions).

contribution to motion sickness. Motion sickness can also occur in animals due to the resonance of transportation vehicles with animals' organs. It has been observed that, during transportation, healthy chickens can become ill and are unable to stand (Ji and Bell, 2008).

The spill of red wine shown in Fig. 1.5 is known as sloshing, and involves the dynamic responses of liquids under excitations. This dynamic response can be amplified if the motions of a liquid container (glass) have a period close to the period of the liquid's (wine) sloshing. Sloshing must be considered for almost any moving vehicle or for structures containing a liquid with a free surface (Faltinsen and Timokha, 2012). This means that, for example, if similar phenomena occur for a chemical tank on a moving truck, the liquids' sloshing motions inside the tank (Fig. 1.6) can exert significant impact on the tank, making the truck unstable or even causing a rollover. It has been reported that 4% of heavy-truck road accidents are directly caused by the sloshing of the liquid cargo within the truck tanks (Romero

Figure 1.5. The sloshing of wine in a glass (photo by Stefan Krause).

Figure 1.6. Sloshing of a liquid inside a tank.

et al., 2005). For onshore tanks, earthquakes may also induce tank liquid sloshing, causing structural damage. An accident of this kind was the damage caused to seven large oil storage units that occurred during the Tokachi-oki earthquake in 2003. Forensic investigations have found that, at the tanks' sites, the earthquake generated ground motions with long peak period of 4–8 s, which is in the range of the tanks' sloshing period of 5–12 s (Hatayama, 2008). For ships carrying LNG (liquefied natural gas), the resonance between liquid cargo sloshing and ship motions can be significant at certain levels to which the liquid cargo is filled, as the large liquid movement creates highly localized impact pressure on tank walls, threatening the structural integrity and stability of the LNG ships. On various cruise ships, sloshing of the water in the swimming pool on the sun deck occurs frequently, often even on a monthly basis. The reader may refer to the online video footage (Youtube, 2007) for an illustration of this. This is mainly caused by the surge (heading) and pitch (rotation around the transverse axis of the ship) motions of the ship that coincide with the natural sloshing frequency of the pool (Ruponen et al., 2009). In addition, during a strong earthquake or storm event, seismic waves and wind can also excite waters in a lake or semi-enclosed sea, causing sloshing with high water surges, known as seiches. Harbors, bays, and estuaries are often prone to small seiches with amplitudes of a few centimeters and periods of a few minutes. The North Sea often experiences a lengthwise seiche with a period of about 36 hours. Geological evidence indicates that the shores of Lake Tahoe may have been hit by seiches and tsunamis as large as 10 m (33 feet) high in prehistoric times, and local researchers have called for the risk to be factored into emergency plans for the region (Brown, 2008). On 26 June 1954, eight fishermen at Lake Michigan were swept away and drowned by a seiche more than 3 m high. Figure 1.7 shows two photos of storm-induced seiches at Canal Park in Duluth, Lake Superior, Minnesota. The two photos were taken just minutes apart.

It is well known that structures loaded repeatedly tend to fail at a lower load level than expected—a phenomenon known as fatigue failure. This type of failure is responsible for most of the material failures in engineering structures. Furthermore,

(a)   (b)

Figure 1.7. (a): Mild seiche at Canal Park in Duluth, MN, and (b): the situation just minutes before the seiche took place (courtesy of Minnesota Sea Grant).

such fatigue damages are often accompanied by unfavorable dynamic excitations relevant to resonance, high frequency loading or repeated significant loadings. A number of accidents due to fatigue failure have become well known. On 3 June 1998, a high-speed intercity train traveling from Hannover to Hamburg derailed in the village of Eschede at a speed of 200 km/h, and crashed into a road bridge after derailment (Fig. 1.8), leading to a loss of 102 lives and 88 injuries. The forensic study followed showed that the accident was caused by a broken wheel tire on the first middle car, in which an undiscovered crack grew to an unacceptable size under repeated fatigue loading.

Environmental loading such as wind or sea waves hit structures repeatedly, and the fatigue caused by this is a typical problem posing risks for structural safety. For example, the sinking of the *MS Estonia* on 28 September 1994, was simply caused by the repeated wave impact on her bow door, leading to fatigue failure of the bow visor locking devices and the formation of opening moments about the deck hinges. This simple failure led to 852 casualties (The Joint Accident Investigation Commission of Estonia, 1994). Other well-known accidents leading to structural failure are the *Ranger I* jackup (Gulf of Mexico, 84 fatalities) and the collapse of the *Alexander Kielland* semi-submersible (North Sea, 123 fatalities), in 1979 and 1980 respectively. The sequence of failure in the *Alexander Kielland* platform accident (lower figure in Fig. 1.9) was: fatigue failure of one brace (shown in Fig. 1.9); overload failure of five other braces; loss of column; flooding into deck; and capsizing (Moan, 2005). For *Ocean Ranger* the accident sequence was: flooding through a broken window into the ballast control room; closed electrical circuit; disabled ballast pumps; erroneous ballast operation; flooding through chain lockers; and capsizing (Moan, 2005). Except for the fatigue cracks that caused the failure, it is noticed that both of these structures were statically determinate platforms with

Figure 1.8. The destruction of rear passenger cars that were pushed into each other and crashed into a road bridge (Eschede accident, 3 June 1998; photo by Nils Fretwurst).

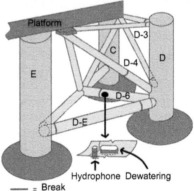

Figure 1.9. Capsizing of *Alexander Kielland* semi-submersible (upper; photo courtesy of Norwegian Petroleum Museum) initiated by a fatigue crack in one brace (lower).

a lack of redundancy. Figure 1.10 shows a member in an offshore jacket structure breaking due to repeated wave loading. Wave-induced ship vibrations are nowadays regarded as an important source contributing to fatigue damage. The vibrations are normally referred to as hull girder vibrations (including whipping and springing), and vibrations at resonance period, so called 2-node mode vibrations, are typically dominant. Hull girder vibrations often occur on ships. In many cases, people on board a ship can easily feel them when the wave height reaches only a few meters. Moreover, due to their low damping and large size, blunt ships may experience more vibrations than slender ships.

Tacoma Narrows Bridge, shown in Fig. 1.11, was opened to traffic on 1 July 1940. It spanned over a mile, the third longest suspension span in the world at that time, with the deck supported by a combination of a cable-supported suspension

Figure 1.10. A break in a primary structural member of an offshore jacket structure due to repeated wave loading. The photo was taken after the jacket structure was decommissioned and transported onshore.

Figure 1.11. The collapse of Tacoma Narrows Bridge due to wind flutter.

structure and steel plate girder. The slenderness of the suspended deck represented a distinct departure from earlier suspension bridge designs, but because of this the bridge had shown vibratory tendencies even during construction. From the beginning of its service, it received many complaints from users because in even a light wind the bridge behaved like a ship riding the waves, with pronounced vertical oscillations, causing the "seasickness" of many passengers in cars (Levy and Salvadori, 2002), thus earning it the nickname "Galloping Gertie." On 7 November 1940, in a wind of 64 km/hour, the bridge twisted so much that the left side of road descended significantly, with the right side rising, and this motion alternating rapidly. The twisting vibrations became more and more significant, finally leading to the total collapse of the bridge, as shown in Fig. 1.11. From an aerodynamic point of view, such violent vibrations are caused by the aero-elastic fluttering due to the feeding of energy when the bridge was subjected to alternative unstable oscillations in strong wind. From a structural engineering point of view, this is a type of self-excited vibration, which is due to the sustained alternating excitations that induce the instability of a system at its own natural or critical frequency (note that this is different to a typical resonance phenomenon). The entire bridge-wind system therefore behaved as if it had an effective negative damping, leading to exponentially growing responses.

Figure 1.12 shows another example of self-excited vibrations of a bridge. While they did not cause a collapse, the large deflection amplitude of the bridge deck can clearly be observed. After the accident on Tacoma Narrows Suspension Bridge, engineers had proposed various mitigation measures to prevent similar accidents from occurring again, such as cutting holes in the web of the underdeck girder or installing curved outriggers to divert the wind (Fig. 1.13), making the wind pass through the holes and thus avoiding the wind fluttering. The Tacoma disaster provided a great impetus to research in the field of aerodynamic stability and structural dynamics, which led to the modifications of the Golden Gate Bridge (Fig. 1.14) and several other significant suspension bridges (White et al., 1972).

As an important dynamic phenomenon, vortex-induced vibrations (VIV) are normally induced on members interacting with external fluid flows, and this produces periodical irregularities (vortices) in the flow, as shown in Fig. 1.15. When the vortices are not symmetrical around the body (with respect to its mid plane), lift forces will be applied on each side of the member, leading to members vibrating perpendicular to the fluid flow. When the vortex-induced frequency coincides with the natural frequency of structural members, the resulting condition is called lock-in. During the lock-in process, the structural member oscillates with increased amplitude but rarely exceeding half of the across wind dimension of the body. Cylinder members, such as subsea pipes or chimneys, are most susceptible to VIV. Figure 1.16 shows the cracks due to fatigue on a tubular member's end (joint). Forensic investigation has found that the buffeting response of the flare boom cannot cause fatigue cracking on this joint (Jia, 2011). Therefore, VIV is most likely to be the reason for the development of these particular fatigue cracks. Another example of VIV-induced vibrations is the "loud singing" of external hand railings of ships during storms or hurricanes. To diminish VIV, it is common to put obstacles around free spans of cylinders. Figure 1.17 shows spiral strakes installed on the upper part of a chimney to diminish VIV.

Figure 1.12. A large deflection amplitude of a bridge deck due to self-excited vibrations can clearly be observed: the left and right figures show the relative vertical position at two time instants (Larsen et al., 2000).

WOULD THIS HAVE SAVED BRIDGE?

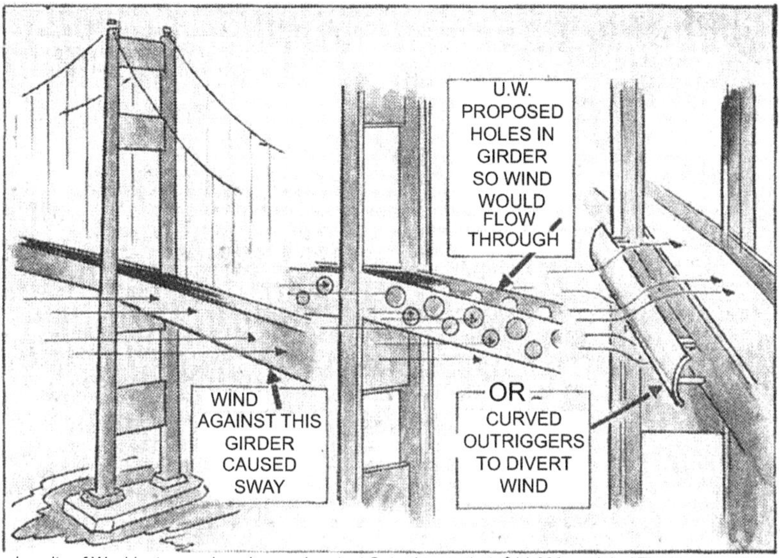

U.W.
PROPOSED
HOLES IN
GIRDER
SO WIND
WOULD
FLOW
THROUGH

WIND
AGAINST THIS
GIRDER
CAUSED
SWAY

OR
CURVED
OUTRIGGERS
TO DIVERT
WIND

university of Washington engineering made a test Saturday on their $14,000 model of The Narrows Bridge, attempting to eliminate the dangerous wind sways which finally caused the real-life structure to collapse yesterday. The sketch at left shows the flat horizontal girder which offered resistance to winds, causing the sway. University recommendations were (center) to drill holes with a torch in the girder, permitting the wind to pass through; or (right) to erect an $80,000 streamlined buffer alongside the girder, to divert winds. Their tests showed the latter materially reduced the vibrations, might have saved the bridge.

Figure 1.13. Proposals to avoid wind flutter for Tacoma Bridge (courtesy of University of Washington Library).

Figure 1.14. The Golden Gate Bridge (photo by Rich Niewiroski Jr.).

Figure 1.15. Vortex produced by fluid passing through a cylinder.

Figure 1.16. Cracks (marked with circle) found on a tubular joint due to wind-induced VIV on a high-rise flare boom in the North Sea (courtesy of Aker Solutions).

Figure 1.17. A spiral strake installed on the upper part of a chimney to diminish VIV (photo by Jing Dong).

Machinery vibrations can cause over-stress and collapse of the main structures, damage to non-structural elements, loosening of fastener(s), collapse of cladding and ceiling, development of cracks on structures, fatigue, accelerated subsidence of foundations, etc. The structural vibrations induced by machinery vibrations can lead to visible motions of building elements or objects, oscillations of suspended light fixtures, or structure- and air-borne noise. Machinery vibrations can also cause fatigue in machinery components, deformation and strength failure, and higher tolerance demands due to unexpected motions of tools and installations, which reduce the production quality and efficiency.

Moreover, continuous exposure to structure-borne or air-borne sound can lead to various types of problems, such as hearing damage if a person is exposed to noise, high cycle fatigue of machinery or structural components, etc. Here, air-borne noise is the sound transmitted through air, while structure-borne sound is that which results from an impact on or a continuous vibration against a part of a structure. Whilst they are sometimes considered to be separate phenomena, air-borne and structure-borne sound are related, in that air-borne sound can cause structure-borne sound and vice versa.

1.3 Utilizing Dynamics

While avoiding resonance disasters is an important concern in the engineering world, vibrations and resonance can also be put to use. Resonant systems can be used to generate vibrations at a specific frequency (e.g., musical instruments), or pick out specific frequencies from a complex vibration containing many frequencies (e.g., filters). For example, many clocks keep time by mechanical resonance in a balance wheel, pendulum, or quartz crystal (Jia, 2014).

Another example is the vibration plate (power vibration plate), which was originally used by Russian scientists to counter muscle atrophy and the reduction of bone density in cosmonauts, and is now used as a piece of fitness equipment to strengthen muscle and reduce weight. As shown in Fig. 1.18, it is essentially

Figure 1.18. A power vibration plate as a piece of fitness equipment.

a flat base that vibrates. An exerciser stands on the plate while it vibrates with a frequency of between 0.4 and 2 Hz. This forces the exerciser's entire body to react to the relatively high frequency vibrations, causing the muscles to contract and stretch in order to maintain balance. By tuning the machine to an appropriate vibration frequency, most of the muscles in the body can be effectively tightened, thus strengthening muscles and reducing weight.

The natural period is an inherent property of any system. Therefore, if one can find a convenient way to measure it, several essential characteristics of the system can be obtained. To demonstrate this, we first go to the London Eye observation wheel, as shown in Fig. 1.19. It has a structural system similar to a bicycle wheel, with its rim stiffened by 16 rotation cables and 64 spoke cables. It is obvious that part of the tension load in each cable will be lost within the lifetime of the structure, requiring a re-tensing in order to maintain the cable tension in accordance with the design requirements. However, it is challenging to directly measure the tension force in each cable. Therefore, engineers used a much more convenient alternative to measure the natural frequency due to the transverse vibration of each cable, from which the tension load can be calculated using the relationship between the natural period and the tension load.

The measurement of important mechanical properties of materials, such as Young's modulus, has traditionally been carried out through a series of mechanical tests by placing the specimen on costly traction-torsion machines. Engineers nowadays have found a less costly and more convenient way to obtain part of the basic mechanical properties: simply hitting the material sample and inducing vibrations in it, as shown in Fig. 1.20. A high precision microphone close to the sample captures vibration signals and transfers them to computers; the signals are then analyzed, and the natural period and internal friction of the sample can be

Figure 1.19. The London Eye observation wheel (left) with the rim supported by tensioned steel cables (right; photo by Christine Matthews). The wheel works like a huge spoked bicycle wheel.

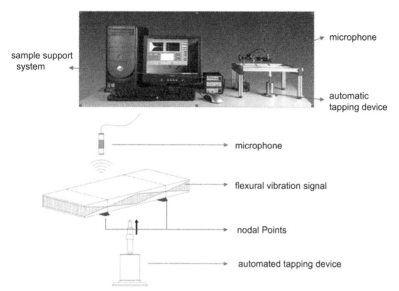

Figure 1.20. Measurement of flexural and torsional eigen frequencies to determine the Young's modulus and shear modulus, together with a measurement of damping through the free decay of the sample's vibrations (courtesy of IMCE Belgium).

obtained. Thereafter, the resonant frequency, together with the dimensions and the weight of the sample, are used to calculate the elastic properties (Young's modulus, Shear modulus and Poisson ratio).

As a counter-measure, masses (often called tuned masses) with their supporting stiffness can be installed to absorb energy of another (primary) structure at their resonant frequency and further dissipate the absorbed energy through the damping of the system. Note that the response of the tuned masses is around 90° out of phase with the response of the primary structure, this difference in phase produces the energy absorption transferred to the tuned masses. Therefore, the resonance response of the structure can be greatly decreased. This is normally referred to as a dynamic absorber (Thomson, 1966). As shown in Fig. 1.21, in a dynamic absorber, the mass

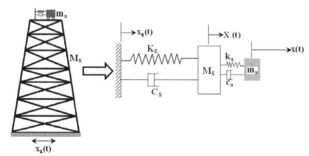

Figure 1.21. Mechanism of a dynamic absorber with mass m_a, stiffness k_a and viscous damping c_a. It is used to mitigate the dynamic responses of the main structure with mass M_s.

and stiffness of it, m_a and k_a, are tuned such that the absorber's natural frequency coincides with the resonance frequency of the main structure.

Similar to the mechanism of a children's swing in Fig. 1.1, Fig. 1.22 shows a type of dynamic absorber—the tuned mass damper (TMD) installed in the Taipei World Financial Center, which has a height of 509.2 m. The TMD weighs 660 tons and is suspended by eight steel cables, arranged in four pairs, from the frame on the 92nd floor as a pendulum system. By adjusting the free cable length, the mass in this pendulum system moves with the building at similar natural period of 6.8 s. Eight primary hydraulic viscous dampers situated beneath the TMD automatically

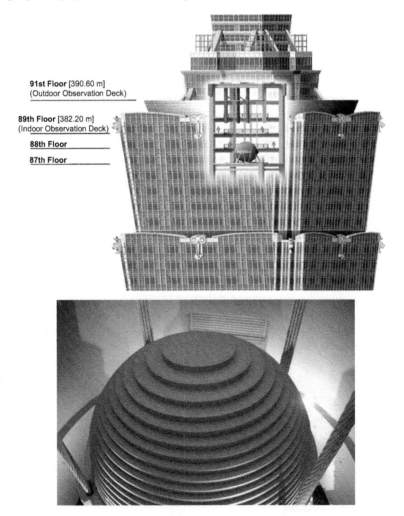

Figure 1.22. A 660-ton pendulum tuned mass damper system installed in the Taipei World Financial Center (with a height of 509.2 m) to mitigate the wind- and earthquake-induced responses of the building. The TMD hangs from the 92nd to the 88th floor.

dissipate energy from vibration impacts. A bumper system of eight hydraulic viscous dampers beneath the TMD absorbs vibration impacts, particularly in major typhoons or earthquakes where movements exceed 1.5 m (www.taipei-101.com.tw). This can greatly decrease dynamic responses due to seismic and wind loadings (Jia, 2017).

Similar to a TMD, a tuned liquid damper (TLD), another type of dynamic absorber, is also a passive damping system in which the damping effects are provided by the motion of liquid in tanks. The moving liquid has a function similar to the moving mass of a TMD, in that gravity is harnessed as a restoring force. Energy is mainly dissipated by using damping baffles to create turbulence in the liquid, as well as through the wave breaking and the impact of liquid on the tank wall. The geometry of the tank that holds the water is determined theoretically to give the desired natural frequency of water motions in accordance with the space in which the tank is to be located. Liquid tanks used as a TLD are typically rectangular or circular, with those of the former shape able to be tuned to two different frequencies in two perpendicular directions. An engineering example of TLD is the water tanks installed on the top of the skyscraper One Rincon Hill in San Francisco, which can hold up to 190 tons of water, as shown in Fig. 1.23. The water level in the tanks is adjusted to achieve a tank sloshing natural frequency close to that of the building structure. Baffles are installed inside the water tanks in order to increase the damping when the water is in motion. In addition to the function as a TLD to mitigate wind- and earthquake-induced responses, the water tanks were also built to hold water for fire fighters. Together with other measures to increase the performance of the building, US\$ 54 is saved per square meter. It is noticed that, compared to a TMD, the TLD has the advantage of low manufacturing and maintenance costs, and it

Figure 1.23. A TLD installed on the top of One Rincon Hill in San Francisco, USA (courtesy of John Hooper, Magnusson Klemencic Associates, USA).

can also serve the purpose of liquid (water, fuel, crude oil or mud, etc.), storage (Lee and Ng, 2010) for emergency, industry, or everyday purposes if fresh water is used (Hitchcock et al., 1997a; Hitchcock et al., 1997b). Furthermore, without adversely affecting the functional use of tanks, TLD tanks can be designed with proper dimensions or reconfigured with internal partitions of existing tanks, which is helpful to cope with physical and architectural requirements (Jia, 2017).

1.4 Dynamics vs. Statics

Over history, the safety and serviceability of structures have basically been measured on the basis of their static behavior, which required adequate stiffness and strength. This was perhaps because the necessary knowledge of dynamics was less accessible to engineers than was that of statics. Nowadays, it is common knowledge that all bodies possessing stiffness and mass are capable of exhibiting dynamic behavior.

The major difference between dynamic and static responses is that dynamics involves the inertia forces associated with the accelerations at different parts of a structure throughout its motion (Jia, 2014). If one ignores the inertia force, the predicted responses can be erroneous. As an example, let's consider a bottom-fixed cantilevered tower subjected to sea wave loadings as shown in Fig. 1.24 (Naess and Moan, 2012). In addition to the static bending moment due to wave loadings applied on the structure, as shown in Fig. 1.24b, the stiffness and mass of the structure will react to the wave loadings and generate internal forces on both the top mass block (Q_i) and the tower (q_i), as shown in Fig. 1.24c. Rather than a single function of mass, the amplitudes of the inertia forces are related to a ratio between stiffness and mass (eigenfrequency), mass, as well as damping, thus resulting in additional dynamic bending action (Fig. 1.24d).

As another example, consider a gravity-based structure (GBS), as shown in Fig. 1.25, that is subjected to the ground motions recorded during the El Centro earthquake, which had a high energy content at vibration periods above 0.2 s (below 5 Hz in Fourier amplitude, shown in Fig. 1.26). The dynamic responses of

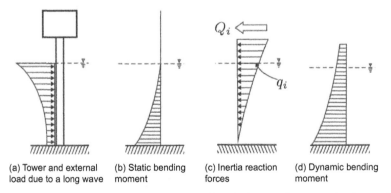

(a) Tower and external load due to a long wave   (b) Static bending moment   (c) Inertia reaction forces   (d) Dynamic bending moment

Figure 1.24. Wave-induced static vs. instantaneous dynamic forces and moments in a bottom-fixed cantilevered tower (Naess and Moan, 2012).

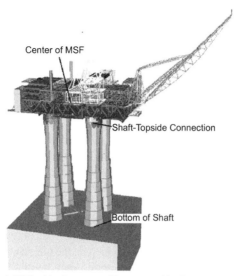

Figure 1.25. A GBS with a heavy topside supported by four concrete shafts (legs).

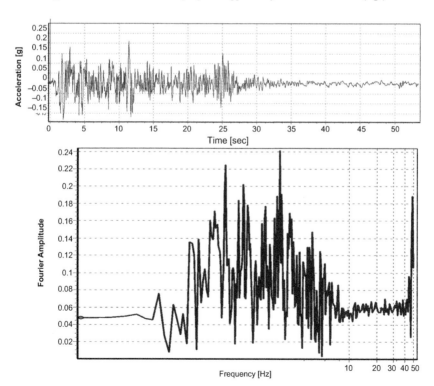

Figure 1.26. Ground motions EW component (upper) recorded during the El Centro earthquake and its Fourier amplitude (lower).

the platform are investigated by varying the thickness of four shafts from half of the reference thickness, to the reference thickness, to twice the reference thickness. It is obvious that the GBS becomes stiff by increasing the shafts' thickness. If a static analysis is performed, under the same seismic excitations the stiffer structure would have lower responses. However, the seismic responses involving dynamic effects may not obey this rule. Figure 1.27 shows the acceleration at the shaft-topside connection. It is clearly shown that the peak acceleration for the reference shaft thickness case is higher than that of the half-thickness case. However, the trend of peak acceleration response variation with the change of stiffness cannot be identified, as the peak acceleration for the double shaft thickness (the stiffest one) is lower than that for other cases with lower stiffness. This indicates the effects of inertia, which are more complex than their static counterpart. The response variation trend can be identified by relating the seismic responses to the dynamic characteristics of both structures and excitations.

Even for dynamic insensitive structures with periods of resonance lower than that of the dynamic loading, dynamics does include the inertia effects due to loading that varies with time, even if this load variation may be quite slow compared to the resonance period of structures. The inertia effects could lead to the fatigue failure of the materials at stress conditions well below the breaking strength of the materials. They may also be responsible for the discomfort of human beings. Figure 1.28 shows an offshore jacket structure subjected to two subsequent sea waves; the jacket has a resonance period of 2.5 s. Figure 1.29 compares the calculated

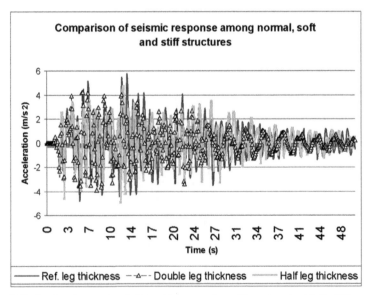

Figure 1.27. Acceleration at the shaft-topside connection with various leg/shaft stiffness (peak acceleration: 4.7 m/s² for double leg thickness, 5.8 m/s² for reference leg thickness, 4.2 m/s² for double leg thickness).

axial force time history at a leg C1 with and without accounting for the dynamic inertia effects. When the dynamic effects are ignored (right figure), the axial force's history entirely follows the variation of the wave and has a period of wave loading (15.6 s) well above the structure's resonance period (2.5 s). However, when the

First wave crest hit Leg C1   Second wave crest hit Leg C1

Figure 1.28. An offshore jacket structure subjected to a wave with a wave height of 31.5 m and a wave peak period 15.6 s (courtesy of Aker Solutions).

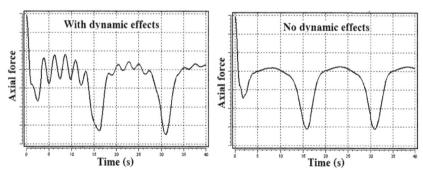

Figure 1.29. Axial force time history on the lower part of leg C1 of the offshore jacket with and without dynamic inertia effects. (The exact magnitude of axial forces is omitted to protect the interests of the relevant parties.)

dynamic effects are accounted for, fluctuations (left figure) of the axial force can be clearly observed as a background noise with the resonance period of the structure (2.5 s). Depending on the magnitude of this background noise, it may undermine the integrity of the structure through fatigue damage. On the other hand, even though some types of loadings (such as high-speed impacts) have a dominant period much shorter than that of the resonance period of structures encountering those loadings, they can excite high frequency eigenmode vibrations of structures, which may be relevant to problems associated with strength, fatigue and noise, etc.

From another angle, dynamic loading often has a different orientation to static loading. For example, the static loading of a structure under the gravity of the Earth is strictly in the direction of the Earth. However, when the structure is subjected to dynamic loading due to, for example, wind, earthquake or sea waves, the direction of resultant loadings changes from downward to a more horizontal orientation. This can result in an entirely different pattern regarding the load level and load path, and this obviously influences the structural design. Therefore, structural engineers are required to have a complete picture of load path and level, and structures designed must have corresponding load resisting systems that form a continuous load path between different parts of the structures and the foundation. The structure shown in Fig. 1.28 represents a typical configuration of the jacket structure and a clear path for load transferring, i.e., the gravity and acceleration loads from topside, the wave load applied on the upper part of the jacket, and the jacket gravity and acceleration loads are all transferred through legs and braces down to the pile foundation at the bottom.

Before concluding this section, it is of great importance to emphasize that dynamics is a rather more complex process than its static counterpart. The natural frequency of a structure can change when a change in its stiffness, mass or damping occurs. What makes dynamics even more complicated is that, strictly speaking, regular harmonic loadings or responses, with a sine or cosine forming at a single frequency, do not exist in the real world, even if they represent a helpful simplification when the dynamics at a single frequency is dominating. This implies that one should always assess whether the vibrations in various frequencies need to be accounted for or not.

1.5 Solving Dynamic Problems (Jia, 2014)

Given the presence of inertia effects as discussed in Section 1.4, dynamic analysis is generally much trickier to solve than its static counterpart. This is mainly because when the inertia term appears in the equilibrium equation (Eq. 1.2), in order to uniquely determine the solution, not only boundary conditions but also initial values are needed, giving the dynamic analysis problem the name "initial boundary value problem." Furthermore, rather than a linear equation, like a static equilibrium has (Eq. 1.3), the additional inertia and damping terms make the equilibrium equation an ordinary second-order differential equation and time dependent, which requires more in-depth knowledge to examine. In addition, if time series responses are needed, a decent time stepping procedure must be employed, rendering the dynamic problems even more complicated.

$$m\,\ddot{x}(t) + c\,\dot{x}(t) + kx(t) = F(t) \tag{1.2}$$

$$kx = F \tag{1.3}$$

where F, k, m and x are the external force on a body, linear stiffness (between the body and the fixed ground), mass, and displacement of the body, respectively; t is the time; the dot over the symbol represents differentiation with respect to t. See Fig. 1.30 for an illustration.

Therefore, if it is possible to calculate responses based on a static analysis, dynamic calculation should always be avoided. However, this is unfortunately not the case for many problems in engineering. The general rule is that if the excitations (loads) have a dominant frequency close to the natural frequency of structures, dynamic analysis has to be adopted. However, even if the load frequency is far from the natural frequency of the structure, the inertia effects may still be important and responsible for certain types of integrity problem (e.g., fatigue), such as the one shown in Fig. 1.29. In addition, high frequency transient loads can also induce significant dynamic responses, such as explosion, car collision, etc., in which the inertia effects of the relevant structures can be rather significant. In these situations, dynamic analysis is also normally required.

It is sometimes convenient to use an amplification factor to simulate the dynamic effects: by scaling the static response with a dynamic amplification factor, only static analysis needs be performed. However, this method lacks a solid theoretical background and has its limitations, and may become seriously erroneous under certain situations.

Before solving a dynamic problem, one needs to classify the vibration problem in terms of whether or not external excitations are presented (forced and free vibrations); whether the excitations are of a deterministic or stochastic type; whether or not the damping is presented (damped or undamped); whether the system can be modeled as a discrete or continuous one; and whether the responses present linear or nonlinear characteristics (Jia, 2014; Jia, 2016).

To perform a dynamic analysis, analysts should fully understand the essential dynamic characteristics of a system or a structure: eigenfrequencies, mode shapes and damping.

In order to find the solutions of vibration responses, the system or the structure must be represented by an idealized model. This model can be either discrete or continuous. The former can be modeled by limited degrees-of-freedoms, while the latter theoretically has infinite degrees-of-freedoms.

For a discrete model, one needs to first construct the governing equations of motions, which can be described in terms of a second-order differential equation with constant coefficients. The information on displacements or rotations (essential boundary conditions) and external forces excitations (natural boundary conditions) needs to be clarified. If time is involved, the initial conditions (boundary conditions in time) also need to be known. After gathering sufficient information on these boundary conditions, the solutions of the equilibrium equations are then unique (Krysl, 2006). One can then solve the equations using decent mathematical treatment.

The system under study can be undamped or damped and with or without external excitations. We pay particular attention to the solutions for forced vibrations, which typically consist of a steady-state term that oscillates and gradually becomes dominant at the forcing frequency, and a transient term at the system's natural frequency that may be important initially but gradually dies out and eventually becomes insignificant due to the presence of damping. Under certain conditions the dominant forced vibrations become rather significant, indicating the occurrence of resonance.

Engineers sometimes need to choose the type of dynamic analysis method to be adopted. Each method has its unique characteristics, merits and limitations, and the various methods also fit different situations in respect of structural and load characteristics, design requirements, limitations of computation tools, and even the skills of analysts, among other factors. Understanding all these factors is essential for choosing the right method: on the one hand, this can increase accuracy; on the other hand, it may also simplify the computation without diminishing reliability. In certain cases, the trade-off may be difficult to judge even for experienced researchers and engineers.

For structures or systems with single or very few degrees-of-freedoms, depending on the types of excitations and responses (duration, shape, deterministic or stochastic, etc.), and their eigenpairs (eigenfrequencies and mode shapes) in comparison with the excitations, based on the pure mathematical formulation of the stiffness, mass and damping of the structures, various types of analytical methods are available for solving the dynamic responses, all of which result in exact solutions.

However, for a structure or a system with multiple or many degrees-of-freedoms, it is almost impossible to perform a dynamic analysis using classical analytical methods. Therefore, approximation methods have to be adopted. Among others, there are two most important types of approximation methods. The first involves approximating the solutions using either a series of solutions or an energy criterion to minimize/control the error, such as the Rayleigh energy method. The second one is essentially a discretization of structures into many sub-domains (elements), and an assembly of these elements expressed in a matrix form for solving, which practically promotes the application of the finite element method.

As illustrated in Fig. 1.30, in complex dynamic analyses for engineering structures, the finite element method, the finite difference method or the modal superposition method, and linear iteration method are the three numerical methods most commonly used in practical computational solid mechanics (Curnier, 1994), solving problems associated with space, time and nonlinearities, respectively, but

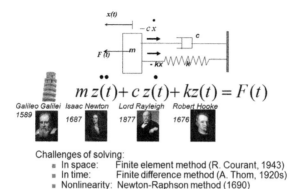

Challenges of solving:
- In space: Finite element method (R. Courant, 1943)
- In time: Finite difference method (A. Thom, 1920s)
- Nonlinearity: Newton-Raphson method (1690)

Figure 1.30. Essentials of applied dynamic analysis (from a presentation by the author at the 11th International Conference on Recent Advances in Structural Dynamics, Pisa, 2013).

sometimes in a combined manner. They are important not only because of their efficiency and generality of application, but also due to the simplicity of their computer implementation.

In the modal superposition method, the coupled equations of motions are transformed into a series of uncoupled/independent equations. Each of these equations is analogous to the equation of motions for a single-degree-of-freedom system, and can be solved in the same manner. The responses are calculated as the linear sum of product between the eigenvectors (constant with time) and the generalized/modal coordinates (varied with time) for each eigenmode. Note that the number of uncoupled equations needing to be solved is equal to the number of eigenmodes to be accounted for. For structures with dynamic responses dominated by the first few eigenmodes, the modal superposition method leads to high computation efficiency. This is more obvious if the structure has a large number of degrees-of-freedoms.

In linear dynamic analysis, the responses of a system/structure are proportional to the loads/excitations to which it is subjected. This enables the utilization of superposition, which brings significant convenience in terms of mathematical treatment, and, in most cases, also ensures the accuracy of the calculation. However, when nonlinearities appear in the system/structure, the stiffness and/or load are dependent on the deformation, and the responses of a system are generally not amenable to any analytical method that can provide exact solutions. A general method for obtaining the exact solution of nonlinear differential equations is not available, and most of the analytical methods that have been developed only yield approximate solutions. Further, the available techniques vary greatly according to the type of nonlinear equation.

Despite the significant efficiency of modal analysis, it generally applies only to linear dynamic problems. Therefore, the nonlinearities involved in a dynamic analysis are theoretically and practically treated with the support of the finite difference method and linear iteration method. The former (typically referred to as the Newmark method) is a step-by-step time integration of the equations of motions, and it can solve for example transient phenomena such as nonlinear vibrations or shock wave propagation. The latter is a generalization of the Newton-Raphson method, which is essentially the application of a linearization in a locally approaching curve between load and deformation, and which is able to overcome numerical challenges introduced by the geometric (e.g., buckling), material (e.g., plasticity), boundary (e.g., contact) and force (e.g., follower forces with change of geometry or hydrodynamic drag load) nonlinearities.

Damping exists in all types of real-world structures or systems. It mainly provides a dissipation of energy. In most cases, it is beneficial to decrease dynamic responses. Viscous damping, which is the most typical type of damping modeling, is only effective at or close to a structure/system's Eigenfrequencies.

1.6 Characteristics of Dynamic Responses

If we take an SDOF spring-mass-damper system under forced excitation as an example, as illustrated in Fig. 1.31, the equation of motions for the system is expressed as:

$$m\,\ddot{x}(t) + c\,\dot{x}(t) + kx(t) = F(t) \qquad (1.4)$$

By exerting an external harmonic force ($F(t) = F_0 \sin(\Omega t)$ with an amplitude of F_0 and an angular frequency of Ω shown in Fig. 1.31, or displacement excitations in a harmonic form on the spring-mass-damper system, an SDOF spring-mass-damper system under forced harmonic excitation is constructed. The governing linear differential equation of motions for this system in case of harmonic force excitations can then be written as:

$$m\,\ddot{x}(t) + c\,\dot{x}(t) + kx(t) = F_0 \sin(\Omega t) \qquad (1.5)$$

Dividing both sides of the equation above by m, this equation is rewritten as:

$$\ddot{x}(t) + \frac{c}{m}\,\dot{x}(t) + \omega_n^2\,x(t) = F_0/m \sin(\Omega t) \qquad (1.6)$$

It is noted that the viscous damping is very important in an oscillating system because it helps to efficiently limit the excursion of the system in a resonance

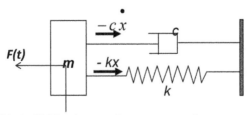

Figure 1.31. An SDOF spring-mass-damper system under an external force $F(t)$.

situation. As a reference, we first define the critical damping c_c, which is the lowest damping value that gives no oscillation responses, i.e., the system does not vibrate at all and decays to the equilibrium position within the shortest time. This represents the dividing line between oscillatory and non-oscillatory motions:

$$c_c = 2\sqrt{km} = 2m\omega_n \qquad (1.7)$$

The actual damping ratio can be specified as a percentage of critical damping:

$$\zeta = \frac{c}{c_c} \qquad (1.8)$$

By realizing that $c = 2\omega_n m\zeta$, the equation of motions for the system finally gives:

$$\ddot{x}(t) + 2\omega_n \zeta \dot{x}(t) + \omega_n^2 x(t) = F_0/m \sin(\Omega t) \qquad (1.9)$$

As the equation above is a second-order non-homogeneous equation, the general solution for it is the sum of the two parts: the complementary solution $x_c(t)$ to the homogeneous (free vibrations) equation and the particular solution $x_p(t)$ to the non-homogeneous equation:

$$x(t) = x_c(t) + x_p(t) \qquad (1.10)$$

The complementary solution exhibits transient vibrations at the system's natural frequency and only depends on the initial condition and the system's natural frequency, i.e., it represents free vibrations and does not contain any enforced responses:

$$x_c(t) = Xe^{-\zeta\omega_n t} \sin\left(\sqrt{1-\zeta^2}\,\omega_n t + \phi\right) \qquad (1.11)$$

It is noticed that this aspect of the vibration dies out due to the presence of damping, leaving only the particular solution exhibiting steady-state harmonic oscillation at excitation frequency Ω. This particular solution is also called the steady-state solution, which depends on the excitation amplitude F_0, the excitation frequency Ω as well as the natural frequency of the system, and it persists indefinitely:

$$x_p(t) = E \sin(\Omega t) + F \cos(\Omega t) \qquad (1.12)$$

By substituting the equation above and its first and second derivatives into Eq. (2.6), one obtains the coefficients E and F as:

$$E = \frac{F_0}{k} \frac{1-(\Omega/\omega_n)^2}{[1-(\Omega/\omega_n)^2]^2 + [2\zeta(\Omega/\omega_n)]^2} \qquad (1.13)$$

$$F = \frac{F_0}{k} \frac{-2\zeta\Omega/\omega_n)}{[1-(\Omega/\omega_n)^2]^2 + [2\zeta(\Omega/\omega_n)]^2} \qquad (1.14)$$

By inserting the expression for coefficient E and F into Eq. (2.9) and rearranging it, one can rewrite the steady-state solution as:

$$x_p(t) = \frac{F_0}{km} \frac{\sin(\Omega t - \varphi)}{\sqrt{[1 - (\Omega/\omega_n)^2]^2 + [2\zeta(\Omega/\omega_n)]^2}} \tag{1.15}$$

where φ is the phase between the external input force and the response output, with the most noticeable feature being a shift (particularly for underdamped systems) at resonance. It can be calculated as:

$$\varphi = \tan^{-1}\left(\frac{2\zeta(\Omega/\omega_n)}{1 - (\Omega/\omega_n)^2}\right) \tag{1.16}$$

It is clearly shown that the steady-state solutions are mainly associated with the excitation force and the natural frequency. Figure 1.32 shows an example of the dynamic responses due to the contribution from both transient and steady-state responses, with a Ω/ω_n ratio of 0.8, a damping value of 0.05 ($\varphi = 0.21$), and $\phi = 0.1$. Phases between the two types of response can be clearly observed.

When mass m in Fig. 1.31 is subjected to harmonic excitations, the magnitude and phase of the displacement responses strongly depend on the frequency of the excitations, resulting in three types of steady-state responses, namely quasi-static, resonance, and inertia dominant responses, which are illustrated in Fig. 1.33.

When the frequencies of excitations Ω are well below the natural frequencies of the structure ω_n, both the inertia and damping term are small, and the responses are controlled by the stiffness. The displacement of the mass follows the time-varying force almost instantaneously. Subject to environmental loading such as wind or ocean wave loading, the majority of land-based structures and fixed offshore

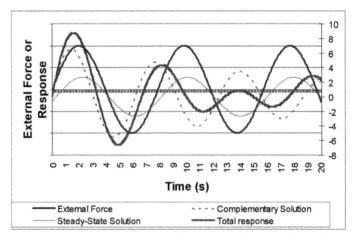

Figure 1.32. Transient and steady-state responses due to external harmonic force excitations applied on a system with $\omega_n = 1.0$, $\Omega = 0.8$, $\zeta = 0.05$, and $\phi = 0.1$.

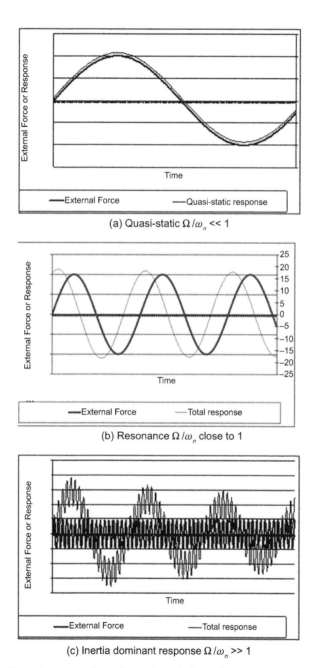

Figure 1.33. Damped responses due to harmonic excitations with the characteristics of (a) quasi-static ($\Omega/\omega_n \ll 1$); (b) resonance (Ω/ω_n close to 1); and (c) inertia dominant responses ($\Omega/\omega_n \gg 1$), for a system with $\omega_n = 1.0$ and viscous damping ratio $\zeta = 0.03$.

structures are designed to reach this condition, as shown in Fig. 1.34. However, earthquake loading is likely to have a dominant frequency higher than the natural frequency of structures.

When the excitation frequencies are close to the natural frequency of the system, the inertia term becomes larger. More importantly, the external forces are almost overcome (controlled) by the viscous damping forces. Resonance then occurs by producing responses that are much larger than those from quasi-static responses, as shown in the circle in the upper figure of Fig. 1.34, and there is a dramatic change of phase angle, i.e., by neglecting the damping, the displacement is 90° out of phase with the force, while the velocity is in phase with the excitation force. In a typical situation in which the damping is well below 1.0, the responses are much larger than their quasi-static counterparts. From an energy point of view, when the frequency of excitations is equal to the natural frequency, the maximum kinetic energy is equal to the maximum potential energy. Almost all engineering structures are designed to avoid this resonance condition. Resonance conditions can occur, for example, when the resonance period (site period) of soil layers at sites due to shear wave transmission is close to the natural period of the structure;

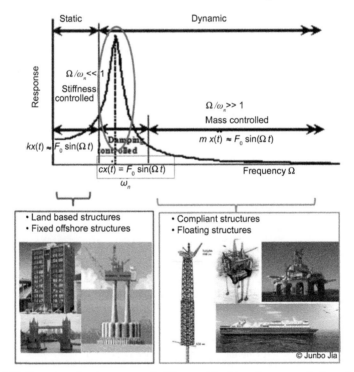

Figure 1.34. Response of various types of offshore and land-based structures in three frequency ranges subjected to external environmental loading with a frequency of Ω, the natural frequency of the structure is denoted as ω_n (from an oral presentation by the author at the 11th International Conference on Recent Advances in Structural Dynamics, Pisa, 2013).

when the resonance period of the surface wave is close to the natural period of the structure; or when significant plasticity develops on structural members during a strong earthquake, leading to a decreased natural frequency of the structure, which may track the decreasing predominant frequency of the ground motions, causing resonance with ground motions (moving resonance) (Jia, 2017; Jia, 2018), etc.

When the excitation frequencies are well above the natural frequency of the system, the external forces are expected to be almost entirely overcome by the inertia force, as the excitations are so frequent that the mass cannot immediately follow them. Transient vibrations are normally more significant than steady-state oscillations. The responses of the mass are therefore small and almost out of phase (phase angle approaches 180°) with the excitation forces, as illustrated in Fig. 1.34. From an energy point of view, this reflects the condition in which the maximum kinetic energy is larger than the maximum potential energy. Offshore compliant structures and floating structures are normally designed to behave "softly" in their motion responses, and therefore have natural frequencies of motion ω_n well below the external wave loading frequency. It is noted that for most engineering structures and typical site conditions, the long period of seismic ground excitations are usually small, except for the ground motions caused by the seismic surface (Rayleigh) waves, which can have dominating long period components of ground motions. Therefore, subject to seismic ground excitations, a large number of offshore and land-based structures are likely to reach this condition.

As a structure typically has—or more precisely, has to be represented/modeled by—a large number of degrees-of-freedoms, in addition to the natural frequency (which is typically the first eigenfrequency of the structure), it has more numbers/orders of eigenfrequencies, and the total number of eigenfrequencies is equal to the number of degrees-of-freedoms of the structure. However, the first few eigenfrequencies, and especially the first one (natural frequency), normally dominate the majority of the total modal mass participating in structural vibrations, and are therefore the most important ones contributing to the dynamic response of the structure. From a modal response point of view, the lower order eigenfrequencies of the structure are normally separated well apart. In this frequency range, with small damping, the modal response will generally be dominated by a single mode with frequency close to the loading frequency and a single mode with natural frequency of the structure. If the loading frequency is lower than the first eigenfrequency of the structure, then the structural response will show two peaks, one at the loading frequency corresponding to quasi-static response, and the other at the natural frequency of the structure contributing to the dynamic response of the structure. As an example, Fig. 1.35 shows ocean wave-induced frequency responses of a welded joint on an offshore jacket structure (Fig. 1.36) in the North Sea. Two peaks can be identified in this frequency response graph: one corresponds to the wave modal frequency (0.08 Hz); the other corresponds to the structure's natural frequency (0.24 Hz). On the other hand, the higher order eigenfrequencies are more closely spaced and modal mass participation of each mode vibration is much lower than that of the first few eigenmodes. This is more obvious for highly redundant structures.

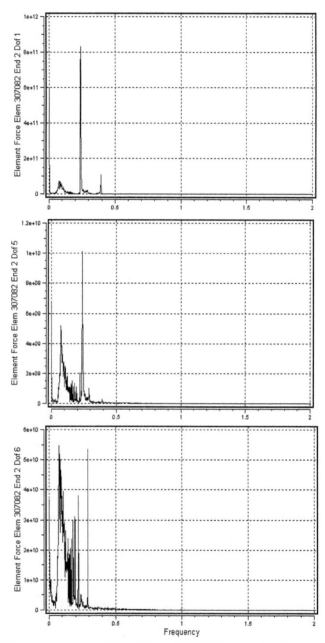

Figure 1.35. Frequency responses of the axial force (DOF 1) [N], in-plane bending moments (DOF 5) [Nm] and out-of-plane bending moments (DOF 6) [Nm] for a weld joint 10803 (Fig. 1.36) at the top of a jacket; the jacket is subject to wave loading corresponding to a sea state with a significant wave height H_s=8.8 m, and a modal wave period T_p= 13.2 s (0.08 Hz).

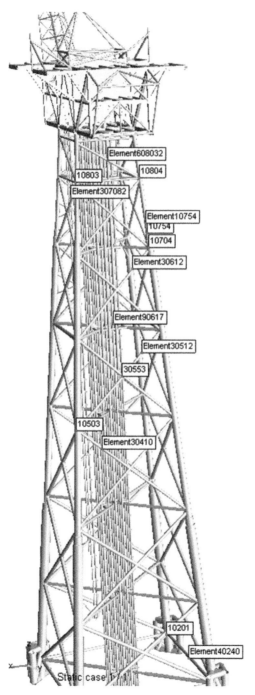

Figure 1.36. The location of the welded joint 10803 on the jacket.

With a dynamic loading in this frequency region, multiple eigenmodes contribute to a similar extent to the modal response, and vibration modes above the loading frequency will be out of phase with those below the loading frequency. Therefore, the net vibration is likely to be less than any of the single mode vibrations in this frequency range (dynamic cancelation).

1.7 Frequency Range of Dynamic Environmental Loading

If likely to be subjected to environmental loading, such as wind, earthquakes, ocean waves, current, and ice, etc., all engineering structures should be designed by accounting for their dynamics with a special consideration of resonance, as this can have relevance for structural performances associated with ultimate strength, fatigue strength and serviceability limits. Each type of loading has different dominant ranges of loading frequency, as shown in Fig. 1.37. In addition, other types of dynamic loading induced by explosion, machinery vibrations, and vehicle- or human-induced excitations may also require a dedicated consideration to solve relevant dynamic problems in the design (Jia, 2006; 2007).

It is noted that the dominant frequency of seismic ground motions is not only dependent on the frequency of seismic waves generated at the source due to fault fractures, but also influenced even more by site conditions associated with soil layers and ground topology. Therefore, their possible peak period has a large range of up to 4 s (Jia, 2017).

Furthermore, Fig. 1.37 also indicates that the difference of dominant frequency range for different types of environmental loading is significant. This can pose a

---

Typical Short Term Environmental Loading Frequency

- **Seismic strong ground motions**
  - Peak period below 4 s
  - Typical dominated period between 0.2 s to 0.5 s
  - Majority of motion energy between 0.3 s and 30 s
- **Wind**
  - Slowly varying mean wind speed: period in order of hours
  - Rapid fluctuating instantaneous wind speed peak period: period 15 s – a few minutes
  - Most of the energy of fluctuating instantaneous wind speed: above 1 s
- **Wave**
  - Slowing varying wave: period in order of hours, e.g., significant wave height and characteristic wave period
  - Rapid fluctuating individual waves: period 3-26 s
- **Current**
  - Varies slowly with time, can be assumed as constant (long period)
- **Ice (level ice) loads**
  - Vertical structures: wide banded
  - Conical structures: narrow banded, above 0.2 s

Figure 1.37. Typical short-term environmental loading frequency/period (from an oral presentation by the author at the 11th International Conference on Recent Advances in Structural Dynamics, Pisa, 2013).

challenge in designing an optimized structure to resist the different types of dynamic loading. For example, for fixed offshore structures or land-based structures, to avoid significant dynamic amplification and/or excessive vibrations, it is usually desirable to design a stiff structure to resist wind and ocean wave loading, so that the natural period of the structure is far below the dominant period range of wind and wave loading, thus avoiding resonance and limiting excessive vibrations. On the other hand, the low dominant period of seismic loading requires a "softer" structure design that has a higher natural period. This contradiction has been encountered in various structural design projects. Sometimes a "balance" between the two needs to be sought, such as the design of Taipei 101. It has a natural period of around 7 s, which is obviously above the dominant period of earthquake loading. Even though this natural period of 7 s is also far below the period of loading due to wind turbulence (fluctuating part of wind), it can induce significant peak acceleration at the top of Taipei 101, causing both human discomfort and structural metal fatigue. To solve this problem, a large tuned mass damper (TMD) weighing 660 tons (Fig. 1.22) was introduced to mitigate sway motion of the building, particularly in major typhoons or earthquakes where movement of the top floor can exceed 1.5 m. The TMD will reduce peak acceleration of the top occupied floor from 7.9 milli-g to 5.0 milli-g due to wind storm with a return period of half a year. Moreover, for a 1000- to 2500-year return period of strong earthquake, the TMD will also be rather effective to mitigate the dynamic response of the structure, and to remain in place and intact after strong seismic ground motions cease and the vibration of the structure terminates. In addition, another two small TMDs are designed to mitigate vibrations at two tip vibration modes at periods of around 1 s.

For slender light-weight structures such as guyed steel stacks, chimneys, slender tips of flare booms or other elevated structures, with two examples shown in Fig. 1.38 and Fig. 1.39, the structural design is governed by the wind loading rather than seismic loading because the structure has a high natural period (compared with the dominant period of earthquake loading) and the wind loading increases

Figure 1.38. Taipei 101 (under license of CC BY-SA 3.0 by Guillom and Peellden).

Figure 1.39. A flare boom with a slender tip (courtesy: Aker Solutions).

with distance from the ground. However, due to the tips of those slender structures normally being much softer (with much lower stiffness) than the structural parts below the tips, they can exhibit significant vibrations during earthquakes, which is referred to as a whipping effect (Jia, 2017). Therefore, the design of the slender tips of structures may be governed by seismic loading and therefore requires dedicated consideration of their seismic resistance.

1.8 Pioneers of Dynamic Analysis

Several great scientists in history need to be mentioned here, as without them and many others, classical dynamics might still today be called "modern" dynamics: Galileo Galilei (1564–1642), who showed that acceleration due to gravity is independent of mass; Isaac Newton (1642–1727), who disclosed the three laws of motion (and, given our purposes here, specifically the second law of motion); Robert Hooke (1635–1703) who developed the law of elasticity; the third Baron Rayleigh (1842–1919), who introduced the concept of modal analysis and viscous damping; Joseph Louis Lagrange (1736–1813), who presented the Lagrange multipliers; and William Rowan Hamilton (1805–1865), who illustrated the Hamiltonian formulation of dynamics (which is essentially a reformulation of Newtonian dynamics).

We should also acknowledge the more recent contributions: Goldstein (Goldstein et al., 2001), Whittaker and Synege (Whittakker, 1988), Timoshenko and Young (Timoshenko and Young, 1948), Den Hartog (Den Hartog, 1930), Griffith (Synge and Griffith, 1959), Nayfeh (Nayfeh, 1973), Crandall and Mark (Crandall and Mark, 1963), Robson (Robson, 1964), Zienkiewicz (Zienkiewicz et al., 2005) and many others who have contributed to the development of dynamic analysis in the last century, and have thus made possible the solving of rather sophisticated dynamic analyses and real engineering vibration problems (Jia, 2014).

Before concluding, it should also be noted that the dynamic, vibration and impact problems elaborated in this edited volume is for a real-time causal system, in which the present responses depend only on the past and present inputs, and not on the future inputs. It is assumed that non-causal systems do not exist in nature.

Acknowledgements

The author wish to acknowledge the permission by Springer and Elsevier to reuse figures and texts from the author's previous book and journal publications.

References

Berg, P.K. and Bråfel, O. 1991. Noise and Vibrations on Board. Joint Industrial Safety Council, Stockholm, Sweden.

Broderick, B.M., Elnashai, A.S., Ambraseys, N.N., Barr, J.M., Goodfellow, R.G. and Higazy, E.M. 1994. The Northridge (California) earthquake of 17 January 1994: Observations, strong motion and correlative response analysis, Engineering Seismology and Earthquake Engineering, Research Report No. ESEE 94/4, Imperial College, London, UK.

Brown, K. 2002. Tsunami! At Lake Tahoe? Science News, Magazine of The Society for Science & the Public. https://www.sciencenews.org/article/tsunami-lake-tahoe.

Crandall, S.H. and Mark, W.D. 1963. Random Vibration in Mechanical Systems. Academic Press, New York, U.S.

Curnier, A. 1994. Computational Methods in Solid Mechanics. Kluwer Academic Publishers, Dordrecht, Netherlands.

Den Hartog, J.P. 1930. Forced vibrations with combined Coulomb and viscous damping. Translations of the ASME 53: 107–115.

Elnashai, A.S. and Sarno, L.D. 2008. Fundamentals of Earthquake Engineering. John Wiley and Sons, Hoboken, NJ, US.

Faltinsen, O.M. and Timokha, A.N. 2012. On sloshing modes in a circular tank. Journal of Fluid Mechanics 695: 467–477.

Goldstein, H., Poole, C.P. and Safko, J.L. 2001. Classical Mechanics. 3rd edn., Addison-Wesley, Boston, MA, USA.

Hatayama, K. 2008. Lessons from the 2003 Tokachi-oki, Japan, earthquake for prediction of long-period strong ground motions and sloshing damage to oil storage tanks. J. Seismol. 12: 255–263.

Hitchcock, P.A., Kwok, K.C.S. and Watkins, R.D. 1997a. Characteristics of liquid column vibration absorbers (LCVA)-I. Engineering Structures 19(2): 126–134.

Hitchcock, P.A., Kwok, K.C.S. and Watkins, R.D. 1997b. Characteristics of liquid column vibration absorbers (LCVA)-II. Engineering Structures 19(2): 135–144.

ISO 2631-1. 1997. Mechanical Vibration and Shock—Evaluation of Human Exposure to whole Body Vibration—Part 1: General Requirements, International Organization for Standardization. 2nd edn., Geneva, Switzerland.

Ji, T.J. and Bell, A. 2008. Seeing and Touching Structural Concepts. Taylor and Francis, Oxon, UK.

Krysl, P. 2006. A Pragmatic Introduction to the Finite Element Method for Thermal and Stress Analysis. World Scientific, London, UK.

Larsen, A., Esdahl, S., Andersen, J.E. and Vejrum, T. 2000. Storebaelt suspension bridge—vortex shedding excitation and mitigation by guide vanes. Journal of Wind Engineering and Industrial Aerodynamics 88: 283–296.

Lee, D. and Ng, M. 2010. Application of tuned liquid dampers for the efficient structural design of slender tall buildings. CTBUH Journal 4: 30–36.

Levy, M. and Salvadori, M. 2002. Why Buildings Fall Down. WW Norton and Company, New York, US.

Moan, T. 2005. Safety of Offshore Structures. No 2005-04, Centre for Offshore Research and Engineering, National University of Singapore, Singapore.

Naess, A. and Moan, T. 2012. Stochastic Dynamics of Marine Structures. Cambridge University Press, Cambridge, UK.

Nayfeh, A.H. 1973. Perturbation Methods. Wiley, New York, U.S.

Robson, J.D. 1964. Random Vibration. Edinburgh University Press, Edinburgh, UK.

Romero, J.A., Hildebrand, R., Martinez, M., Ramirez, O. and Fortanell, J.A. 2005. Natural sloshing frequencies of liquid cargo in road tankers. International Journal of Heavy Vehicle System 2(2): 121–138.

Ruponen, P., Matusiak, J., Luukkonen, J. and Ilus, M. 2009 Experimental study on the behavior of a swimming pool onboard a large passenger ship. Marine Technology 46(1): 27–33.

Synge, J.L. and Griffith, B.A. 1959. Principles of Mechanics. McGraw-Hill, New York, U.S.

Jia, J.B. and Ulfvarson, A. 2006. Dynamic Analysis of Vehicle-deck Interactions. Ocean Engineering 33(13): 1765–1795.

Jia, J.B. 2007. Investigations of vehicle securing without lashings for Ro-Ro ships. Journal of Marine Science and Technology 12(1): 43–57.

Jia, J.B. 2008. An efficient nonlinear dynamic approach for calculating the wave induced fatigue damage of offshore structures and its industry applications for the lifetime extension purpose. Applied Ocean Research 30(3): 189–198.

Jia, J.B. 2011. Wind and structural modeling for an accurate fatigue life assessment of tubular structures. Engineering Structures 33(2): 477–491.

Jia, J.B. 2012. Seismic Analysis for Offshore Industry: Promoting State of the Practice toward State of the Art, ISOPE 2012. 438–447, Rhodes.

Jia, J.B. 2014. Essentials of Applied Dynamic Analysis. 424 pp., Springer, Heidelberg, Germany.

Jia, J.B. 2016. The effect of gravity on the dynamic characteristics and fatigue life assessment of offshore structures. Journal of Constructional Steel Research 118(1): 1–21.

Jia, J.B. 2017. Modern Earthquake Engineering–Offshore and Land-based Structures. 848 pp., Springer, Heidelberg, Germany.

Jia, J.B. 2018. Soil Dynamics and Foundation Modeling–Offshore and Earthquake Engineering. 740 pp., Springer, Heidelberg, Germany.

Jia, J.B. and Ringsberg, J.W. 2009. Numerical dynamic analysis of nonlinear structural behaviour of ice-loaded side-shell structures. International Journal of Steel Structures 9(3): 219–230.

The Joint Accident Investigation Commission of Estonia, Finland and Sweden. 1997. Final Report on the Capsizing on 28 September 1994 in the Baltic Sea of the Ro-Ro passenger vessel MV ESTONIA.

Thomson, W.T. 1966. Vibration Theory and Applications. George Allen and Unwin, London, UK.

Timoshenko, S. and Young, D.H. 1948. Advanced Dynamics. 1st edn, McGraw-Hill Book Company, New York, U.S.

Whittaker, E.T. 1988. A Treatise on the Analytical Dynamics of Particles and Rigid Bodies. Cambridge Mathematical Library, Cambridge, UK.

White R.N., Gergely P. and Sexsmith, R.G. 1972. Structural Engineering, Volume 1, Introduction to Design Concepts and Analysis. John Wiley and Sons, NY, U.S.

Youtube. 2007. Crazy "wave pool" aboard Sun Princess, http://www.youtube.com/watch?v=AJCurMmkNTY.

Zienkiewicz, O.C., Taylor, R.L. and Zhu, J.Z. 2005. The Finite Element Method: Its Basis and Fundamentals. 6th edn, Butterworth-Heinemann, Oxford, UK.

2

Nonlinear Structural Responses Associated with Hydrocarbon Explosions

Jeom Kee Paik,[1,*] *Sang Jin Kim*[2] *and Junbo Jia*[3]

2.1 Introduction

While in service, ships and offshore structures are subjected to various types of actions and action effects, which are usually normal, but they are sometimes extreme and even accidental, as shown in Fig. 2.1. Hydrocarbon explosions and fires are two of the most typical types of accidents associated with offshore installations that develop oil and gas.

Explosions are a major type of accident on offshore platforms that develop offshore oil and gas, which are flammable. These explosions occur because hydrocarbons are often released from flanges, valves, seals, vessels, or nozzles of offshore installations and may be ignited by sparks. When hydrocarbons are combined with an oxidizer (usually oxygen or air), they can explode by ignition. Combustion occurs if temperatures increase to the point at which the hydrocarbon

[1] Department of Naval Architecture and Ocean Engineering and The Korea Ship and Offshore Research Institute (The Lloyd's Register Foundation Research Centre of Excellence) at Pusan National University, Korea; and Department of Mechanical Engineering at University College, London, UK.
[2] The Korea Ship and Offshore Research Institute (The Lloyd's Register Foundation Research Centre of Excellence) at Pusan National University, Korea.
[3] Aker Solutions, Bergen, Norway.
* Corresponding author

Figure 2.1. Various types of extreme and accidental events involving ships and offshore installations (Paik, 2015).

molecules react spontaneously with an oxidizer. A blast or a rapid increase in pressure results from such an explosion. Offshore structures subjected to the impact of overpressure from explosions can be significantly damaged, and catastrophes may result, with casualties, asset damage, and marine pollution.

Successful engineering and design should meet not only functional requirements but also Health, Safety, Environment & Ergonomics (HSE&E) requirements. Functional requirements address operability in normal conditions, and HSE&E requirements represent safe performance and integrity in accidental and extreme conditions. Normal conditions can usually be characterized by a solely linear approach, but more sophisticated approaches must be applied to accidental and extreme conditions that involve highly nonlinear responses, as shown in Fig. 2.2 (Paik et al., 2014; Paik, 2015). The risk-based approach is known to be the best method for successful design and engineering to meet HSE&E requirements against accidental and extreme conditions.

In industry practices, prescriptive (predefined or deterministic) methods are often applied for risk assessment and management (FABIG, 1996; API, 2006; ABS,

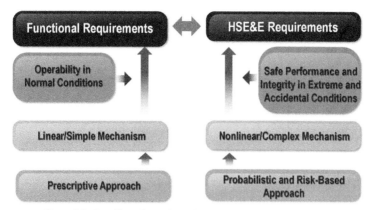

Figure 2.2. Paradigm change in engineering and design (Paik, 2015).

2013; DNVGL, 2014). However, application of a fully probabilistic approach for quantitative risk assessment and management is highly desirable (Czujko, 2001; Vinnem, 2007; NORSOK, 2010; Paik and Czujko, 2010; Paik, 2011; ISO, 2014; LR, 2014).

Within the framework of risk assessment and management, the characteristics of actions and action effects are identified by taking advantage of advanced engineering models associated with nonlinear structural mechanics. This chapter describes the nonlinear structural mechanics associated with hydrocarbon explosions. Current rules and industry practices for risk assessment are surveyed, and advanced procedures and recommended practices are investigated. For nonlinear structural response analysis due to explosions, the blast pressure actions must be defined. Therefore, both blast pressure actions and action effects are described.

2.2 Fundamentals—Theory

2.2.1 Profile of Blast Pressure Actions

Figure 2.3 represents a typical profile of the blast pressure actions caused by hydrocarbon explosions, which are generally characterized by four parameters: (a) rise time until the peak pressure, (b) peak pressure, (c) pressure decay type beyond the peak pressure, and (d) pressure duration time. The peak pressure value often approaches some two to three times the collapse pressure loads of structural components under quasi-static actions. However, the rise time of blast pressure actions is very short, only a few milliseconds. The duration (persistence) of blast pressure actions is often in the range of 10–50 ms. It is necessary to define the structural consequences (damage) of blast pressure actions within the quantitative risk assessment and management.

When the rise and duration times of blast pressure actions are very short, the blast pressure response is often approximated to an impulsive type of action characterized by only two parameters, the equivalent peak pressure P_e and the

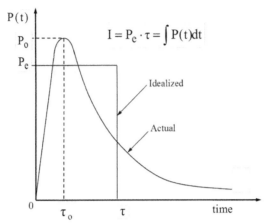

Figure 2.3. Typical profile of blast pressure action and its idealization (Paik and Thayamballi, 2007).

duration time τ, as long as the corresponding impulse is identical (Paik and Thayamballi, 2007). The two parameters may be defined so that the actual and idealized impulses of the impact pressure action are equal, namely,

$$I = P_e \tau = \int P(t) dt \qquad (2.1)$$

where I is the impulse of the impact pressure action, t is the time, P_e is the effective peak pressure, and τ is the duration time of P_e. Taking P_e as the same as P_o (peak pressure value) may be unduly pessimistic for obvious reasons, and thus P_e is sometimes obtained by multiplying a relevant knockdown factor to P_o. Once the impulse I and the effective peak pressure value P_e are defined, the duration time τ can be determined from Eq. (2.1).

In analytical methods for prediction of structural damage due to blast pressure action, P_o and τ can be dealt with as parameters of influence. In computational models, however, the actual profile of blast pressure action is directly applied to simulate the nonlinear structural responses.

2.2.2 Thermodynamics of Hydrocarbons

1. Definitions of Physical Parameters

This section presents the definitions of the physical parameters associated with hydrocarbon explosions. When a single type of gas is involved, the number of moles of a species is defined as follows (FLACS, 2016):

$$n_i = \frac{m_i}{M_i} \qquad (2.2)$$

where n_i is the number density, m_i is the mass, and M_i is the molecular weight of a mixture of species.

The mole fractions are defined as follows:

$$X_i = \frac{n_i}{\sum\limits_{i=1}^{N} n_i} \tag{2.3}$$

where X_i is the mole fraction.

Mass fractions are defined as follows:

$$Y_i = \frac{m_i}{\sum\limits_{i=1}^{N} m_i} \tag{2.4}$$

where Y_i is the mass fraction.

The fuel-oxidant ratio is defined as follows:

$$\left(F/O\right) = \frac{m_{fuel}}{m_{oxygen}} \tag{2.5}$$

where F/O is the fuel-oxidant ratio, m_{fuel} is the mass of fuel, and m_{oxygen} is the mass of oxygen.

The equivalence ratio is then defined as follows:

$$\Phi = \frac{\left(F/O\right)_{actual}}{\left(F/O\right)_{stoichometric}} = \frac{\left(m_{fuel}/m_{oxygen}\right)_{actual}}{\left(m_{fuel}/m_{oxygen}\right)_{stoichometric}} \tag{2.6}$$

where Φ is the equivalence ratio.

When several gases are mixed in explosions, the mole fraction is defined by

$$X_i = \frac{Y_i/M_i}{\sum\limits_{i=1}^{N} Y_i/M_i} \tag{2.7}$$

The mass fraction is defined as follows:

$$Y_i = \frac{X_i M_i}{\sum\limits_{i=1}^{N} X_i M_i} \tag{2.8}$$

The ideal gas law for a mixture is given by

$$p = \rho R T \tag{2.9}$$

where p is the absolute pressure, ρ is the density, R is the gas constant of a mixture, and T is the absolute temperature.

For a perfect gas, Dalton's law is represented by

$$p = \sum_{i=1}^{N} p_i = \frac{R_u T}{V} \sum_{i=1}^{N} n_i \tag{2.10}$$

where R_u is the universal gas constant and V is the volume.

The isentropic ratio is defined as follows:

$$\gamma = \frac{c_p}{c_v} \tag{2.11}$$

where γ is the isentropic ratio and c_p and c_v are the specific heat capacities at constant pressure and volume, respectively.

The speed of sound is defined by

$$c \equiv \sqrt{\gamma R T} = \sqrt{\gamma \frac{\rho}{p}} \tag{2.12}$$

where c is the speed of sound.

The relation between blast pressure-density-temperature is given as follows:

$$\left(\frac{p}{p_0} \right) = \left(\frac{\rho}{\rho_0} \right)^{\gamma} = \left(\frac{Y}{Y_0} \right)^{\gamma/(\gamma-1)} = \left[1 + \frac{(\gamma-1)}{2} \left(\frac{u}{c} \right)^2 \right]^{-\gamma/(\gamma-1)} \tag{2.13}$$

where p_0 is the ambient pressure, ρ_0 is the initial density, Y_0 is the initial mass fraction, and u is the flow velocity.

2. Stoichiometric Reaction

Combustion is a burning process by which a fuel is oxidized with an oxidant (usually air), producing heat and light. The chemical process of reaction can be given as follows (FLACS, 2016):

$$C_{nc} H_{nh} O_{no} + \alpha O_2 \rightarrow nc CO_2 + b H_2 O + Q \tag{2.14}$$

When both the fuel and oxidant disappear entirely after the reaction is completed, the process is termed a stoichiometric reaction. The stoichiometric amount of oxidant on a molar basis can be defined by

$$\alpha = nc + \frac{nh}{4} - \frac{no}{2} \tag{2.15}$$

where n_c is the number of carbons, n_h is the number of hydrogens, and n_o is the number of oxygens.

2.2.3 Governing Equations for Fluid Flow (Dispersion and Explosion)

The conservation of mass is given by

$$\frac{\partial}{\partial t}\left(\beta_v \rho\right) + \frac{\partial}{\partial x_i}\left(\beta_i \rho u_i\right) = \frac{\dot{m}}{V} \tag{2.16}$$

where β_v is the volume porosity, β_i and u_i are the area porosity and mean velocity in the ith direction, respectively, and $\dot{m}$ is the mass rate.

The momentum equation is given by

$$\frac{\partial}{\partial t}\left(\beta_v \rho u_i\right) + \frac{\partial}{\partial x_j}\left(\beta_j \rho u_j u_i\right) = -\beta_v \frac{\partial p}{\partial x_i} + \frac{\partial}{\partial x_j}\left(\beta_j \sigma_{ij}\right) + F_{o,i} + \beta_v F_{w,i} + \beta_v \left(\rho - \rho_0\right) g_i$$

$$F_{o,i} = -\rho \left|\frac{\partial \beta}{\partial x_i}\right| u_i \left|u_i\right| \tag{2.17}$$

where σ_{ij} is the stress tensor, $F_{o,i}$ and $F_{w,i}$ are the obstruction and wall friction forces, respectively, and g_i is the gravitational acceleration in the ith direction.

The transport equation for enthalpy is given by

$$\frac{\partial}{\partial t}\left(\beta_v \rho h\right) + \frac{\partial}{\partial x_j}\left(\beta_j \rho u_j h\right) = \frac{\partial}{\partial x_j}\left(\beta_j \frac{\mu_{eff}}{\sigma_h} \frac{\partial h}{\partial x_j}\right) + \beta_v \frac{Dp}{Dt} + \frac{\dot{Q}}{V} \tag{2.18}$$

where h is the specific enthalpy, μ_{eff} is the effective viscosity, σ_h is the Prandtl-Schmidt number of specific enthalpy (typically $\sigma_h = 0.7$), and $\dot{Q}$ is the heat rate.

The transport equation for fuel mass fraction is given by

$$\frac{\partial}{\partial t}\left(\beta_v \rho Y_{fuel}\right) + \frac{\partial}{\partial x_j}\left(\beta_j \rho u_j Y_{fuel}\right) = \frac{\partial}{\partial x_j}\left(\beta_j \frac{\mu_{eff}}{\sigma_{fuel}} \frac{\partial Y_{fuel}}{\partial x_j}\right) + R_{fuel} \tag{2.19}$$

where Y_{fuel} is the mass fraction of fuel, σ_{fuel} is the Prandtl-Schmidt number of fuel (typically $\sigma_{fuel} = 0.7$), and R_{fuel} is the fuel reaction rate.

The transport equation for turbulent kinetic energy is given by

$$\frac{\partial}{\partial t}\left(\beta_v \rho k\right) + \frac{\partial}{\partial x_j}\left(\beta_j \rho u_j k\right) = \frac{\partial}{\partial x_j}\left(\beta_j \frac{\mu_{eff}}{\sigma_k} \frac{\partial k}{\partial x_j}\right) + \beta_v P_k - \beta_v \rho \varepsilon \tag{2.20}$$

where k is the turbulent kinetic energy, σ_k is the Prandtl-Schmidt number of turbulent kinetic energy (typically $\sigma_k = 1.00$), P_k is the gauge pressure of kinetic energy, and ε is the dissipation of turbulent kinetic energy.

The transport equation for the dissipation rate of turbulent kinetic energy is given by

$$\frac{\partial}{\partial t}(\beta_v \rho \varepsilon) + \frac{\partial}{\partial x_j}(\beta_j \rho u_j \varepsilon) = \frac{\partial}{\partial x_j}\left(\beta_j \frac{\mu_{eff}}{\sigma_\varepsilon} \frac{\partial \varepsilon}{\partial x_j}\right) + \beta_v P_\varepsilon - C_{2\varepsilon} \beta_v \rho \frac{\varepsilon^2}{k} \tag{2.21}$$

where σ_ε is the Prandtl-Schmidt number of the dissipation rate of turbulent kinetic energy (typically $\sigma_\varepsilon = 1.30$), P_ε is the gauge pressure of the dissipation rate of kinetic energy, and $C_{2\varepsilon}$ is the constant in the $k - \varepsilon$ equation (typically $C_{2\varepsilon} = 1.92$).

2.2.4 Turbulence Model (k-ε Model)

In industry practice, the $k - \varepsilon$ model is often applied to model turbulence in association with hydrocarbon explosions. In this model, two additional transport equations are solved: one for turbulent kinetic energy and one for the dissipation of turbulent kinetic energy.

$$\frac{\partial}{\partial t}(\rho k) + \frac{\partial}{\partial x_i}(\rho k u_i) = \frac{\partial}{\partial x_j}\left(\frac{\mu_t}{\sigma_k} \frac{\partial k}{\partial x_j}\right) + 2\mu_t E_{ij}^{\ 2} - \rho \varepsilon \tag{2.22}$$

where E_{ij} is the component of rate of deformation.

The turbulence model for the dissipation rate of turbulent kinetic energy is given by

$$\frac{\partial}{\partial t}(\rho \varepsilon) + \frac{\partial}{\partial x_i}(\rho \varepsilon u_i) = \frac{\partial}{\partial x_j}\left(\frac{\mu_t}{\sigma_\varepsilon} \frac{\partial \varepsilon}{\partial x_j}\right) + C_{1\varepsilon} \frac{\varepsilon}{k} 2\mu_t E_{ij}^{\ 2} - C_{2\varepsilon} \rho \frac{\varepsilon^2}{k} \tag{2.23}$$

where $C_{1\varepsilon}$ is the constant in the $k - \varepsilon$ equation (typically $C_{1\varepsilon} = 1.44$).

2.2.5 Wind Boundary (Dispersion)

In hydrocarbon explosions, structural responses are affected by wind boundaries, which may reproduce the properties of the atmospheric boundary layer near the Earth's surface. In industry practice, the concept of a characteristic length scale is often applied in association with buoyancy effects on the atmospheric boundary layer (Monin and Obukhov, 1954).

$$L = \frac{\rho_{air} c_p T_{air} (u^*)^3}{\kappa g H_s} \tag{2.24}$$

where L is the Monin-Obukhov length scale, ρ_{air} and T_{air} are the absolute pressure and temperature of the air, respectively, u^* is the friction velocity, κ is the Von Karman constant (typically $\kappa = 0.41$), and H_s is the sensible heat flux from the surface. Table 2.1 indicates the Monin-Obukhov lengths and stability, which are an interpretation of the Monin-Obukhov lengths with respect to atmospheric stability.

Table 2.1. Monin-Obukhow lengths and stability.

Monin-Obukhov length (m)	Stability
Small negative, $-100 < L$	Very unstable
Large negative, $-10^5 < L < -100$	Unstable
Very large, $|L| > 10^5$	Neutral
Large positive, $10 < L < 10^5$	Stable
Small positive, $0 < L < 10$	Very stable

2.2.6 Combustion Model

An explosion may be escalated by ignition of a premixed cloud of fuel and oxidant. However, a steady non-turbulent premix of fuel and oxidant may burn with a laminar burning velocity before escalation.

$$S_L^0 = S_L^0 \left(fuel, \Phi \right) \tag{2.25}$$

The fuel and the equivalence ratio F affect the laminar burning velocity, which is zero, or mixtures with fuel contents below the lower flammability limit (LFL) or above the upper flammability limit (UFL) will not burn. In a hydrocarbon explosion, the flame accelerates and becomes turbulent. The turbulent burning velocity is much greater than the laminar one because the reactants and products are much better mixed. In numerical models of combustion, the correlations are used for both laminar and turbulent burning velocities that originate from experimental work.

In industry practice, a hypothesis is applied in which the reaction zone in a premixed flame is thinner than the practical grid resolutions. In this case, the flame needs to be modeled where the flame zone is thickened by increasing the diffusion with a factor b and reducing the reaction rate with a factor $1/b$. In this regard, the flame model is often called the β-model.

1. Flame Model

The diffusion coefficient D for fuel comes from the transport equation for fuel,

$$D = \frac{\mu_{eff}}{\sigma_{fuel}} \tag{2.26}$$

where D is the diffusion coefficient.

A dimensionless reaction rate W is defined by adjusting D and W as follows:

$$W^* = \frac{W}{\beta} = W \frac{l_{LT}}{\Delta g} \tag{2.27a}$$

$$D^* = D\beta = D \frac{\Delta g}{l_{LT}} \tag{2.27b}$$

where W is the dimensionless reaction rate, β is the transformation factor in the β-model, l_{LT} is the mixing length in the β-model, and D is the diffusion coefficient.

2. Burning Velocity Model

As far as a weak ignition source is associated with a combustible cloud under quiescent conditions, the initial burning process may be laminar. In this case, the front of the flame may be smooth, and the propagation of flame is governed by thermal and/or molecular diffusion processes. Immediately after the initial stage, the flame surface is wrinkled by instabilities from various sources (e.g., ignition, flow dynamics, Rayleigh-Taylor) where the speed of flame increases and becomes quasi-laminar. Depending on the flow conditions, a transition period may occur, and eventually, the turbulent burning regime is reached.

It is obvious that the laminar burning velocity depends on the type of fuel, the fuel-air mixture, and the pressure. For a mixture of fuels, the laminar burning velocity is estimated as the volume-weighted average. The laminar burning velocity is given as a function of the pressure as follows:

$$S_L = S_L^0 \left(\frac{P}{P_0}\right)^{\gamma_P} \tag{2.28}$$

where γ_P is the pressure exponent for the laminar burning velocity, which is a fuel-dependent parameter.

In the quasi-laminar regime, the turbulent burning velocity is given by

$$S_{QL} = S_L \left(1 + \chi \min\left(\left(\frac{R_{\text{flame}}}{3}\right)^{0.5}, 1\right)\right) \tag{2.29}$$

where R_{flame} is the flame radius and χ is the fuel-dependent constant.

2.2.7 Numerical Models for Nonlinear Structural Responses

The equations of the dynamic equilibrium are solved numerically (Paik, 2018). Implicit and explicit approaches are relevant. For the explicit scheme associated with the time integration of the dynamic equations of motion, the displacements at time $t + \Delta t$ are calculated from the equilibrium of the structure at time t when the effect of a damping matrix is neglected:

$$[m][\ddot{w}]^t = [F]^t - [S]^t \tag{2.30}$$

where $[m]$ is the mass matrix of the structure, $[w]^t$ is the vector of nodal displacements and rotations at time t, $[\ddot{w}]^t$ is the corresponding acceleration vector, $[F]^t$ is the vector of the external nodal forces, and $[S]^t$ is the vector of the internal forces-moments equivalent to the internal stresses at time t.

The vector $[S]^t$ varies depending on the configuration of the structure with the displacements at time t, the stresses, and the material constitutive models. For a linear elastic response, $[S]^t$ is given by

$$[S]^t = [K][w]^t \tag{2.31}$$

where $[K]$ is the constant in the time stiffness matrix.

The nodal point displacements at the next time step, $t + \Delta t$, are obtained by substituting an approximation for the acceleration vector into the above equation. The most common approximation used is that obtained by using the central difference operator, given by

$$\left[\ddot{w}\right]^t = \frac{[w]^{t+\Delta t} - 2[w]^t + [w]^{t-\Delta t}}{\Delta t^2} \tag{2.32}$$

By substituting Eq. (2.31) into Eq. (2.29), the displacements at time $t + \Delta t$ are calculated. In practical problems, $[m]$ is often a diagonal matrix. In this case, the equation is uncoupled, and therefore the structural responses at time $t + \Delta t$ are computed easily, where the inverse of any coefficient matrix of the system is not necessary. This is the main advantage of the use of the implicit time integration method. The major disadvantage is that relatively very small solution time increments must be used to obtain a stable and reliable solution.

For the implicit time integration scheme, the displacements at time $t + \Delta t$ are obtained from the equilibrium of the structure as follows:

$$[m]\left[\ddot{w}\right]^{t+\Delta t} + [K]^t \Delta[w]^t = [F]^{t+\Delta t} - [S]^t \tag{2.33}$$

where $[K]^t$ is the tangent stiffness matrix of the structure at time t and $\Delta[w]^t = [w]^{t+\Delta t} - [w]^t$.

Several implicit schemes are available to approximate the acceleration $\left[\ddot{w}\right]^{t+\Delta t}$ in Eq. (2.32). One is the trapezoidal rule, which is given by

$$\left[\dot{w}\right]^{t+\Delta t} = \left[\dot{w}\right]^t + \frac{\Delta t}{2}\left(\left[\ddot{w}\right]^t + \left[\ddot{w}\right]^{t+\Delta t}\right) \text{ and } \left[w\right]^{t+\Delta t} = \left[w\right]^t + \frac{\Delta t}{2}\left(\left[\dot{w}\right]^t + \left[\dot{w}\right]^{t+\Delta t}\right) \tag{2.34}$$

By substituting Eq. (2.33) into Eq. (2.32), the equilibrium equation is transformed into the following equation:

$$\left([K]^t + \frac{4}{\Delta t^2}[m]\right)\Delta[w]^t = [F]^{t+\Delta t} - [S]^t + [m]\left(\frac{4}{\Delta t}[\dot{w}]^t + [\ddot{w}]^t\right) \tag{2.35}$$

Equation (2.34) can be solved for the displacement increments, $\Delta[w]^t$, where the inversion of a matrix is required with a time step that is larger than the one required for the explicit solution scheme.

2.3 Current Rules and Industry Practices

2.3.1 American Bureau of Shipping (ABS)

The American Bureau of Shipping (ABS, 2013) specifies guideline for evaluation of the risk of explosion that consists of two steps, preliminary and detailed risk assessment, as shown in Fig. 2.4, and the analysis of nonlinear structural responses is a key task. The ABS procedure applies three steps of the analysis: (i) screening, (ii) strength-level analysis, and (iii) ductility-level analysis, similar to API (2006), considering that the profile of blast pressure loads is idealized as shown in Fig. 2.5.

- **Screening analysis** is the simplest approach with which to assess the structural response under a blast event. This method applies an equivalent static load and evaluates the response by means of accidental limit state-based design checks. The equivalent static load is the peak overpressure in accordance with the strength level associated with the blast scaled by a dynamic load amplification factor.

- **Strength-level analysis** is a linear-elastic analysis of an equivalent static load corresponding to the blast overpressure, taking into account the effect

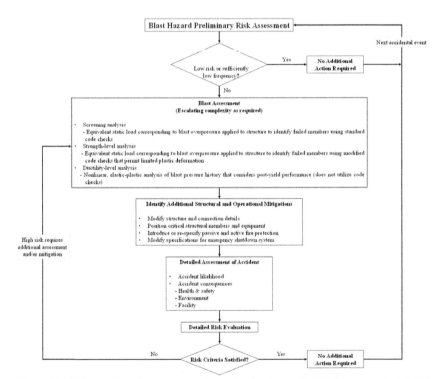

Figure 2.4. ABS procedure for assessment of structural safety against blast pressure loads (ABS, 2013).

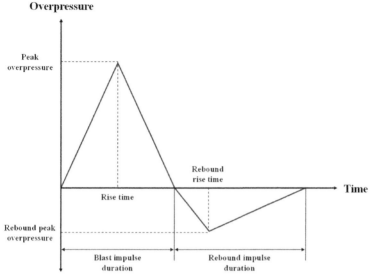

Figure 2.5. Profile of blast pressure loads (ABS, 2013).

of plasticity. The overpressure peak is represented by a dynamic load amplification factor.

- **Ductility-level analysis** takes into account the effects of geometric and material nonlinearities as the most refined approach.

2.3.2 American Petroleum Institute (API)

A prescriptive method is suggested by the American Petroleum Institute (API, 2006) to define the blast loads and assess the structural responses. Figure 2.6 shows the API procedure. Three kinds of models are relevant.

- **Empirical models** with overpressure correlated to experimental data associated with accuracy and applicability limited by the model database.
- **Phenomenological models** with overpressure characterized by incorporating physical principles into empirical observations (i.e., interpreting observations so that they are consistent with fundamental theory).
- **Numerical models** with overpressure defined by solving the appropriate relationships for gas flow, combustion, and turbulence, that typically make use of computational fluid dynamics (CFD) principles.

The numerical models are more refined than other models, but they require more time and effort. In this regard, the API procedure uses a prescriptive model associated with the nominal value of overpressure for specific areas of the structures. The API procedure is composed of four steps: (i) selection of the concept type; (ii) establishment of the conditioning factors to apply; (iii) determination of nominal overpressures; and (iv) application of safety factors to account for data uncertainties.

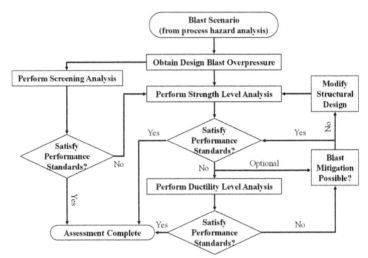

Figure 2.6. API procedure for assessment of structural safety against blast pressure loads (API, 2006).

Tables 2.2 and 2.3 indicate the nominal overpressures by the types of offshore installations and modification factors associated with the project parameters, respectively. For other cases not indicated in Table 2.3, design explosion loads from the explosion load exceedance curves with risk acceptance level are used. 10^{-3} to $10^{-4}/y$ of risk levels are generally recommended to be used depending on the performance criteria.

Methods for structural response assessment against blast pressure actions proposed by API are similar to those of ABS. The structural assessment is performed by a screening check, strength level analysis, and ductility level analysis, in that order.

Table 2.2. Nominal overpressures by the type of offshore installation (API, 2006).

Blast prone areas	Nominal overpressure in offshore installation type (bar)				
	Integrated production/drilling		Bridge linked production/drilling (multiple platforms)	Production only	
	Single platform	TLP/Wet trees		Single jacket	Mono-hull FPSO*
Wellhead/drill deck	2.50	2.50	2.00	-	-
Gas separation facilities	2.00	1.00	1.50	1.50	1.00
Gas treatment/ compression facilities	1.50	1.00	1.00	1.00	1.00
Turret (internal)	-	-	-	-	3.00
FPSO* main deck	-	-	-	-	2.00
TLP moon pool	-	2.00	-	-	-
TLP deck box	-	2.50	-	-	-
Other	1.00	0.75	1.00	1.00	0.50

*FPSO: Floating, Production, Storage and Offloading unit.

Table 2.3. Load modifiers associated with project parameters (API, 2006).

Project parameters		Nominal blast load modifiers*
Item	Range/rate/quantity	
Production rate	Less than 50,000 bbl/day	0.90
	50,000 to 100,000 bbl/day	1.05
	More than 100,000 bbl/day	1.10
Gas compression pressure	Less than 100 bar	1.00
	100 to 200 bar	1.05
	More than 200 bar	1.10
Gas composition	Normal	1.00
	Onerous	1.10
	More onerous	1.35
Production trains	1	0.90
	2	0.95
	4	1.10
Module footprint area	Less than 75,000 sqft	0.90**
	75,000 to 150,000 sqft	1.00
	More than 150,000 sqft	1.10
Confinement	3 sides of more open	0.85
	1 to 2 sides open	0.95
	All sides closed	1.25
Module length to width aspect ratio	Less than 1.0	0.90
	1.0 to 1.7	1.05
	More than 1.7	1.10

* Load modifier should not be applied to wellheads/drilling decks, moonpools, and FPSO main deck.
** For small and very congested platforms ($\sim$ 10,000 sqft), the load modifier of 0.9 should not be applied to reduce the nominal explosion overpressure for module area.

2.3.3 Det Norske Veritas and Germanischer Lloyd (DNVGL)

Det Norske Veritas and Germanischer Lloyd (DNVGL, 2014) suggests a deterministic method for prediction of design explosion loads in terms of overpressure and pulse duration. The design loads are subdivided depending on the conditions of confinement and congestion. Table 2.4 summarizes the typical design explosion values. However, a specific analysis with the use of actual details is also recommended if accurate predictions are needed because the explosion overpressures depend on numerous variables.

For the structural design of offshore structures to protect against explosions, DNV-RP-C204 (DNVGL, 2010) proposes the use of nonlinear dynamic finite analysis or simple calculation methods based on single and/or multiple degree of freedom (SDOF and/or MDOF) analogies with idealized design blast loads. DNVGL (2010) classifies the analysis models that depend on the failure mode that a designer wishes to check. Figure 2.7 and Table 2.5 show the failure modes for two-way stiffened panel and recommended analytical models.

Table 2.4. Nominal explosion design values proposed by DNVGL (2014).

Type of offshore installations	Working areas	Design blast overpressure (barg)	Pulse duration (s)
Drilling rig	Drill floor with cladded walls	0.1	0.2
	Shale shaker room with strong walls, medium sized	2.0	0.3
Mono-hull FPSO	Process area, small	0.3	0.2
	Process area, medium-sized with no walls or roof	1.0	0.2
	Turret in hull, STP/STL room with access hatch	4.0	1.0
Mono-hull FPSO (Large)	Process area, large with no walls or roof	2.0	0.2
Production platform (Sumi-sub)	Process area, large with no or light walls, 3 storeys, grated mezzanine and upper decks	2.0	0.2
Production platform (Fixed)	Process area, medium-sized, solid upper and lower decks, 3 storeys, 1 or 2 sides open	1.5	0.2
Integrated production and drilling	Process area and drilling module each medium sized on partly solid decks, 3 storeys, 3 sides open	1.5	0.2
	X-mas tree/wellhead area, medium sized with grated floors	1.0	0.2

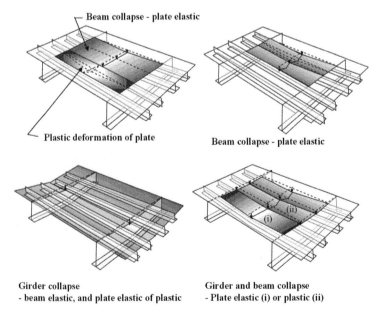

Figure 2.7. Failure modes for two-way stiffened panel for adoption of analysis models (DNVGL, 2010).

Table 2.5. Analytical models according to failure modes suggested by DNVGL (2010).

Failure modes	Simplified analysis models	Comments
Elastic-plastic deformation of plate	SDOF	-
Stiffener plastic - plate elastic	SDOF	Elastic, effective flange of plate.
Stiffener plastic - plate plastic	SDOF	Effective width of plate at mid-span. Elastic, effective flange of plate at ends.
Girder plastic - stiffener and plating elastic	SDOF	Elastic, effective flange of plate with concentrated loads (stiffener reactions). Stiffener mass included.
Girder plastic - stiffener elastic - plate plastic	SDOF	Effective width of plate at girder mid span and ends. Stiffener mass included.
Girder and stiffener plastic - plate elastic	MDOF	Dynamic reactions of stiffeners → loading for girders.
Girder and stiffener plastic - plate plastic	MDOF	Dynamic reactions of stiffeners → loading for girders.

2.3.4 Fire and Blast Information Group (FABIG)

The Fire and Blast Information Group (FABIG) Technical Note 4 specifies the necessity of establishing base, lower, and upper cases for the definitions of blast loads because realistic pressures cannot be obtained without the majority of the piping and structure congestion included in the geometry model for explosion simulations (FABIG, 1996).

Both SDOF and MDOF (finite-element method [FEM] in this section) can be used for structural response analysis, taking into account an idealized explosion load, as shown in Fig. 2.3 (FABIG, 1996). FABIG also advises that dynamic effects such as a strain-rate effect should be considered when the structural response analysis under explosion is performed (FABIG, 1996).

2.3.5 International Standards Organization (ISO)

The International Standards Organization (ISO) specifies the international standard (ISO 19901-3), which suggests specific requirements for the design of topside structures against fires and explosions, as shown in Fig. 2.8 (ISO, 2014). ISO proposes worst-case explosion actions with a fully detailed structure for escape routes and safe areas. Probabilistic approaches used to assess explosion actions are suggested as follows (ISO, 2014):

- Worst-case gas clouds containing stoichiometric mixes, for which it is certain or at least highly probable that the resulting actions are conservative;
- A distribution of gas clouds with associated probabilities, for which the resulting actions and their probabilities can be presented as a series of curves that show a range of overpressures with associated probabilities.

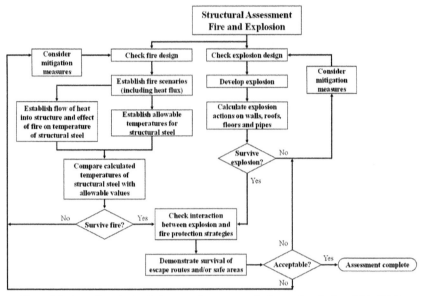

Figure 2.8. Procedure for detailed structural assessment for fires and explosions (ISO, 2014).

- In other areas, the design of explosion actions from explosion exceedance curves with acceptable risk levels are used (ISO, 2014). The risk levels proposed by ISO are as follows:
  - ° Strength level explosion (SLB): an explosion with a probability of exceedance of around $10^{-2}/y$;
  - ° Ductility level explosion (DLB): an explosion with a probability of exceedance of around $10^{-4}/y$.

This standard offers different action effect analysis methods depending on the probability of exceedance. For SLB, SDOF and/or linear finite-element analysis (FEA) are sufficient. Nonlinear FEA is also recommended for DLB.

2.3.6 Lloyd's Register (LR)

The LR guideline recommends the use of a probabilistic approach to determine the design explosion loads (LR, 2014). It suggests CFD simulation of gas dispersion and explosion scenarios that consider various parameters.

This guideline suggests the use of a design chart such as a pressure-impulse (P-I) curve with the design load for the structural response assessment. A simple explosion design load can be determined where the design accidental load (DAL) is defined by the risk acceptance (i.e., frequency cut-off) criterion (LR, 2014).

2.3.7 NORSOK (Standards Norway)

NORSOK Z003 adopts a probabilistic approach to the determination of explosion loads. Figure 2.9 shows the schematics of a procedure for calculating explosion risk (explosion loads) (NORSOK, 2010). It considers the most influential factors regarding gas release (rate and direction), wind (speed and direction), ignition source, gas cloud (size, location, and concentration), and frequency/probability of each parameter in the definition of explosion loads, including the steps of gas dispersion and explosion (NORSOK, 2010).

This standard provides three different applications of probabilistic accidental loads to the structural response analysis as follows (NORSOK, 2010):

- Use the design explosion load calculated by both the pressure and impulse exceedance curves based on acceptance criteria;
- Evaluate the structural response based on the load-frequency relation, such as with a P-I diagram;
- Directly apply the calculated explosion-time history from each explosion scenario to the structural response analysis.

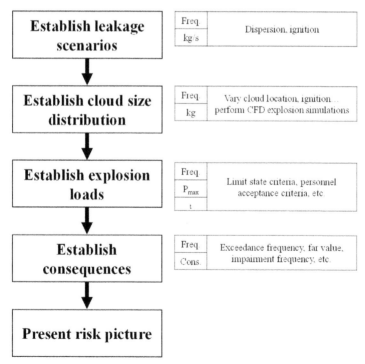

Figure 2.9. Schematics of the procedure for calculation of explosion risk (NORSOK, 2010).

In another standard, NORSOK-N004, proposed by NORSOK (2004), processes for structural analysis and design against explosion events are noted in detail. The methods for structural response assessment of partial structures suggested by NORSOK (2004) are similar to the analysis proposed by DNVGL (2010), as shown in Fig. 2.7 and Table 2.5. NORSOK (2010) recommends the proper use of SDOF, MDOF, linear FEM, and/or nonlinear FEM.

2.4 Recommendations for Advanced Engineering Practice

2.4.1 Recommended Methods

To define explosion loads, either deterministic or probabilistic models may be used.

- Deterministic models for load prediction
  - Empirical models
  - Phenomenological models
  - CFD (FLACS) models

- Probabilistic models of explosion loads
  - Model proposed by NORSOK (2001)
  - Probabilistic explosion load assessment with the help of quantitative risk analysis

The following methods are also available for analysis of nonlinear structural responses under blast pressure loads associated with hydrocarbon explosions:

- SDOF
  - Static and quasi-static analysis with the dynamic amplification factor
  - Dynamic analysis (linear and nonlinear systems)

- Conservation of momentum and energy method
  - Impulsive analysis

- Design chart
  - P-I diagram based on FEA and SDOF

- FEM
  - Nonlinear static analysis to assess static resistance and the static failure modes
  - Nonlinear dynamic analysis with dynamic effects to assess the time-dependent response

In a probabilistic approach to define explosion loads, the following parameters may be considered:

- Location of the leak source
- Direction of the gas jet
- Flow rate of the leak
- Performance of barrier element

The profile of the blast pressure loads may be simplified as either a simplified triangular (or rectangular) pressure pulse considering the defined blast loads or a detailed pressure-time history calculated by CFD simulation. Nonlinear FEMs will be used to analyze the nonlinear structural responses:

- Dynamic responses to pressure-time histories (detailed or simplified, triangular loads)
- Nonlinear aspects of the structural response

In determining the explosion design loads, not only overpressure-related loads, but also drag force and drag force impulse, must be considered. The design loads are then defined in terms of four kinds of explosion load (overpressure, overpressure impulse, drag force, and drag force impulse) exceedance curves, with a $10^{-4}/y$ level of the risk acceptance criterion.

Three approaches are relevant for the computing actions and action effects of offshore installations against explosions, although the use of CFD and nonlinear FEM is strongly recommended for refined computations.

- SDOF
  - ° Application of an idealized explosion load

- Nonlinear FEM
  - ° Application of an idealized design load
  - ° Application of an actual explosion load using an interface between CFD and FEM

- Design chart
  - ° P-I diagram based on FEA and SDOF

Figure 2.10 presents the accidental limit state design procedure for explosion actions and action effects suggested by ISSC (2015), in which three methods are considered for the explosion load assessment, and four kinds of structural analysis approach with details for the structural design of topside structures under explosions are introduced depending on the design stages, as indicated in Table 2.6. Figure 2.11 presents an advanced procedure for the quantitative explosion risk assessment and management proposed by Paik et al. (2014).

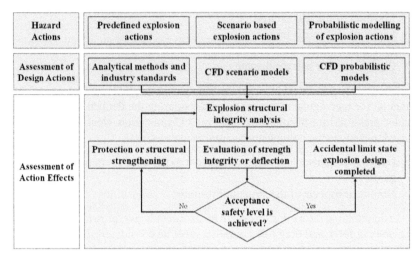

Figure 2.10. Accidental limit state design procedure for explosion actions and action effects (ISSC, 2015).

2.4.2 Comparisons between Recommended Practices

1. Definition of Explosion Loads

ABS, API, and DNVGL suggest a deterministic approach to the definition of design explosion loads (API, 2006; ABS, 2013; DNVGL, 2014), whereas LR and NORSOK propose a probabilistic method (NORSOK, 2010; LR, 2014). In contrast, FABIG (1996) recommends the use of explosion loads from predefined best- and worst-case explosion simulations. ISSC (2015) has issued good guidelines, including all possible and practicable approaches for the definition of hydrocarbon explosion actions.

2. Structural Assessment

For the structural response analysis, ABS and API use a stepwise analysis with screening, linear analysis, and nonlinear FEA (API, 2006; ABS, 2013). Others apply the linear, nonlinear dynamic finite analysis or simple calculation methods based on the SDOF or MDOF analogies with idealized design blast loads.

3. Comparison of Applied Methods

Table 2.7 summarizes a comparison of methods for the definition of explosion loads, application to structural assessment, and structural analysis methods. Most methods adopt an idealized explosion load obtained by a deterministic, predefined, or probabilistic approach. Simplified structural analysis methods with an idealized structural model are often recommended. However, the idealized approaches to

Table 2.6. Choice of design approach for topside structures under gas explosion loadings (ISSC, 2015).

Design stage	Analysis method	Dynamic behavior	Nonlinear behavior	Acceptance criteria	Structural model
Basic	SDOF method	- Intrinsic capability (or by DAF* from response charts) - Enhanced yield stress (strain rate effect × 1.2)	- Intrinsic capability - Enhanced yield stress (full plastic section × 1.12) - Strain hardening (ultimate tensile strength/1.25)	- Ductility ratio	- Member by member - Plate only or Stiffened plate idealized as beam
Basic	Linear static FE analysis	- Intrinsic incapability and considered by DAF - Enhanced yield stress (strain rate effect × 1.2)	- Intrinsic incapability and partially considered by modified code check - Enhanced yield stress (full plastic section, × 1.12) - Strain hardening (ultimate tensile strength/1.25)	- Yield strength with modified code check (utilization factor × 1.5 for ASD**)	- Framed - Plate only - Stiffened plate (idealized stiffeners)
Detail	Nonlinear static FE analysis	- Intrinsic incapability and considered by DAF - Enhanced yield stress (strain rate effect, × 1.2)	- Intrinsic capability	- Strain limit (or ductility ratio)	- Framed - Plate only - Stiffened plate (idealized stiffeners)
Detail	Dynamic nonlinear FE analysis	- Intrinsic capability	- Intrinsic capability	- Strain limit (or ductility ratio)	- Framed - Plate only - Stiffened plate (idealized stiffeners)
Detail	Dynamic nonlinear FE analysis	- Intrinsic capability	- Intrinsic capability	- Strain limit (or ductility ratio)	- All structures

* DAF: Dynamic Load Amplification Factor

** ASD: Allowable Stress Design

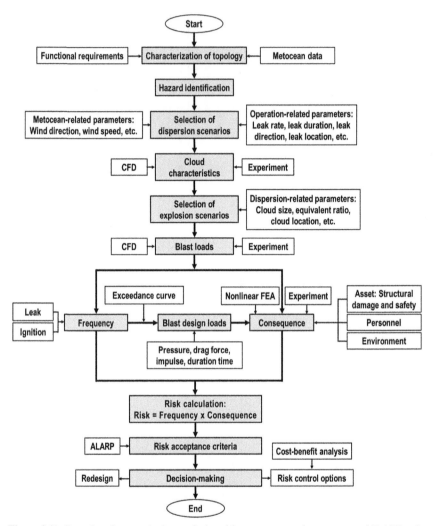

Figure 2.11. Procedure for quantitative explosion risk assessment and management (ALARP = As Low As Reasonably Practicable risk) (Paik et al., 2014).

explosion loads and/or structural analysis yield incorrect results compared with realistic models with respect to the actual explosion loads and the entire structural model.

2.5 Applied Example

In this section, an applied example of structural response analysis subjected to the explosion is introduced.

Table 2.7. Comparison of methods for explosion load definition, structural analysis, and application of explosion load to structural analysis.

	Approach to definition of explosion load	Application of explosion loads to structural assessment	Method for structural consequence analysis
ABS (2013)	Deterministic	Idealized explosion load	Screening analysis → linear FEM → NLFEM
API (2006)	Deterministic/probabilistic	Idealized explosion load	Screening analysis → linear FEM → NLFEM
DNVGL (2010; 2014)	Deterministic	Idealized explosion load	SDOF or MDOF
FABIG (1996)	Predefined (lower and upper cases)	Idealized explosion load	SDOF or FEM
ISO (2014)	Probabilistic	Idealized explosion load of worst-case/design load	SDOF or FEM
LR (2014)	Probabilistic	Idealized explosion load	P-I chart
NORSOK (2004; 2010)	Probabilistic	Idealized explosion load/actual explosion load	SDOF, MDOF or FEM
Czujko (2001)	Deterministic/probabilistic	Idealized explosion load	SDOF, analytical method, design chart or FEM
Vinnem (2007)	Probabilistic	Idealized explosion load	FEM
Paik and Czujko (2010); Paik (2011)	Probabilistic	Idealized design explosion load	SDOF, NLFEM or design chart
Czujko and Paik (2015)	Probabilistic	Actual load	FEM
ISSC (2015)	Predefined, scenario based/probabilistic	Idealized explosion load/actual explosion load	SDOF, static or dynamic FEM

Note: FEM includes linear and nonlinear FEM in this table.

2.5.1 Assessment of Explosion Loads

When the structural analysis with the quantitative explosion risk assessment approach is applied as shown in Fig. 2.11, the explosion loads by the probabilistic method should be defined. In the assessment of explosion loads, gas explosion simulations are performed after gas dispersion simulations. Sometimes, however, the dispersion simulation can be skipped when the gas explosion scenarios are previously defined.

1. Gas Dispersion Simulation

Three-dimensional gas dispersion simulations are needed to investigate characteristics of gas clouds, which are used for gas explosion simulations. Gas dispersion simulations with probabilistic dispersion scenarios can identify the position and concentration of gas clouds.

Figure 2.12 illustrates the relationship between the maximum flammable (actual) and equivalent gas clouds based on the gas dispersion simulations. The equivalent gas cloud has the perfect mixture of fuel and oxygen: no fuel or oxygen remain after combustion.

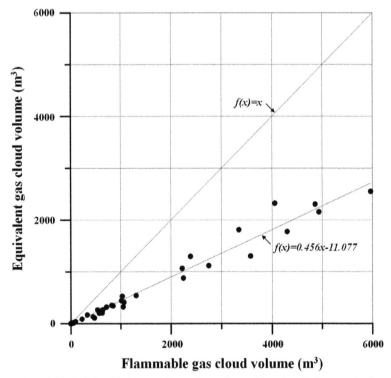

Figure 2.12. Relationship between the maximum flammable and equivalent gas clouds.

2. Gas Explosion Simulation

With the results of the gas dispersion scenarios, the explosion scenarios are defined for gas explosion simulations. Previously described explosion scenarios can also be used.

Using the gas explosion simulations, the characteristics of explosion load, such as overpressure, impulse, drag force, and duration time, are investigated. Figure 2.13 shows an example of gas explosion simulation results that show the effect of gas cloud volume on maximum overpressure. The explosion load by gas explosion simulations with or without dispersion simulations is applied to the structure directly (actual explosion load) or indirectly (idealized explosion load).

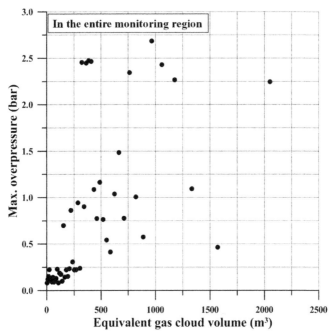

Figure 2.13. Effect of equivalent gas cloud volume on maximum overpressure.

2.5.2 Applying Explosion Loads to the Nonlinear Structural Response Analysis

There are two methods to apply the explosion load to the structural analysis, an idealized or an actual explosion load.

1. Idealized Explosion Loads

An idealized explosion load can be defined with several parameters, including peak positive pressure, peak positive pressure duration time (rising and decaying times), peak negative pressure, and negative pressure duration time, as shown in Fig. 2.5.

Deterministic approach

In the deterministic approach, it is not necessary to perform both gas dispersion and explosion simulations because the explosion load is defined by rules and recommended practices, as described in Section 2.3.

Probabilistic approach

In quantitative explosion risk assessment, which uses a probabilistic approach, the idealized design explosion load is defined with the characteristics of the explosion loads of many explosion scenarios, as introduced in Section 2.4, and their frequency.

With the consequence (explosion load) and frequency, the exceedance curve for the definition of the design load is generated. The design load can then be defined with the As Low As Reasonably Practicable (ALARP) risk level. Figure 2.14 shows an example of the definition of maximum panel pressure, which is a parameter of design load in the explosion exceedance curve with an ARALP level of 10^{-4}/y, which is generally adopted for explosion risk assessment. The curve can be individually generated for factors such as panel pressure, overpressure, drag force, and duration time to define the idealized explosion load.

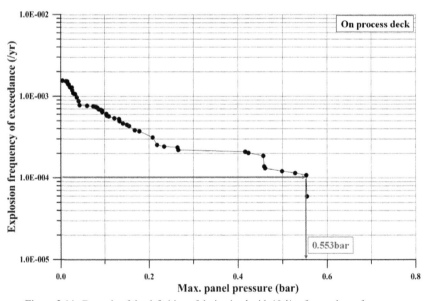

Figure 2.14. Example of the definition of design load with 10^{-4}/y of exceedance frequency.

2. Actual Explosion Loads

When actual explosion loads are applied, an interface program is needed to transfer explosion loads from the CFD simulation to nonlinear FE analysis.

FLACS2DYNA is one of the interface programs. Figure 2.15 presents the concept of the FLACS2DYNA interface program. It deals with both the monitoring point and the control volume in FLACS. The pressure loads on the shell elements are mapped from the nearest monitoring points or the centers of the control volumes by the interface (Kim et al., 2012).

Figure 2.16 shows a mapping view of the explosion loads between the FE model and monitoring points or control volume. Each actual explosion profile at each location can be transferred to the structure by the FLACS2DYNA interface program.

Figure 2.17 illustrates examples of the distribution of the actual and idealized overpressure time histories on specific scenario. The idealized load is uniformly distributed on an area. The figure shows large differences between the actual and idealized explosion loads.

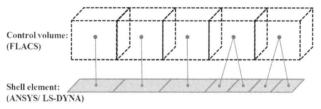

Figure 2.15. Concept of the FLACS2DYNA interface program (Kim et al., 2012).

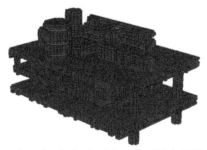

Figure 2.16. Mapping view of explosion loads between CVs in CFD and elements in FEM.

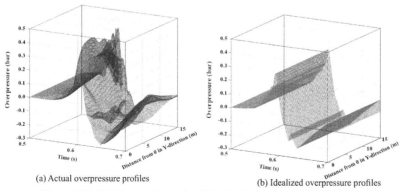

(a) Actual overpressure profiles   (b) Idealized overpressure profiles

Figure 2.17. Distribution of actual and idealized overpressure-time histories.

2.5.3 Nonlinear Structural Response Analysis

After defining the explosion load by the deterministic or probabilistic approach, the nonlinear structural response analysis under the actual or idealized explosion load is performed.

Nonlinear FEA is generally used for structural response analysis, and factors (geometry modeling, element type, strain rate effect, boundary condition, etc.), that affect the structural responses under impact loads should be considered to obtain more accurate results in the nonlinear FE analysis. Figure 2.18 shows a generated FEM with shell elements.

Figures 2.19 through 2.22 show examples of structural response by nonlinear FE analysis under explosion loads to compare the responses by the actual and idealized explosion loads. Figures 2.19 and 2.20 present the deflection distributions of the blast wall and decks, respectively, and Fig. 2.21 illustrates the total displacement. These figures show that the actual load application with nonuniform distributions causes the torsional moment, whereas the application of the idealized uniformly distributed loads could not capture this behavior.

The structure subjected to the impact load is usually assessed by a plastic strain. Thus, the plastic strain should also be investigated as shown in Fig. 2.22.

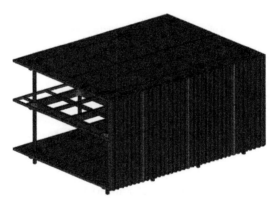

Figure 2.18. Example of the FEM for the topside structure with blast wall on FPSO.

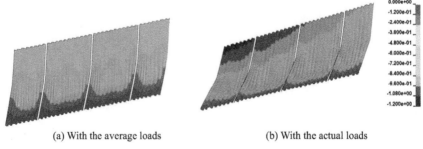

(a) With the average loads   (b) With the actual loads

Figure 2.19. Deflection idealized of blast walls at 0.68 s plotted by an amplification factor of 5 (in m).

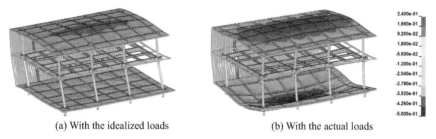

(a) With the idealized loads   (b) With the actual loads

Figure 2.20. Deflection distribution of decks at 0.68 s plotted by an amplification factor of 5 (in m).

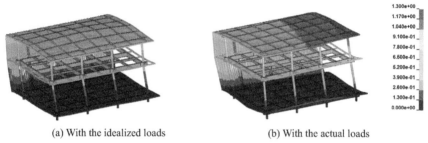

(a) With the idealized loads   (b) With the actual loads

Figure 2.21. Total displacement distribution at 0.68 s plotted by an amplification factor of 5 (in m).

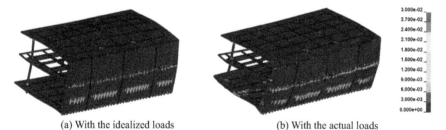

(a) With the idealized loads   (b) With the actual loads

Figure 2.22. Plastic strain distribution at 1.0 s, with deformations plotted by an amplification factor of 5.

2.6 Concluding Remarks and Further Studies

Risk is defined as either the product or a composite of (i) the probability or likelihood of occurrence of any accident or limit state that leads to severe consequences such as human injuries, environmental damage, and loss of property or financial expenditure, and (ii) the resulting consequences (Paik and Thayamballi, 2007). The resulting consequences are associated with nonlinear structural responses, so the analysis of nonlinear structural responses is a key task within the framework of quantitative risk assessment and management. This chapter described procedures for the nonlinear structural response analysis due to explosions. The definition of explosion loads is also described because it is required for analysis of structural responses.

In the conventional design of structures for explosion loads, it is usually assumed that the explosion loads are distributed uniformly among the individual structural members. However, actual explosion loads are not uniformly distributed, as illustrated in Fig. 2.23. The structural responses as calculated with uniform or actual explosion pressure loads can differ greatly. The assumption of uniform loads can result in overestimation of structural damage in some cases and underestimation in others. Therefore, it is important to use the actual load distributions for accurate response analyses of structures.

A variety of influencing parameters are involved in the nonlinear structural responses associated with explosions. Some important factors include the blast load profile, strain rate, and temperature. The blast load profiles of explosions are the main factors to be considered in a structural integrity analysis. In general, idealized pressure loads are uniformly distributed. Four kinds of general idealization are available for blast loading shapes to different pulse shapes, as illustrated in Fig. 2.24: the rise time until the peak pressure is reached, peak pressure, the decaying shape after the peak pressure is reached, and duration time. Among the options for analysis of these parameters, the symmetric triangular load approach is often adopted for dynamic structural analysis in considering hydrocarbon explosion accidents. Other approaches are used to analyze solid explosions, such as those caused by TNT associated with detonation.

The uncertainties associated with load profile idealization can significantly affect the nonlinear structural responses. In this regard, the use of actual load profiles is recommended in addition to actual loads with non-uniformly distributed overpressure that are directly obtained from CFD simulations without any modifications.

Material properties are also major factors in the structural analysis of dynamic events. Structural analyses that make use of nonlinear FEM should consider the dynamic properties of the materials used. The material properties such as yield stress and fracture strain should be considered along with the dynamic effects, which are called strain rate effects. The duration time of explosions is extremely

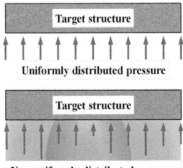

Figure 2.23. Uniform (upper) and non-uniform (lower) distribution of pressure loads in an explosion event.

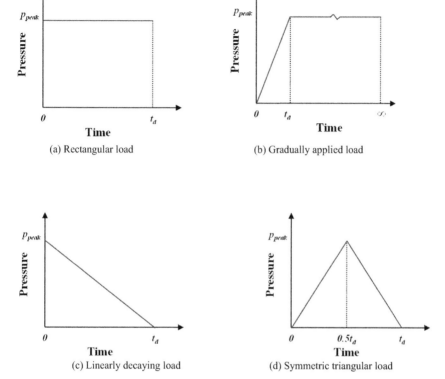

Figure 2.24. Different idealizations of blast loading shapes for structural analyses under explosion loads.

short—several milliseconds—thus, the effect of temperature is usually neglected for nonlinear structural response analysis because much more time is required to transfer the gas cloud temperature to the steel temperature due to heat and flame.

Abbreviations and Terminologies used

Actual gas cloud size	Flammable gas cloud size obtained from CFD simulation or experiment
Actual loads	Overpressures or drag forces obtained from CFD simulation or experiment
ALARP	As low as reasonably practicable risk
ANSYS/LS-DYNA	A computer code for the nonlinear structural response analysis
Average loads	Average values of actual overpressures or drag forces
CFD	Computational fluid dynamics
Control volume	Mathematical abstraction employed in the process of creating mathematical models in CFD simulation

DAF	Dynamic load amplification factor
ER	Equivalent gas concentration ratio
Equivalent gas cloud size	Gas cloud size equivalent to the stoichiometric condition
FEM	Finite element method
FLACS	A computer code for simulating gas dispersion and explosion
FLACS2DYNA	A computer code for transferring the results of FLACS simulations to the input data of ANSYS/LS-DYNA
HSE&E	Health, safety, environment & ergonomics
Monitoring panel or point	A predefined area or point of interest where the computational results are monitored
Porosity	Volumetric measure of void spaces in the range of 0 to 1.0 in which 1 indicates an empty space and 0 indicates a solid condition
SDOF method	Single degree of freedom method using an analytical approach

Nomenclature

A	Effective area under the pressure load
b_f	Breadth of flange
C	Cowper-Symonds coefficient
$C_{1\varepsilon}, C_{2\varepsilon}$	Constant in the $k - \varepsilon$ equation
C_{A1-12}	Monitoring points at elevation level A
C_{B1-12}	Monitoring points at elevation level B
C_d	Drag coefficient
c	Speed of sound
c_p	Specific heat capacity at constant pressure
c_v	Specific heat capacity at constant volume
D	Diffusion coefficient
E	Component of rat of deformation
F	Applied blast force at instant of maximum dynamic reaction
F/O	Fuel-oxidant ratio
F_o	Obstruction friction force
F_w	Wall friction force
g	Gravitational acceleration
H	Hydrogen
H_s	Sensible heat flux from the surface

h	Specific enthalpy
h_w	Height of web
k	Turbulent kinetic energy
L	Monin-Obukhov length scale
l_{LT}	Mixing length in the β-model
M	Molecular weight of a mixture
m	Mass
$\dot{m}$	Mass flow
m_{fuel}	Mass of fuel (gas)
m_{oxygen}	Mass of oxygen in the fuel (gas)
n	Number of density
O	Oxygen
P	Gauge pressure
P_B	Monitoring panel on blast wall
P_M	Monitoring panel on mezzanine deck
P_P	Monitoring panel on process deck
P_U	Monitoring panel on upper deck
p_0	Ambient pressure
p_{air}	Absolute pressure
p_{peak}	Peak (maximum) overpressure
Q	Heat
$\dot{Q}$	Heat rate
q	Cowper-Symonds coefficient
R	Gas constant of a mixture
R_B	Maximum resistance to blast loading
R_{flame}	Flame radius
R_{fuel}	Fuel reaction rate
R_u	Universal gas constant
r_{inner}	Inner radius of column
r_{outer}	Outer radius of column
S_L	Laminar burning velocity
T	Natural period
T_{air}	Absolute temperature
t_d	Specific time

t_f	Thickness of flange
t_w	Thickness of web
u	Velocity of the fluid
u^*	Friction velocity
u_i	Mean velocity in the ith direction
V	Volume
W	Dimensionless reaction factor
X	Mole fraction
Y	Mass fraction
Y_0	Initial mass fraction
y_{el}	Deflection at elastic limit
y_m	Maximum total deflection
β	Transformation factor in the β-model
β_i	Area porosity in the ith direction
β_v	Volume porosity
γ	Isentropic ratio
γ_P	Pressure exponent for the laminar burning velocity
ε	Dissipation of turbulent kinetic energy
$\dot{\varepsilon}$	Strain rate
ε_f	Critical fracture strain under quasi-static load
ε_{fd}	Critical fracture strain under dynamic load
κ	Von Karman constant
μ	Dynamic viscosity
μ_{eff}	Effective viscosity, $\mu_{eff} = \mu + \mu_t$
μ_t	Dynamic turbulent viscosity
ρ	Density
ρ_0	Initial density
σ	Prandtl-Schmidt number
σ_{ij}	Stress tensor
σ_Y	Yield stress under quasi-static load
σ_{Yd}	Yield stress under dynamic load
Φ	Equivalence ratio
χ	Fuel-dependent constant

References

ABS. 2013. Accidental Load Analysis and Design For Offshore Structures, American Bureau of Shipping, TX, USA, Design.

API. 2006. Design of Offshore Facilities Against Fire and Blast Loading, API-RP2FB, American Petroleum Institute, WA, USA.

Czujko, J. 2001. Design of Offshore Facilities to Resist Gas Explosion Hazard: Engineering Handbook, CorrOcean ASA, Oslo, Norway.

Czujko, J. and Paik, J.K. 2015. A new method for accidental limit states design of thin-walled structures subjected to hydrocarbon explosion loads. Ships and Offshore Structures 10(5): 460–469.

DNVGL. 2010. Design Against Accidental Loads, DNV-RP-C204, Det Norske Veritas, Oslo, Norway.

DNVGL. 2014. Safety Principles And Arrangements, DNV-OS-A101, Det Norske Veritas, Oslo, Norway.

FABIG. 1996. Explosion resistant design of offshore structures, Technical Note 4, Fire and Blast Information Group, Berkshire, UK.

FLACS. 2016. User's manual for FLame ACceleration Simulator (FLACS) version 10.1, Gexcon AS, Bergen, Norway.

ISO. 2014. Petroleum and natural gas industries—specific retirements for offshore structures—Part 3: topside structure, ISO 19901-3. International Standards Organization, Geneva, Switzerland.

ISSC. 2015. Committee V.1: Guidelines on the use of accidental limit states for the design of offshore structures, International Ship and Offshore Structures Congress, Rostock, Germany.

Kim, S.J., Sohn, J.M., Kim, C.K., Paik, J.K., Katsaounis, G.M. and Samuelides, M. 2012. Computational modelling of interaction between CFD and FEA simulations under gas explosion loads, Proceedings of the International Conference on Ship and Offshore Technology (ICSOT): Developments in Fixed and Floating Offshore Structures, May 23–24, Busan, Korea.

LR. 2014. Guideline for the calculation of probabilistic explosion loads, Report No. 104520/R1, Lloyd's Register, Southampton, UK.

Monin, A.S. and Obukhov, A.M. 1954. Basic laws of turbulent mixing in the surface layer of the atmosphere. Tr. Akad. Nauk SSSR Geofiz. 24: 163–187.

NORSOK. 2001. Risk and Emergency Preparedness Analysis, NORSOK-Z013, Norway Standard, Lysaker, Norway.

NORSOK. 2004. Design of Steel Structures, NORSOK-N004, Norway Standard, Lysaker, Norway.

NORSOK. 2010. Risk and emergency preparedness assessment, NORSOK-Z003. Norway Standard, Lysaker, Norway.

Paik, J.K. 2011. Explosion and fire engineering on FPSOs (Phase III): nonlinear structural consequence analysis, Report No. EFEF-04, The Korea Ship and Offshore Research Institute, Pusan National University, Busan, Korea.

Paik, J.K. 2015. Making the case for adding variety to Goal-Based Standards, The Naval Architect, The Royal Institution of Naval Architects, UK, January 22–24.

Paik, J.K. 2018. Ultimate Limit State Analysis and Design of Plated Structures, 2nd Edition, John Wiley & Sons, Chichester, UK.

Paik, J.K. and Czujko, J. 2010. Explosion and fire engineering on FPSOs (Phase II): definition of design explosion and fire loads, Report No. EFEF-03, The Korea Ship and Offshore Research Institute, Pusan National University, Busan, Korea.

Paik, J.K. and Thayamballi, A.K. 2007. Ship-Shaped Offshore Installations: Design, Building and Operation, Cambridge University Press, Cambridge, UK.

Paik, J.K., Czujko, J., Kim, S.J., Lee, J.C., Seo, J.K., Kim, B.J. and Ha, Y.C. 2014. A new procedure for the nonlinear structural response analysis of offshore installations in explosions. Transactions of The Society of Naval Architects and Marine Engineers 122: 1–33.

Vinnem, J.E. 2007. Offshore Risk Assessment—Principles, Modelling and Application of QRA Studies, Springer, Stavanger, Norway.

3

Stochastic Dynamic Analysis of Marine Structures

Bernt J. Leira* and *Wei Chai*

3.1 Introduction

Stochastic models have been successfully developed and are being applied for representation of a number of a different environmental processes also including the corresponding load processes. Particular examples of such models are those which have been developed in relation to wave, wind and earthquake processes. In the following, focus is on the wave process together with the associated loading processes.

The main features of associated structural response analysis methods are basically the same for both deterministic and stochastic load models. However, practical implementation of the relevant analysis procedures for the two categories may differ somewhat. This applies in particular to the evaluation of probabilistic response properties in the case of stochastic loading for cases where the structural behavior is of a non-linear character.

The two main categories of methods for dynamic response analysis are classified as time domain versus frequency domain approaches. The former is particularly relevant for the purpose of studying non-linear effects associated with both loads and structural behavior. Time domain analysis in connection with stochastic loading is generally based on generating a number of sample functions for the load processes and subsequently computing the associated sample time histories for the response quantities of interest.

Department of Marine Structures, Norwegian University of Science and Technology, Otto Nielsens veg 10, 7491-Trondheim, Norway.
* Corresponding author

For some categories of marine structures (e.g., floating wind turbines or vessels with DP and other propulsion systems), the effects of actuators and control forces with respect to the dynamic response need to be taken into account. For such cases, the time domain approach is much applied. Further details of the corresponding terms in the dynamic equilibrium equation associated with floating structures are given, e.g., in Fossen (2002).

Structural response analysis in the time-domain is based on a stepwise integration of the dynamic equilibrium equation, see, e.g., Clough and Penzien (1975), in addition to Newland (1993). In the case of non-linear structural behaviour, the incremental form of this equilibrium equation is quite commonly applied. Classical references in relation to assessment of numerical stability related to step-wise time integration in structural dynamic analyses are, e.g., Newmark (1959), Belytscho and Shoeberle (1975), and Hughes (1976; 1977). More recent studies are summarized, e.g., in Bathe (1996). Elaboration of suitable time integration methods for systems with large displacements and constraint conditions can be found, e.g., in Krenk (2008).

The frequency domain approach is most relevant in connection with linear (or linearized) models of load and structural behavior, and it is generally superior in terms of computation time as well as manual processing time. Further details of frequency domain analysis of multi-degree-of-freedom are found, e.g., in Newland (1993), and Jia (2014).

For evaluation of the so-called quasistatic response, dynamic effects are neglected. This implies that the response is obtained by inverting the stiffness matrix which represents a numerical model of the structure. This type of analysis can, e.g., be relevant if the frequencies associated with the loading are much lower than the natural frequencies of the structure.

Before entering a more detailed description of procedures for stochastic analysis, application of deterministic analysis methods also deserve some further comments. The benefits of deterministic methods are that they are much easier to apply due to a simplified modelling of the sea surface as a monochromatic oscillation with a given amplitude and a given period (or alternatively as a Stokes wave with a given amplitude and a given period). The weakness of such methods is, e.g., that for dynamic structures, the choice of period may have a very strong influence on the computed response characteristics. Still, it is found that deterministic analysis is still quite widespread in use. This applies in particular to studies which are performed during early design stages, partly due to the superior speed of calculation and also limited amount of post-processing of the computed response.

Historically, estimation of extreme dynamic response for offshore structures has accordingly in many cases been based on regular wave analysis. The characteristics of such "representative" waves (i.e., wave height and wave period) are then determined in some way or another by consideration of the full probabilistic description. However, fully stochastic dynamic response analyses are increasingly being utilized and this applies in particular during the verification phase of the projects. The so-called contour-line method, which is described in more detail in the following chapters, has also been introduced in the design standards and codes, see, e.g., N-003 Actions and action effects (Edition 2, September 2007). This approach is based on a stochastic description of the wave environment and the dynamic response properties. Regarding design criteria in general, high-level prescriptions are typically found in NORSOK (1997; 2007; 2012; 2013; 2015) and ISO (2004; 2006; 2006a; 2006b; 2009; 2010; 2012; 2013; 2013a; 2014; 2014a, 2015; 2016; 2016a; 2016b; 2017). More specific design details are frequently found in the rules, standards and guidelines of the Classification Societies, see, e.g., DNVGL (1992; 1992a; 2010; 2015; 2015a; 2015b; 2015c; 2015d; 2015e; 2016; 2016a; 2016b; 2017; 2017a; 2017b; 2017c), as well as the API set of documents, API (2007; 2009; 2011; 2011a; 2014; 2014a; 2015; 2015a; 2015b; 2015c).

The concept of stochastic dynamic response is clearly based on an extension of classical dynamic response analysis where the dynamic load is represented by a deterministic time history. By means of the concept of "samples" of a stochastic process, the connection between the two approaches is made quite transparent. For each sample function of the load process, a classical dynamic response analysis can be performed by well-established numerical schemes, see, e.g., Newmark (1959), Belytschko and Schoeberle (1975), Hughes (1976; 1977), Bathe (1996), Krenk (2008) and Jia (2014) for an overview of these procedures. By repeating the response analysis for a number of sample functions, the statistical properties of the corresponding response processes can be established.

For analysis and design of marine structures, a number of different mechanical limit states must furthermore be considered. These limit states mainly belong to one of the following categories of criteria: Serviceability limit state (SLS), Ultimate limit state (ULS), Fatigue limit state (FLS) and Accidental limit state (ALS). In the present text, focus is on the Ultimate and Fatigue limit states (i.e., ULS and FLS).

For these limit states, efficient numerical methods which at the same are sufficiently accurate are in demand for calculation of extreme load and response, as well as for estimation of fatigue damage. The so-called contour-line methods represent a step in this direction, and this approach is described in more detail as part of the following chapter.

There are typically different rules for bottom-fixed versus floating (including compliant) systems as exemplified by the NORSOK and DNV set of rules which are applied for the Norwegian Continental Shelf, see, e.g., NORSOK (1997; 2007; 2012; 2013; 2015) and DNVGL (1992; 1992a; 2010; 2015; 2015a; 2015b; 2015c; 2015d; 2015e; 2016; 2016a; 2016b; 2017; 2017a; 2017b; 2017c). Similar differences between bottom-fixed and floating systems are also found in the ISO and API set of rules, see, e.g., ISO (2004; 2006; 2006a; 2006b; 2009; 2010; 2012; 2013; 2013a; 2014; 2014a; 2015; 2016; 2016a; 2016b; 2017) and API (2007; 2009; 2011; 2011a; 2014; 2014a; 2015; 2015a; 2015b; 2015c).

3.2 Fundamentals—Theory

3.2.1 General

As discussed above, stochastic models of the wave environment and the corresponding hydrodynamic loading will form the basis for the ensuing static and dynamic response analysis. The ultimate goal of all methods for analysis of structural response is to establish statistical models that can be applied for design purposes, also allowing that computation of proper measures for the structural reliability level can be made. As a first step, short- and long-term statistical models of the wave environment and the associated structural response need to be established, which are elaborated in the next paragraphs.

3.2.2 Stochastic Models of the Wave Environment

Due to the non-stationary nature of most environmental processes, the corresponding modeling typically consists of two main building blocks which represent the so-called "short-term" and "long-term" behavior. The short-term modeling is associated with "stationary" conditions which correspond to a given set of values of the characteristic environmental parameters.

Models related to the "long-term" representation of environmental processes will typically involve joint statistical models of several characteristic parameters. As an example, the wave climate is typically represented by the significant wave height and a characteristic period (e.g., peak period or zero-crossing period). The wind climate is similarly characterized by the mean wind and turbulence intensity. (The expression "long-term" can in the present context also refer to a sequence of environmental conditions within a limited time window, and can hence reflect, e.g., seasonal environmental characteristics which are relevant for marine operations.)

Knowledge of the joint statistical properties of two or several simultaneous environmental parameters will accordingly play an important role for many activities at sea. This pertains both to the open ocean and coastal areas. In particular, the bivariate probability distribution of significant wave height and characteristic period is highly relevant for a number of applications, see, e.g., Bitner-Gregersen and Guedes Soares (1997), Bitner-Gregersen and Guedes Soares (2007).

The significant wave height characterizes the intensity of the sea states, while the mean period or the peak period is relevant for assessing the possibility of exciting the natural periods of a given structure. Hence, the joint distribution of significant wave height and characteristic period is required in order to address several issues which are of key importance for design of marine structures. Furthermore, it represents a key issue in connection with planning of marine operations.

An example of a joint probability density function (pdf) of significant wave height (H_s) and peak period (T_p) based on a data set given in Bitner-Gregersen and Guedes Soares (1997) is shown in Fig. 3.1. The correlation between the two parameters is clearly reflected in the shape of the pdf and its associated level curves.

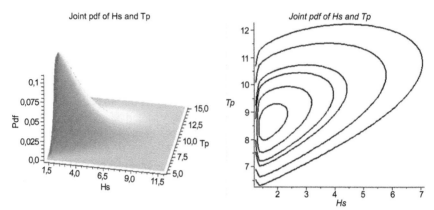

Figure 3.1. Example of joint pdf of wave height and peak period based on data given in Bitner-Gregersen and Guedes Soares (1997) (Pdf contour levels are at [0.01, 0.02, 0.04, 0.05, 0.07, 0.095]).

A number of other statistical models have also been applied in order to model the joint behavior of these variables. Ochi (1978) has adopted a bivariate Lognormal distribution, which implies an exponential transformation of the bivariate normal distribution. This model has the great advantage of simplicity but is not always quite accurate in the upper "low probability" range.

To ensure a good fit to the data, Haver (1985) has chosen separate models for the significant wave height (H_s) and the peak period (T_p). The combination of a Weibull and a Lognormal distribution was applied in order to model the marginal distribution of significant wave height. The conditional distribution of the peak period was fitted by a Lognormal distribution. A regression equation was proposed for the parameters of the conditional distributions of the peak period as a function of significant wave height in order to be able to extrapolate the parameters at the low probability end.

Mathiesen and Bitner-Gregersen (1990) applied a three-parameter Weibull distribution to model the marginal distribution of significant wave height. This was combined with the conditional Lognormal distribution for wave periods. They compared this model with the bivariate Lognormal and with a bivariate Weibull distribution. It was concluded that the approach with the conditional distribution of wave period provided the best fit to data.

Athanassoulis et al. (1994) have proposed an approach that combines some degree of flexibility with a certain simplicity and parsimony in the number of required parameters. They use the Plackett bivariate model to fit bivariate distribution functions to the data. This model, even though not being completely general, allows the specification of any two marginal distributions and accounts for the dependence structure by means of a parameter related to the correlation between the variables.

Ferreira and Guedes Soares (2001) decided to use the bivariate normal distribution to model the data after transformation and they have chosen the

Box-Cox transformation to make the "background variables" close to normally distributed. This transformation has been applied in several univariate and bivariate autoregressive models of wave time series. Prince-Wright (1995) applied an extended version of this transformation to the data. In Soares and Guedes Soares (2007) it was also concluded that the Box-Cox model may be a good choice for many applications, as it represents a compromise between accuracy of the fit and simplicity with respect to the number of parameters involved.

The models discussed above apply to the joint occurrence of two parameters such as the significant wave height and the characteristic period. However, the literature contains less information when additional parameters are involved. Most of these models use a conditional approach that results in a rapid increase with respect to the amount of data required as a function of the problem dimension.

Other studies on joint modeling have also been performed, see, e.g., Nerzic and Prevosto (2000) in connection with the Plackett model and by Prince-Wright (1995) in relation to the Johnson transformations (see, e.g., Johnson and Kotz (1972)). In Fouques et al. (2004) a joint model for several metocean parameters is presented which includes the significant wave height, the mean wave period, the mean wind velocity and sea state persistence parameters. Seasonal models for the different parameters are also introduced. Multivariate extreme-value models for significant wave height and peak period have been considered by Jonathan et al. (2010).

3.2.3 Linear versus Nonlinear Structural Response Analysis

3.2.3.1 General

As a general observation, there are main differences between analysis methods for linear versus non-linear physical mechanism. This applies not the least to analysis of marine structures for which there are a number of sources of non-linearity. These can broadly classified into three different types:

  i) Hydrodynamic non-linearities
 ii) Non-linearities associated with fluid-structure interaction
iii) Purely structural non-linearities

The present section contains a broad outline of these three categories. A brief summary of linear load calculation for marine structures is first given.

Actions on fixed large volume structures are typically calculated on the basis of diffraction theory. The wave potential is expressed as the linear superposition of the incident and the scattered wave systems. For simple geometrical shapes analytical solutions of the linear diffraction problem may be used. In the case of structures with a large extension in one direction as compared to the others, strip theory may be applied which is based on simplified two-dimensional calculations, see, e.g., Faltinsen (1990).

For general structures consisting of several large volume components, boundary-element (i.e., panel methods) or finite fluid elements should be used. It is important that the results from boundary element methods should be carefully checked with respect to convergence of the solution for increasingly refined discretization of the surface geometry.

For irregular waves, a linear superposition of wave forces for different frequencies is generally performed. For each frequency, the load amplitude is then proportional to the wave amplitude which is determined from the relevant spectral density for the sea state.

Wave actions on structures composed of large volume parts and slender members may be computed by a combination of wave diffraction theory and Morison's equation (see the next section). Parts of the structure may be modelled by a combination of boundary elements to represent the potential hydrodynamic actions and beams to represent the viscous drag actions. The modifications of velocities and accelerations as well as surface elevation due to the large-volume parts should be taken into account when using Morison's equation.

For large-volume structures with significant motion (e.g., for ships and floating structures in general) the radiation wave system generated by the moving structure should be added to the incident and scattered wave systems. The radiation wave system is determined by diffraction analysis and gives rise to added mass and wave damping actions.

3.2.3.2 Hydrodynamic Non-Linearities

Wave diffraction solutions do not include viscous actions. Viscous effects may be important for body members that are relatively slender (i.e., relative to the wave length) or have sharp edges. Viscous actions on the hull, possible mooring system or thrusters and risers need to be considered.

For slender structures, the wave force on an incremental length is generally computed by means of the so-called Morison's equation, see, e.g., Faltinsen (1990). This force consists of two terms, which generally are referred to as the drag term and the inertia term. The drag term is a quadratic function of the water particle velocity, or in many cases it is expressed as a quadratic function of the relative velocity between the water and the structural element. The inertia term is proportional to the acceleration of the fluid, and is frequently also expressed in terms of the relative acceleration between the fluid and the structure. The force coefficients which are applied for the two different terms are referred to as the drag coefficient, C_D, and the inertia coefficient, C_M, respectively. The values of these coefficients will generally depend strongly on the flow regime under consideration.

The external action vector of the complete structure is subsequently obtained by integrating the force per incremental length over each of the slender structural members. When using the relative velocity formulation for structures with large motion, the force should be calculated for zero as well as maximum current velocity. Furthermore, calculations both with a low and high value of the drag coefficient should be performed, depending upon whether the corresponding drag force primarily induces damping or excitation.

Higher order terms in the potential theory or finite wave height kinematics used in conjunction with Morison's equation, cause mean and time-variant sum and difference frequency action in irregular waves. These actions may cause significant response if resonance occurs.

Similar effects may also be present for (large volume) structures, see, e.g., Faltinsen (1990) for a more thorough description. As the higher order terms will give rise to both sum frequency and difference frequency effects (i.e., corresponding to different linear wave components with different amplitudes and frequencies), both subharmonic and superharmonic response contributions need to be considered.

Difference (low) frequency actions associated with wave-body nonlinear interaction, may be important for global motions and positioning systems when the action synchronise with fundamental periods of vibration for the system or parts of the system. Model tests are frequently performed in order to confirm the numerical calculations for cases where the associated response level is significant.

Sum (high) frequency actions (causing springing) may be important for the response of restrained modes of tension-leg platforms, ships and jack-ups. Such effects are typically of particular importance for the fatigue limit state.

Other non-linear hydrodynamic effects such as wave in deck and green water on deck are of crucial importance for certain types of structures. Phenomena such as slamming loads and whipping response also need to be accounted for in order to achieve adequately designed marine structures.

Steep and high waves which are acting on structural components extending above the still water level may cause nonlinear transient actions. Structural responses to these actions may be dynamically amplified and cause increased extreme response (e.g., so-called ringing). Such transient nonlinear actions may be important for structures consisting of large diameter shafts if also dynamic amplification can occur. Available analysis methods seem generally only to be suited for screening analysis of the ringing problem, and quantification based on model tests is typically required. It may then generally be difficult to distinguish impact/slamming from higher order inertia (ringing) effects.

Finite surface effects which resulted in somewhat amplified dynamic response levels were observed, e.g., in connection with design of the Troll Gravity Based concrete platform. These effect were found both in model test results and numerical calculations, see, e.g., Fergestad et al. (1994).

3.2.3.3 Non-Linearities Associated with Fluid-Structure Interaction

Some of the non-linear hydrodynamic effects which were discussed in the previous Section were strictly also associated with the response of the structure (e.g., the relative velocity term in the Morison's equation). Here, some additional and highly non-linear effects due to fluid-structure interaction are briefly considered. The focus is on current-induced vibrations which must also be taken into account. These effects are observed in the form of vortex-induced vibration and/or instability caused by varying orientation of the structure in relation to the wave and current direction (i.e., the so-called galloping phenomenon).

Actions by vortex shedding may be of significant importance in the design of slender structures that may respond in resonance modes to the cyclic vortex action. This is particularly the case when the damping level is small. The onset of vortex-induced-vibrations (VIV) generally occurs when the flow velocities due to current and waves exceed certain critical levels. The excitation may be characterized by the motion amplitude and/or the forces on the member. The possibility of vortex shedding of larger parts of the structure due to current combined with wave excitation should also be considered.

VIV will typically also increase the mean in-line drag coefficient. In a simplified way, this may be accounted for by multiplying the drag coefficient for a stationary cylinder with an amplification factor. The value of this factor will depend upon the ratio of the transverse motion and the member diameter.

Vortex shedding may especially contribute to fatigue. The cumulative effects of resonant actions during construction, transportation and operation need to be included in the calculations. The effects of vortex shedding may be reduced by introducing devices to prevent or reduce the vortex intensity, or by changing the vibration properties of the structure, e.g., natural period and damping level.

For slender and flexible structures where the surrounding flow-field is influenced by shielding and wake effects, assessment of the fluid-structure interaction may become a challenging task, see, e.g., Fu et al. (2016). For such cases, the analysis of possible vortex-induced-response and possibly galloping behavior is even more complex.

Models and procedures for calculation of loads and load effects are found in a number of design standards and guidelines, examples are NORSOK STANDARD N-003 Actions and Action effects (2007) which is most relevant for the Norwegian Continental Shelf, and ISO19902 (2008) Petroleum and natural gas industries—Fixed steel offshore structures which is mainly intended for worldwide applications. For particular cases with complex fluid-structure interaction which are not covered by existing standards or research literature, specialized calculations and subsequent verification by model tests and possibly full-scale measurements may be required.

3.2.3.4 Non-Linear Structural Behaviour

Geometric as well as material non-linearities need to be properly represented for stochastic analysis of marine structures. The former can be present even for moderate excitation levels. As an example, the analysis of mooring and riser systems must take into account the variation of tension stiffening effects as a function of the dynamic response displacements.

Material non-linearity, e.g., due to plastic deformations becomes increasingly important for the ultimate and the accidental states. Examples of the effects of material non-linear behavior can be found, e.g., in Jia (2014).

Structural behavior can also be interpreted to cover the foundation of the structure and the corresponding soil properties if these are relevant. An example of non-linear soil behavior with a strong influence on the extreme response level was observed in connection with design of the Troll Gravity Based platform, see, e.g., Leira et al. (1994).

3.2.3.5 Structural Response Analysis Methods

If both the stochastic loading and the applied structural model are of a linear nature, the corresponding dynamic response analysis will also generally be linear. The structural properties are then characterized by stiffness, damping and mass matrices that do not depend on the extended state vector of the structure (i.e., the displacement, velocity and acceleration vectors which correspond to a specific set of degrees-of-freedom for a given numerical model). In the following, algorithms for stochastic response analysis both in the time and frequency domain are reviewed. The latter is most widely applied for cases where linear properties hold both for the loading mechanism and the structural properties.

3.2.4 Frequency versus Time Domain Dynamic Response Analysis

3.2.4.1 Dynamic Equilibrium Equation

When the structure is discretized by means of the finite element method the relationship between the loading and response stochastic processes is expressed as an equilibrium equation on matrix form. A fairly general expression in the time domain is given by:

$$\int_0^{+\infty} \mathbf{M}(t-\tau)\ddot{\mathbf{r}}d\tau \ + \ \int_0^{+\infty} \mathbf{C}(\mathbf{r},\dot{\mathbf{r}},t-\tau)\dot{\mathbf{r}}d\tau \ + \ \mathbf{K}(\mathbf{r})\mathbf{r}(t) = \mathbf{Q}(t) \tag{3.1}$$

The lower integration limit is 0 due to the principle of causality. The convolution integrals are due to the mass and damping matrices being frequency dependent. Nonlinearities are accounted for by allowing the damping matrix to be a function of displacement and velocity, and with the stiffness matrix expressed as a function of the displacement vector. In addition to the equilibrium equation, initial displacement and velocity vectors at time t = 0 must be specified. The mass matrix includes both structural and hydrodynamic contributions, with the latter being frequency dependent while the pure structural contribution typically is taken to be constant. Similarly, the damping matrix includes a frequency dependent hydrodynamic term and a constant structural contribution modelled, e.g., by means of Rayleigh damping. Frequently, the dependence on the displacement vector is neglected. The structural stiffness matrix also includes contributions due to possible surface-piercing members such as pontoons as well as contributions due to tethers and anchor lines.

For the special case that the frequency dependence of the mass and damping matrices can be neglected, the two convolution integrals in the equilibrium equation disappear. This results in a simpler product form for the inertia and damping terms. For a stepwise time-integration of the system equations, it is furthermore convenient to express the equilibrium equation on incremental form (i.e., in terms of increments of the displacement, velocity and acceleration vectors together with increments of the load vector).

3.2.4.2 Short-Term Response-Statistics for Gaussian Processes

In connection with formulation of mechanical failure functions which are applied for evaluation of the structural reliability, probability distributions of local response maxima and extreme values are highly relevant. Furthermore, for the fatigue limit state the probability distribution of stress cycles is required. This distribution is closely related to the probability distribution of local maxima. These topics are addressed in the following.

The probability distribution of local maxima for a scalar Gaussian process was derived by Rice (1944), Longuet-Higgins (1952), and Cartwright and Longuet-Higgins (1956), and is referred to as the Rice distribution. The shape of this distribution depends strongly on the so-called bandwidth parameter, ε, which is defined as

$$\varepsilon = \sqrt{1 - \frac{\dot{\sigma}_x^2}{\sigma_x^2 \ddot{\sigma}_x^2}} \tag{3.2}$$

where σ_x^2 is the variance of the response process, $\dot{\sigma}_x^2$ is the variance of the associated velocity process and $\ddot{\sigma}_x^2$ is the variance of the response acceleration process.

For a so-called wide-band process (for which the bandwidth parameter approaches 1.0), the Rice distribution approaches the Gaussian distribution in the limit. The associated mean value is zero, which implies that positive and negative local maxima are equally likely. For a so-called narrow-band process (for which the bandwidth parameter approaches 0.), instead the Rayleigh distribution applies for the local maxima. The corresponding distribution function is then given by

$$F_S(s) = 1 - \exp\left(-\frac{s^2}{2\sigma_x^2}\right) \tag{3.3}$$

where s is the magnitude of the local maximum. The associated density function becomes

$$f_S(s) = \frac{s}{\sigma_x^2} \exp\left(-\frac{s^2}{2\sigma_x^2}\right) \tag{3.4}$$

The distribution of local maxima can also be obtained by means of the so-called Powell approximation by utilization of the so-called up-crossing rate for the level s, which is denoted by $v_x^+(s)$. The so-called zero-crossing rate is obtained by setting $s = 0$, i.e., $v_x^+(0)$. The distribution of local maxima based on the Powell approximation is then expressed as:

$$F_{S,Powell}(s) = 1 - \left(\frac{v_x^+(s)}{v_{x,max}^+}\right) \tag{3.5}$$

where $v_{x,max}^+$ is the maximum possible value of the up-crossing rate. For a Gaussian process the up-crossing rate is expressed as

$$v_x^+(s) = \frac{\dot{\sigma}_x}{2\pi\sigma_x} \exp\left(-\frac{s^2}{2\sigma_x^2}\right) \tag{3.6}$$

and the maximum value occurs just for the level $s = 0$. Accordingly, the resulting cumulative distribution of local maxima based on the Powell approximation is expressed as

$$F_{S,Powell}(s) = 1 - \left(\frac{v_x^+(s)}{v_{x,max}^+}\right) = 1 - \exp\left(-\frac{s^2}{2\sigma_x^2}\right) \tag{3.7}$$

which is identical with the Rayleigh distribution, i.e., the narrow-band limit of the Rice distribution.

Based on the distribution function for local maxima, the corresponding extreme value distribution for a given duration T can also be obtained. The number of local maxima for a narrow-band process during this period can be estimated based on the zero-crossing frequency as: $N = v_x^+(0)T = \frac{\dot{\sigma}_x T}{2\pi\sigma_x}$. By further assuming that the local maxima are statistically independent, the cumulative distribution function of the extreme value during T (i.e., which is referred to as $X_{E,T}$) is obtained as

$$F_{X_{E,T}}(x_{E,T}) = \left(F_s(x_{E,T})\right)^N = \left(1 - \frac{v_x^+(x_{E,T})}{v_x^+(0)}\right)^N = \left(1 - \exp\left(-\frac{x_{E,T}^2}{2\sigma_x^2}\right)\right)^N \tag{3.8}$$

The corresponding density function is readily obtained by differentiation. A plot of this density function is shown in Fig. 3.2 below for increasing values of the exponent N (in the range from 50 to 5000). The x-axis in the figure corresponds to the normalized variable, i.e., $z = \frac{x_{E,T}}{\sigma_x}$. The figure clearly shows the increase of the mean value for increasing values of the exponent N.

We also note that when N approaches infinity, the distribution function in Eq. (3.8) can be rewritten as:

$$\lim_{N \to \infty}\left(F_{X_{E,T}}(x_{E,T})\right) = \lim_{N \to \infty}\left(F_s(x_{E,T})\right)^N = \lim_{N \to \infty}\left(1 - \frac{Tv_x^+(x_{E,T})}{Tv_x^+(0)}\right)^N = \lim_{N \to \infty}\left(1 - \frac{Tv_x^+(x_{E,T})}{N}\right)^N$$

$$= \exp\left(-Tv_x^+(x_{E,T})\right) = \exp\left(-\frac{\dot{\sigma}_x T}{2\pi\sigma_x}\right)\exp\left(-\frac{x_{E,T}^2}{2\sigma_x^2}\right) \tag{3.9}$$

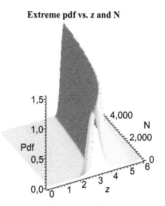

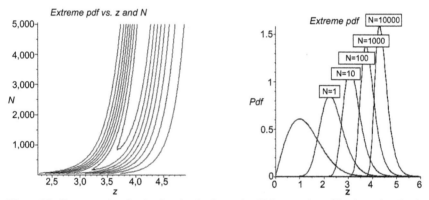

Figure 3.2. Extreme value density function for increasing N (i.e., number of local maxima for the parent Gaussian process).

For high levels, this expression can subsequently be approximated by a Gumbel distribution function of the following form:

$$F_{x_{E,T}}\left(x_{E,T}\right) = \exp\!\left(-\left(\exp\left(-\alpha\left(x_{E,T}-u\right)\right)\right)\right)$$

(3.10)

where α and u are the two parameters (i.e., constants) of the Gumbel distribution function. The proper expressions for these parameters are now obtained by setting the second exponential term for each of the two distribution functions equal to each other, i.e.,

$$\left(\exp\left(-\alpha\left(x_{E,T}-u\right)\right)\right) = \left(\left(\frac{\dot{\sigma}_x T}{2\pi\sigma_x}\right)\exp\left(-\frac{x_{E,T}^2}{2\sigma_x^2}\right)\right) = \left(N\exp\left(-\frac{x_{E,T}^2}{2\sigma_x^2}\right)\right)$$

(3.11)

By taking the logarithm of both sides of this equation we obtain:

$$\left(-\alpha\left(x_{E,T}-u\right)\right) \approx \left(\ln(N)\right) - \frac{x_{E,T}^2}{2\sigma_x^2}$$

(3.12)

The Gumbel density function has a peak at $x_{E,T} = u$, and hence we can also perform a Taylor series expansion of the quadratic term around this value:

$$\frac{x_{E,T}^2}{2\sigma_x^2} \approx \frac{1}{2\sigma_x^2}\left(u^2 + 2u(x-u)\right) = \frac{1}{2\sigma_x^2}\left(2ux - u^2\right) \tag{3.13}$$

By inserting this expression in Eq. (3.12) above, this gives two equations for the two constants α and u. These equations are solved to give:

$$\alpha = \frac{u}{\sigma_x^2} \quad \text{and} \quad \alpha u = \left(\frac{u}{\sigma_x^2}\right)u = \left(\frac{u^2}{\sigma_x^2}\right) = \ln(N) + \frac{1}{2}\left(\frac{u^2}{\sigma_x^2}\right) \tag{3.14}$$

The last equality in the second of these equations then gives $u = \sigma_x\sqrt{2\ln(N)}$.

This value is just equal to the so-called characteristic (or most likely) largest value. The corresponding expected largest value for the Gumbel distribution is obtained from the expression that applies to that particular distribution by inserting the above values for u and α:

$$E[x_{E,T}] = u + \frac{0.5772}{\alpha} = \sigma_x\left(\sqrt{2\ln(N)} + \frac{0.5772}{\sqrt{2\ln(N)}}\right) \tag{3.15}$$

where $\gamma = 0.5772$ is the Euler constant.

3.2.4.3 Frequency Domain Analysis

In the following, it is assumed that the structural matrices and the hydrodynamic loading have been linearized. From the theory of multi-degree-of-freedom systems under random loading, the response spectral density matrix can then be expressed in terms of the load spectral matrix as:

$$\mathbf{S}_r(\omega) = \mathbf{H}(\omega)\mathbf{S}_Q\mathbf{H}^{*T}(\omega) \tag{3.16}$$

where

$$\mathbf{H}(\omega) = \left[\mathbf{K} + i\omega\mathbf{C}(\omega) - \omega^2\mathbf{M}(\omega)\right]^{-1} \tag{3.17}$$

is the *virtual frequency response function* of the structure-fluid system. In turn, the load matrix can be expressed by the one-dimensional sea elevation spectral density, yielding:

$$\mathbf{S}_r(\omega) = \mathbf{H}(\omega)\mathbf{F}(\omega)\mathbf{H}^{*T}(\omega)S_\eta(\omega) = \mathbf{B}(\omega)S_\eta(\omega) \tag{3.18}$$

where $\mathbf{F}(\omega)$ is referred to as the *hydrodynamic transfer function*. The matrix $\mathbf{B}(\omega)$ is designated the *transfer function* of the total system relating the response spectral density matrix to the scalar wave spectral density.

These two matrices also account for directionality effects related to hydrodynamic loading and structural response. In addition to the mean direction of the incoming wave system, the directional spreading of wave energy around this direction is important. This is expressed in terms of the wave spreading function $\Psi(\theta)$ where θ is the wave direction. The load amplitudes (per unit wave height) will typically also depend on the direction of the incoming wave. Furthermore, the structural response may exhibit a strong sensitivity to load direction.

By introducing the direction and frequency-dependent loading $\mathbf{Q}(\omega,\theta)$, the directional integration can be expressed on the following form:

$$\mathbf{F}(\omega) \;=\; \int_{\theta} \mathbf{Q}(\omega,\theta)\mathbf{Q}^{*\mathrm{T}}(\omega,\theta)\Psi(\theta)\mathrm{d}\theta \tag{3.19}$$

where the integration itself can be performed either on element or system level.

Accurate computation of the direction and frequency dependent load vector requires in general utilization of advanced and time-consuming numerical schemes such as the sink-and source method. The same applies to the frequency-dependent hydrodynamic contributions to the damping and mass matrices. Marine structures with complex geometries, caused, e.g., by surface pontoons, demand careful modelling and fine-meshed spatial discretization if high precision is aimed at. However, at a preliminary design stage simplifications can be introduced in order to increase the computational efficiency. As an example, simplified and approximate three-dimensional hydrodynamic models can be assembled from a sequence of two-dimensional approximations. Furthermore, analytical or semi-analytical expressions for axi-symmetric three-dimensional components such as pontoons and shafts can be utilized.

Due to dynamic amplification effects, there will be a significant sensitivity of the response with respect to variation in peak period of the extreme sea state. Accordingly, a more consistent approach is to establish a so-called long-term statistical distribution. Probability distributions of local response maxima are then computed for a number of sea-states. A set of significant wave heights and peak periods are then specified, together with corresponding relative frequencies of occurrence from the so-called scatter diagram applicable for the relevant site. The long-term distribution of local response maxima is subsequently obtained as a weighted combination of the distributions for all these sea-states. The response level corresponding to a given return period is finally estimated by specifying the corresponding probability of exceeding local maxima at different levels, and then inverting the long-term distribution. This is further described in Section 3.2.6 and 3.2.7 below.

A particular feature of some categories of marine structures is a significant sensitivity of structural response to the main wave direction and the degree of directional spreading. This implies that the relative frequency of occurrence for a range of wave directions and corresponding spreading parameters must be considered when computing the long-term distribution.

3.2.4.4 Time Domain Analysis

Simplistic time domain analysis is based on application of regular waves with given amplitude and period. The period is subsequently varied in order to maximize the structural response. This approach allows to take into account structural nonlinear behavior. However, the dynamic amplification effects are less accurately represented.

By means of stochastic analysis a proper distribution of wave energy across the relevant frequency band according to Eq. (3.20) below. The time domain analysis is based on simulation of a number of samples of the stochastic load process. The samples can be generated for instance by means of Monte Carlo simulation methods, see, e.g., Borgman (1969), Shinozuka (1972) and Hammersley and Handscomb (1964). For each of these sample functions, the corresponding load vector time series is computed, and subsequently a deterministic type of response analysis is performed.

In this type of approach, the sea-elevation process can be approximated by a discrete sum as:

$$\eta(\mathbf{x}, t) = \sum_{k=1}^{N_1} \sum_{l=1}^{N_2} A_{kl} \left\{ \cos \omega_k t - \kappa(\omega_k) \cdot (x \cos \theta_l + y \sin \theta_l) + \phi_{kl} \right\} \tag{3.20}$$

where

$$A_{kl} = \sqrt{2S_\eta(\omega_k, \theta_l) \Delta \omega \Delta \theta} \tag{3.21}$$

Here, ϕ_{kl} are random independent phase angles distributed uniformly between 0 and 2π. Furthermore,

$$\Delta \omega = \frac{\omega_{up}}{N_1} \quad \text{and} \quad \omega_k = (k-1)\Delta \omega \tag{3.22}$$

$$\Delta \theta = \frac{(\theta_{up} - \theta_{low})}{N_2} \quad \text{and} \quad \theta_l = \theta_{low} + (l-1)\Delta \theta \tag{3.23}$$

where ω_{up} is the upper limit of the frequencies to be included in the analysis; θ_{up} and θ_{low} designate the upper and lower limits for the summation with respect to wave direction.

The double summation can be efficiently carried out by means of FFT-techniques. Evaluation of the load vector for the relevant structure can be performed based on the same type of expression simply by inserting the frequency and direction dependent (complex-valued) vector transfer functions between each harmonic component of the sea-elevation process and the corresponding harmonic load vector component, i.e.:

$$Q(t) = \sum_{k=1}^{N_1} \sum_{l=1}^{N_2} A_{kl} Q(\omega_k, \theta_l) \{\cos \omega_k t - \kappa(\omega_k) \cdot (x \cos \theta_l + y \sin \theta_l) + \phi_{kl}\} \quad (3.24)$$

where for each component of the load vector a double summation of the same type as for the sea-elevation process is performed.

For each generated sample of hydrodynamic load time series, corresponding response time series are computed by step-by-step time integration of the equations of motion.

3.2.4.5 Time versus Frequency Domain Analysis

Summarizing the main features of the two types of response analysis procedures, the benefits of the frequency domain approach are as follows:

- Frequency dependency of hydrodynamic coefficients for damping, added mass and excitation forces are easily incorporated.
- Estimation of extreme values for design purposes is straightforward due to available closed-form analytical expressions.
- The computational effort is significantly reduced as compared to time domain analysis.

However, on the negative side, this type of analysis is not well-suited for incorporation of nonlinearities related to structural behaviour or hydrodynamic modelling. The benefits of the time domain approach can be summarized as:

- Nonlinear effects related, e.g., to nonlinear material behaviour, geometric stiffness, finite surface wave effects and viscous loading can be incorporated in a direct manner.
- Simulation in the time domain provides insight into the physical behaviour of the structure. Instantaneous deformation patterns and time variation of response can easily be visualized.

Among the negative aspects of this procedure are: Increased effort related to estimation of extreme response statistics, and difficulties related to implementation of frequency dependent damping and mass coefficients. In general, both types of analyses could be performed in order to assess the importance of various modelling assumptions in relation to the computed response.

3.2.5 Stochastic Analysis in the Probability Domain

3.2.5.1 Introduction

For structural response analysis in connection with stochastic loading additional tools are also available for a more direct evaluation of specific probabilistic features of the response. Examples of such "probability domain" methods are the so-called covariance analysis and moment equations. A complete solution method in terms of probability density functions can be achieved by solution of the Fokker-Planck equation. For a classical reference related to the latter approach, see, e.g., Risken (1989). A quite comprehensive overview of different types of solution methods is provided by Kumar and Narayanan (2006). Challenges related to higher-dimensional formulations of this equation is discussed by Masud and Bergman (2005). The following text is mainly based on Chai et al. (2015; 2016).

3.2.5.2 *Fokker-Planck Equation*

The Fokker-Planck equations describe the evolution of stochastic systems. In this section, derivation of this equation is first considered. The probability density function of the dynamic response vector process $\mathbf{x}(t)$ of the system can be obtained by solving this equation system as will be illustrated by a specific example.

The probability density function (PDF) of the process $\mathbf{x}$ at time t, $p(\mathbf{x},t)$, can be obtained by the basic equation, i.e., the well-known Chapman-Kolmogorov equation:

$$p(\mathbf{x},t) = \int p(\mathbf{x},t \mid \mathbf{x}',t')p(\mathbf{x}',t')d\mathbf{x}' \tag{3.25}$$

where p($\mathbf{x}$,t|$\mathbf{x}'$,t') is the transition probability density from state x at time t to x' at time t'.

We next introduce the following quantities for $j = 1,2,3 \ldots\infty$

$$\mu_j(\mathbf{x}') = E[(\mathbf{x}-\mathbf{x}')^j]$$

which implies that μ_j represents the j-th moment of the increment x–x'. The probability density function of the stat x at time t can then be expressed as:

$$p(\mathbf{x},t) = \sum_{j=0}^{\infty} \frac{1}{j!} \frac{1}{2\pi} \int \int \exp(-iu(\mathbf{x}-\mathbf{x}')) \cdot (iu)^j \, du \, \mu_j(\mathbf{x}')p(\mathbf{x}',t')d\mathbf{x}' \tag{3.26}$$

Furthermore, it is found that

$$p(\mathbf{x},t) = p(\mathbf{x}',t') + \sum_{j=0}^{\infty} \frac{1}{j!}(-\frac{\partial}{\partial \mathbf{x}})^j \left[\mu_j(\mathbf{x})p(\mathbf{x},t) \right] \tag{3.27}$$

Dividing Eq. (3.27) by Δt and passing to the limit $\Delta t \rightarrow 0$, the following expression can be obtained:

$$\frac{\partial p(\mathbf{x},t)}{\partial t} = \sum_{j=0}^{\infty} \frac{1}{j!}(-\frac{\partial}{\partial \mathbf{x}})^j \left[K_j(\mathbf{x})p(\mathbf{x},t) \right] \tag{3.28}$$

in which

$$K_j(\mathbf{x}) = \lim_{\Delta t \to 0} \frac{\mu_j(\mathbf{x})}{\Delta t}, \quad j = 1,2,\ldots. \tag{3.29}$$

where $K_j(\mathbf{x})$ represents the intensity coefficient and Eq. (3.29) is called the *Kramers-Moyal expansion*.

Assuming that the Markov process x(*t*) is continuous and the Kramers-Moyal expansion can be truncated with the higher-order intensity $K_3, K_4 \ldots$ being equal to zero, the distribution $p(\mathbf{x}, t)$ of the diffusion process x(*t*) follows the Fokker-Planck equation which is written as, see, e.g., Chai et al. (2016):

$$\frac{\partial p(\mathbf{x},t)}{\partial t} = -\frac{\partial}{\partial \mathbf{x}}\big[K_1(\mathbf{x})p(\mathbf{x},t)\big] + \frac{1}{2}\frac{\partial^2}{\partial \mathbf{x}^2}\big[K_2(\mathbf{x})p(\mathbf{x},t)\big] \tag{3.30}$$

The transition probability can then be found as the solution of the equation:

$$\frac{\partial p(\mathbf{x},t|\mathbf{x}_0,t_0)}{\partial t} = -\frac{\partial}{\partial \mathbf{x}}\big[K_1(\mathbf{x})p(\mathbf{x},t|\mathbf{x}_0,t_0)\big] + \frac{1}{2}\frac{\partial^2}{\partial \mathbf{x}^2}\big[K_2(\mathbf{x})p(\mathbf{x},t|\mathbf{x}_0,t_0)\big] \tag{3.31}$$

Furthermore, there is a close connection between the Fokker-Planck equation and the stochastic differential equation (SDE). For the one-dimensional case, the Fokker-Planck equation is given as:

$$\begin{aligned} \frac{\partial p(x,t|x_0,t_0)}{\partial t} &= -\frac{\partial}{\partial x}\Big[K_1(x)p(x,t|x_0,t_0)\Big] + \frac{1}{2}\frac{\partial^2}{\partial x^2}\Big[K_2(x)p(x,t|x_0,t_0)\Big] \\ &= -\frac{\partial}{\partial x}\Big[a(x,t)p(x,t|x_0,t_0)\Big] + \frac{1}{2}\frac{\partial^2}{\partial x^2}\Big[b^2(t)p(x,t|x_0,t_0)\Big] \end{aligned} \tag{3.32}$$

in which $a(x, t)$ represents the drift coefficient and $b^2(t)$ is the square of the diffusion coefficient for the one-dimensional stochastic differential equation: $dx = a(x, t)dt + b(t)d\mathrm{W}(t)$.

For *n*-dimensional stochastic differential equations, $n > 1$, the Fokker-Planck equation is written as:

$$\begin{aligned} \frac{\partial p(\mathbf{x},t|\mathbf{x}_0,t_0)}{\partial t} &= -\frac{\partial}{\partial \mathbf{x}}\Big[K_1(\mathbf{x})p(\mathbf{x},t|\mathbf{x}_0,t_0)\Big] + \frac{1}{2}\frac{\partial^2}{\partial \mathbf{x}^2}\Big[K_2(\mathbf{x})p(\mathbf{x},t|\mathbf{x}_0,t_0)\Big] \\ &= \sum_{i=1}^{n} -\frac{\partial}{\partial x_i}\Big[\mathbf{a}(\mathbf{x},t)p(\mathbf{x},t|\mathbf{x}_0,t_0)\Big] \\ &\quad + \frac{1}{2}\sum_{i=1}^{n}\sum_{j=1}^{n}\frac{\partial^2}{\partial x_i \partial x_j}\Big[(\mathbf{b}(t)\cdot\mathbf{b}^T(t))_{ij}\, p(\mathbf{x},t|\mathbf{x}_0,t_0)\Big] \end{aligned} \tag{3.33}$$

Based on the derivation above, the Fokker-Planck equations for the one-dimensional and multidimensional SDE are obtained. Alternatively, the Fokker-Planck equation can also be derived by other approaches as described in Ochi (1978).

3.2.5.3 Example of Application

In this section, the stochastic response of the roll motion of an ocean surveillance ship in random beam seas is studied. The GZ curve for this model and the relevant parameters for the vessel are given in Chai et. al. (2015).

The random stationary sea state is specified by the modified Pierson-Moskowitz spectrum, which is widely used for fully developed sea states. The wave spectrum is given as:

$$S_{\xi\xi}(\omega) = \frac{5.058g^2 H_s^2}{T_p^4 \omega^5} \exp(-1.25\frac{\omega_p^4}{\omega^4}) \tag{3.34}$$

in which H_s denotes the significant wave height, ω_p is the peak frequency at which the wave spectrum $S_{\xi\xi}(\omega)$ has its maximum value and T_p is the corresponding peak period.

A sea state with $H_s = 4.0$ m and $T_p = 11.0$ s is selected for the subsequent study. The selected wave spectrum and the roll excitation moment per unit wave height $|F_{roll}(\omega)|$ are presented in Chai et. al.(2015). Subsequently, the roll excitation moment spectrum $S_{MM}(\omega)$ and the relative roll excitation moment spectrum $S_{mm}(\omega)$ can be obtained.

After determining the target spectrum $S_{mm}(\omega)$, which is shown in Fig. 3.3, the parameters α, β, γ in the second-order linear filter should be determined in order to establish the 4D dynamic system. In this regard, the least-square scheme is available as a part of the curve fitting algorithms in MATLAB® and the fitting result is shown in Fig. 3.3. It can be readily seen that the filtered spectrum is reasonable in terms of bandwidth, peak frequency and peak value.

In particular, for the roll motion cases, the transfer function between the roll excitation moment and the roll response in the SDOF model is narrow banded and

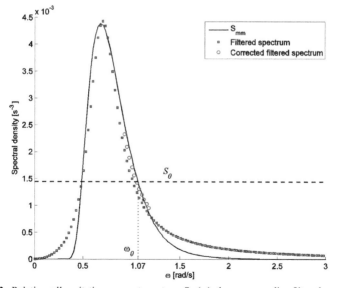

Figure 3.3. Relative roll excitation moment spectrum $S_{mm}(\omega)$, the corresponding filtered spectrum S_{2nd} and the corrected filtered spectrum (part) for the sea state with $H_s = 4.0$ m and $T_p = 11.0$ s, and spectrum of the equivalent Gaussian white noise, S_0 (from Chai et al. (2016)).

peaked near the natural roll frequency ω_0 due to the light roll damping. Therefore, in Fig. 3.3, for the selected ship model, the obvious discrepancies between the spectrum generated by the second-order linear filter and the target spectrum $S_{mm}(\omega)$ in the low-frequency and high-frequency regions would not impact the subsequent roll response to a significant extent. However, a slight discrepancy between the above two spectra in the critical frequency region near ω_0 can be observed. The fitting accuracy in the critical region is crucial for evaluating the roll response since the latter is sensitive to variation of the external excitation in this frequency region. Therefore, a constant, c, should be introduced as a correction factor for the filtered spectrum in order to decrease the discrepancy in the critical region. Then, the filtered spectrum is changed into:

$$S_{2nd}(\omega) = \frac{1}{2\pi} \frac{(c \cdot \gamma)^2 \omega^2}{(\alpha - \omega^2)^2 + (\beta\omega)^2} \tag{3.35}$$

In this work, for the selected sea state and ship model, the correction factor c is taken to be 1.07 by considering the mean difference between the two spectral densities in the critical region. In addition, the corrected (or modified) spectrum in the critical region is also presented in Fig. 3. The 4D dynamic system is established after the spectrum fitting has been performed, and subsequently the joint probability density function (PDF) of the roll angle process and the roll velocity process can be obtained directly by the 4D PI method. The joint PDF for the selected sea state calculated by the 4D PI method is shown in Fig. 3.4.

The marginal PDF of the roll angle process and the marginal PDF of the roll velocity process obtained by the 4D Path Integration method and the corresponding empirical estimations evaluated by Monte Carlo simulation are plotted in Figs. 3.5 and 3.6, respectively.

The Gaussian distributions of the marginal PDFs in Figs. 3.5 and 3.6 are obtained by using the variances evaluated by the straightforward Monte Carlo

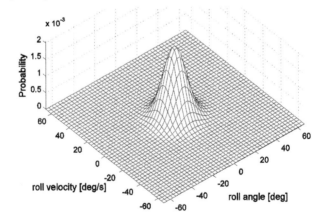

Figure 3.4. Joint PDF of the roll response obtained by the 4D PI method for the sea state with $H_s = 4.0$ m and $T_p = 11.0$ s (from Chai et al. (2016)).

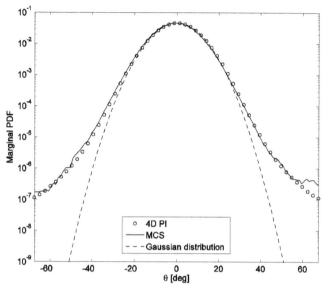

Figure 3.5. Marginal PDF of the roll angle process for the sea state with $H_s = 4.0$ m and $T_p = 11.0$ s (from Chai et al. (2016)).

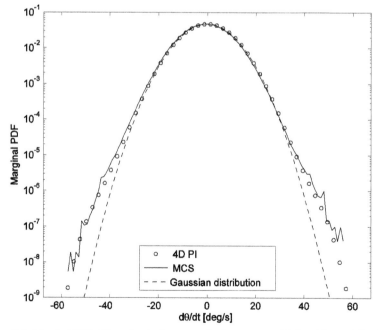

Figure 3.6. Marginal PDF of the roll velocity process for the sea state with $H_s = 4.0$ m and $T_p = 11.0$ s (from Chai et al. (2016)).

simulation. Actually, they are the marginal PDFs of the 4D Gaussian PDF, $p(x^{(0)}, t_0)$, which serves as the initial PDF in the 4D Path Integration procedure. It is shown that the Gaussian distributions in Figs. 3.5 and 3.6 give reasonable approximations of the statistics for the small amplitude motions. However, for the high-level responses, the distributions of the roll angle process and the roll velocity process are very different from the normal distribution, which underestimates the corresponding low probability levels. Moreover, it is shown in Figs. 3.5 and 3.6 that the comparisons of the marginal PDFs obtained by the 4D Path Integration (PI) method and Monte Carlo simulation demonstrate the high-level accuracy of the 4D PI method.

The upcrossing rate calculated by the 4D PI method and the Rice formula for the selected sea state and the corresponding empirical estimation of the upcrossing rate as well as the 95% confidence interval obtained by 4D Monte Carlo simulation are shown in Fig. 3.7. For Monte Carlo simulation, long time domain simulations are required to obtain the upcrossing rate for high response levels. It can be readily seen that the 4D PI technique yields quite accurate and reliable result, even in the high roll response region.

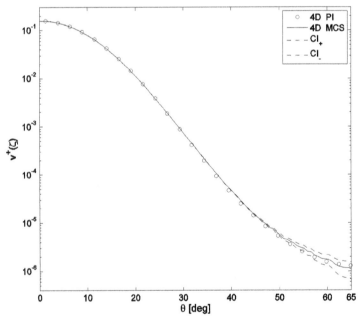

Figure 3.7. Upcrossing rate obtained by the 4D PI method and the empirical mean upcrossing rate evaluated by the 4D Monte Carlo simulation for the sea state with $H_s = 4.0$ m and $T_p = 11.0$ s (from Chai et al. (2016)).

3.2.6 Long-Term Response Properties

Long-term response properties are typically obtained by means of frequency- or time-domain response analyses for a number of short-term conditions for which stationarity assumptions are supposed to hold.

The long-term distribution of response amplitudes is obtained by considering that for a random point in time the sea state parameters (contained in a vector $\mathbf{X}$) are themselves random variables with joint probability density function $f_x(\mathbf{x})$. The relevant types of joint density function for significant wave height and peak period were discussed above. The corresponding probability distribution of the response (i.e., r) is obtained by weighting the conditional distribution of local maxima by multiplying with the joint density function of sea state parameters and subsequently integrating:

$$F_{R,L}(r) = \int_x F_R(r|\mathbf{x})f_x(\mathbf{x})w(\mathbf{x})d\mathbf{x} \tag{3.36}$$

Here, $w(\mathbf{x})$ is a weighting factor which accounts for the relative number of response peaks in each sea state, and $F_s(r|\mathbf{x})$ is the short-term conditional response distribution for a specific sea state corresponding to a given outcome of the vector $\mathbf{X}$. The conditional distribution of local maxima is typically taken to be of the Rice, Rayleigh or Weibull type.

A similar expression also holds for the complement of the cumulative long-term distribution. This complementary distribution corresponds to the probability of exceeding a specific response levels. This probability is obtained by replacing the short-term distributions of the response amplitudes by their complements in Eq. (3.41). A very convenient and generally applied approximation to the long-term distribution $F_{R,L}(r)$ is provided by the Weibull model.

Estimation of extreme values (e.g., 10-year and 100-year wave amplitudes and wave heights) can be performed based on the long-term distribution by application of the proper probability level. Alternatively, extreme response levels can also be estimated based on the corresponding extreme value distributions for each of the "short-term" conditions (i.e., for all the different sea states). The short- and long-term distributions in Eq. (3.36) are then replaced by the corresponding short- and long-term extreme-value distributions, i.e., $F_{S,E}(r_E|\mathbf{x})$ and $F_{L,E}(r_E)$, where the subscript E refers to extreme value:

$$F_{L,E}(r_E) = \int_x F_{S,E}(r_E|\mathbf{x})f_x(\mathbf{x})d\mathbf{x} \tag{3.37}$$

The resulting extreme response which results by application of this expression will typically agree quite well with that obtained by application of the long-term distribution.

A more correct formulation of the long-term extreme-value distribution based on direct application of the upcrossing-rate can be expressed as:

$$F_{L,E}\left(r_E\right) = \exp\left\{-\left(T\int_x v^+\left(r_E\,|\,\mathbf{x}\right)f_x\left(\mathbf{x}\right)d\mathbf{x}\right)\right\}$$ (3.38)

It is anticipated that for most cases the difference between results obtained by application of Eq. (3.42) versus Eqs. (3.42) and (3.43) typically is negligible. Methods for computation of the upcrossing rate for Gaussian as well as non-Gaussian processes are found, e.g., in Wen and Chen (1989)], Hagen and Tvedt (1991), and Beck and Melchers (2004).

Analysis of accumulated fatigue damage in a structural component based on the SN-curve approach is generally based on the following equation for the expected damage:

$$E\left[D(T)\right] = \frac{N(T)}{\bar{a}}E\left[\left(\Delta\sigma\right)^m\right] = \frac{N(T)}{\bar{a}}\int_0^\infty \left(\Delta\sigma\right)^m f_{\Delta\sigma}\left(\Delta\sigma\right)d\left(\Delta\sigma\right)$$ (3.39)

where the probability density function of the stress-ranges, i.e., $f_{\Delta\sigma}(\Delta\sigma)$, may apply to one specific sea state under consideration or may correspond to the long-term cycle distributions. In the former case, summation across the range of possible sea states is required. The probability density function for the stress range is frequently taken to be of the Weibull type, both within a short-term and long-term framework. The quantities $(\bar{a}, m)$ are constants which define a particular SN-curve, and N(T) is the expected number of stress cycles (or local maxima) during the time period T. For S-N curves with one and two slopes, analytical formulas for the resulting fatigue damage are found, e.g., in Almar-Naess et al. (1999).

3.2.7 Long-Term Extreme Response Analysis Based on Contour Analysis

3.2.7.1 Introduction

For evaluation of extreme structural response and fatigue damage accumulation, environmental load parameters (e.g., wind, wave and current characteristics) are typically represented by a sequence of (short-term) piecewise stationary process intervals. This type of stochastic process is frequently referred to as a Borges process or sometimes as an FBC-process (see the work by Ferry Borges and Castanheta (1971)). Response analysis is then greatly facilitated for each of these intervals due to stationarity. The corresponding statistical response distribution for each short term condition is subsequently weighted by the probability of occurrence for each environmental condition. This allows the long-term response distribution to be established as discussed above.

In order to minimize the number of analyses for estimation of extreme response by application of such a procedure, so-called design contour methods are frequently applied. Extreme environmental conditions (corresponding to given return periods) are then identified based on the associated probability distributions. As a second step, response analyses are performed for a selection of these extreme conditions.

The highest response level among these conditions is subsequently applied for design purposes. Hence, the term "design contour" refers to the collection of relevant environmental conditions which correspond to a specific return period (e.g., 100 years).

Inasmuch as the environmental parameters refer to different types of processes (e.g., wind and waves), the design contour represents a solution to the so-called problem of "load combinations". General methods for identification of load combinations in relation to continuous stochastic processes are first highlighted in the present paper. The particular case of multiple FBC-processes with known probability amplitude distribution is subsequently addressed. The case where all process components have identical basic time intervals is first considered. FBC-processes with widely different basic time intervals are next investigated. A methodology is outlined which enables to establish the environmental design contour also for this case.

The present section intends to highlight and further extend the methodology behind construction of design contours which are to be applied for structures where multiple response components need to be addressed simultaneously. A procedure for transformation of non-Gaussian processes into normalized Gaussian components is described. Fractiles of the extreme-value distribution for the normalized components can then be obtained and transformed back into non-Gaussian components. The developed formulation is applied for analysis of an example cross-section in a flexible riser with interacting curvature and tension responses. This leads to the concept of so-called contour tubes.

3.2.7.2 Environmental Contours

A distinction should be made between combination of loads versus combination of load effects. Clearly, the combination of load components at a specific ratio to each other will in general imply a different ratio between the associated load effects. In codified design, combination of different types of loading are typically specified in terms of return periods for the different environmental processes. As an example, for offshore structures the dominant load component is frequently specified to have a return period of 100 years, while the secondary component is specified to have a return period of 10 years (when the process components are assumed to be uncorrelated).

A further distinction should be made between cases where the relationship between the load-effects and environmental parameters are known and cases where these relationships have not yet been obtained. A further distinction can be made between cases for which the capacity surface (mechanical limit state function) for a given structural component is available and cases for which it is not given. If a limit state function is specified, a load-effect combination needs to be analyzed which frequently involves dynamic effects.

In the present paper, focus is on continuous–time processes. For cases where the process components are also continuous-valued and in addition the limit state function is known, the combined load effect is frequently analyzed by means of the so-called up-crossing rate (or more generally the out-crossing rate for multidimensional formulations). For linear combinations, upper bound expressions can be derived which involve the up-crossing rate for each of the component processes, see, e.g., Madsen et al. (1986) and Melchers (1999).

Simplified methods for definition of relevant "point values" which are assumed to cover the most critical load combinations have also been introduced. One of these is the celebrated Turkstra rule which selects the expected extreme value of one component, which is then combined with the expected instantaneous values of the other components, see Turkstra (1970). Evaluation of a sequence of such combinations (which is equal to the number of components) is then required.

A second type of simplified method is the so-called "Square-root-of-sum-of-squares" rule (SRSS-rule). The expected extreme values for all the components are then squared and added together, and the square-root of the resulting sum is subsequently computed.

The simplified load combination methods do not explicitly take into account the particular distribution functions which apply for the involved components (except possibly for computation of the associated expected values). For process components which are discrete instead of continuous-valued the up-crossing rate can still be applied. However, for this case the analysis can be made somewhat simpler by utilization of the step-wise behavior of the sample functions. This is achieved by means of the FBC-process representation which was mentioned above. The particular type of distribution functions and the characteristic time interval for each process component can then be properly accounted for. An example of a FBC process together with the basic time interval is shown in Fig. 3.8.

In general the time interval will be different for the different process components. In some cases the lengths of the time scales are widely different as for example in connection with the joint representation of wind and snow parameters, see, e.g., Næss and Leira (1999).

For cases where the limit state function is not specified, a range of environmental conditions that are relevant for analysis can still be identified. This is based on consideration of the multi-dimensional joint probability density and the corresponding distribution functions for the process components. Iso-probability

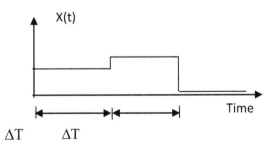

Figure 3.8. Example of FBC process with characteristic time interval.

surfaces can then be computed which correspond to a specified exceedance probability (or equivalently a specified return period). In the following, a brief review of the much-applied contour methods for identification of such relevant design events is first given. The connection with the FBC process is also highlighted.

Design contours

Environmental processes such as wind and wave characteristics (e.g., significant wave height and peak period) are generally of a non-stationary character. A simplified representation is typically applied where these processes are modeled according to the step-wise representation in Fig. 3.8. The statistical properties of the "step-levels" of the basic components are generally of a non-Gaussian character. However, they can still be represented as being transformations of processes which have Gaussian distributed step levels, and they are frequently referred to as "translation processes".

For such processes transformation into the auxiliary normalized Gaussian processes is provided by the Rosenblatt transformation, see, e.g., Madsen et al. (1986) and Melchers (1999). For two process components this transformation is expressed as:

$$\Phi(u_1(t)) = F_{x1}(x_1(t))$$

$$\Phi(u_2(t)) = F_{x2|x1}(x_2(t)|x_1(t))$$

(3.40)

where the second equation involves the conditional distribution function of x_2 given x_1. The associated partial derivatives (i.e., the elements of the Jacobian matrix) are expressed as:

$$J_{ij}(x_i) = \frac{\partial u_i}{\partial x_j} = \begin{cases} 0 & \text{for } i < j \\[2ex] \dfrac{f_i(x_i | x_1, ..., x_{i-1})}{\phi(u_i(x_i))} & \text{for } i = j \\[3ex] \dfrac{\dfrac{\partial F_i}{\partial x_j}(x_i | x_1, ..., x_{i-1})}{\phi(u_i(x_i))} & \text{for } i > j \end{cases}$$

(3.41)

In the case of uncorrelated basic components only the diagonal terms will be non-zero, and the expressions simplify into the following:

$$\partial u_i / \partial x_i = J_{ij}(x_i) = f_i(x_i)/\varphi(u_i(x_i))$$

(3.42)

Other possible types of transformations also exist, somewhat depending on the type of statistical information which is available. As an example, the Nataf transformation can be applied if only the marginal distributions and the pairwise correlation coefficients are known, see Nataf (1962), Der Kiureghian and Liu (1986).

Having performed the transformation into normalized components, the corresponding cumulative distribution for the distance from the origin to a specific point will be independent of the direction in the transformed space (i.e., due to the isotropic properties of the transformed processes). This implies that the iso-probability levels will correspond to concentric circles. The probability of exceeding a given value of the radius (R) in any direction is given by the following expression:

$$p_f(R) = 1 - \Phi(R) = \Phi(-R) \tag{3.43}$$

This probability of exceedance can also be interpreted in terms of a specific return period in the following manner: Designating the number of events (i.e., number of repetitions of the basic time interval) which corresponds to the given return period by N, the probability of exceeding the corresponding radius value is expressed as:

$$p_f(R) = \Phi(-R) = 1 - (1/N) \tag{3.44}$$

Further details and example applications (for the case with identical time scales for the environmental components) are found in Winterstein et. al. (1993), Haver et al. (1998), Johannesen et. al. (2001), Baarholm et al. (2001).

3.2.7.3 Two-Dimensional Wave Height—Wave Period Contour

We next specialize to the 2D contour which represents the joint characteristics of the significant wave height (i.e., H_s) and the peak period (i.e., T_p). Both of these environmental parameters are assumed to have the same basic time interval which is taken to be 3 hours.

A joint probabilistic model is fitted to data which represents simultaneous observation of H_s and T_p. The joint probability density function is expressed on conditional form as:

$$p(H_s, T_p) = p(H_s)p(T_p|H_s) \tag{3.45}$$

The first factor in this expression represents the marginal density function for the significant wave height. It is typically given by the following combined lognormal and Weibull distributions, see, e.g., Haver et al. (1985):

$$p(H_s) = \begin{cases} \dfrac{1}{\sqrt{2\pi}\kappa H_s} \exp\left[-\dfrac{(\ln H_s - \theta)^2}{2\kappa^2} \right] ; & H_s \leq 3.25 \text{ m} \\[4mm] \beta \dfrac{H_s^{\beta-1}}{\zeta^\beta} \exp\left[-\left(\dfrac{H_s}{\zeta} \right)^\beta \right] ; & H_s > 3.25 \text{ m} \end{cases} \tag{3.46}$$

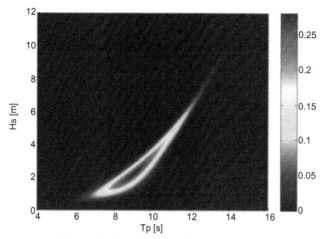

Figure 3.9. Joint probability density function for significant wave height (vertical axis) and peak period (horizontal axis).

Here, θ and κ^2 are the mean value and the variance, respectively, of $\ln(H_s)$; ζ and β are the corresponding Weibull parameters.

The conditional density of the peak period for given values of Hs is expressed by the following lognormal model:

$$p\left(T_{char} \mid H_s\right) = \frac{1}{\sqrt{2\pi}\sigma T_{char}} \exp\left[-\frac{\left(\ln T_{char} - \mu\right)^2}{2\sigma^2} \right] \tag{3.47}$$

where $T_{char} = T_p$ and the parameters $\mu(H_s)$ and $\sigma^2(H_s)$ are respectively the conditional expected value and conditional standard deviation of the variable $\ln(Tp)$ expressed as function of H_s. It is noted that a Gamma distribution could also be considered as being a relevant alternative for the significant wave height, see, e.g., Fouques et al. (2004).

The joint density function of significant wave height and peak period is shown in Fig. 3.9 below. The corresponding two-dimensional contours (based on the expressions above) which represent the significant wave height and peak period for the site where the present platform is located are displayed in Fig. 3.10, see Lindstad (2013). Contours for return periods of 1 year, 10 years and 100 years are shown.

3.2.7.4 Contour Response Tubes

When applying the contour approach for estimation of extreme load effects, the structural response needs to be computed for a number of sea states along the environmental contour. For the case of a single response component, a single value of the estimated extreme response for each sea state is obtained. For the case of several response components, an extreme response curve (which corresponds to a surface if more than two response components are involved) can be constructed.

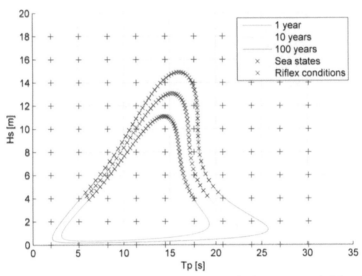

Figure 3.10. Contour lines corresponding to 1-year, 10-year and 100 year return periods based on the joint density function shown in Fig. 3.9.

There are several ways of carrying out such a construction. The required input will be the joint distribution of the response processes (including correlation properties) which can be estimated, e.g., based on numerical simulation and response analysis in the time domain.

One option is to apply so-called convex hulls which represent envelopes of the response vector process, see Ottesen and Aarstein (2006). In the example study below a modified approach of this type is adopted.

Schematic illustrations of contour response tubes with one-dimensional and two-dimensional cross-sections are given in Fig. 3.11a and Fig. 3.11b respectively. It is seen that for both cases a convenient representation of the response variation is obtained.

For three-dimensional cross-section each sea state can, e.g., give rise to a two-dimensional ellipsoidal surface. Cross-sections in even higher dimensions would then yield higher dimensional ellipsoids (or other shapes) which would be quite difficult to visualize. However, simplifications can be achieved by focusing on the two most dominant response quantities in each case if such can be identified (which will usually be the case).

Construction of the cross-section of the response tube is described in more detail in the example of application which follows next. Basically, the cross-section is three-dimensional but it was found that a 2D representation based on the two dominant response processes was adequate. This greatly simplifies the visualization of the contour response tube.

(a) "One-dimensional" response tube

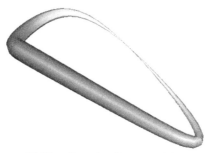

(b) "Two-dimensional" response tube

Figure 3.11. Example of contour response tubes. (a) "One-dimensional" response tube (b) "Two-dimensional" response tube.

3.2.7.5 Example of Multidimensional Response

As an example of application, we consider the dynamic response of a flexible riser configuration. The 2D response corresponding to the tension and bending moment at a critical cross-section is considered. The model used in this study is a flexible riser suspended from a semisubmersible in the Northern part of the North Sea at a water depth of 100 m, see Fig. 3.12.

For each sea-state at the rectangular grid which is indicated by red crosses in Fig. 3.10, a dynamic response analysis is performed in the time domain by application of the riser analysis computer program Riflex, Sintef (2012). The response quantities which are in focus are the tension and curvature at the upper end of the riser. At present, a simplistic (or naïve) model is applied with a moment free hinge instead of the bending stiffener which is applied in reality. Accordingly, the present analysis would correspond to a pre-study where the detailed design of the bending stiffener at the upper end will be performed as part of the next design phase.

Based on the simulated sample functions of the tension and curvature response processes, probability distributions are fitted to the local maxima and extreme values of the respective responses. The curvature has two components, one in the vertical and one in the horizontal plane.

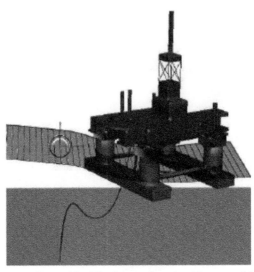

Figure 3.12. Semi-submersible and flexible riser configuration (Courtesy of Statoil and Marintek, 2013).

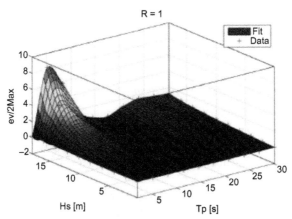

Figure 3.13. Scale parameter of extreme value distribution (Gumbel type) for tension at upper end of flexible riser as function of sea state parameters.

For the extreme value distribution, repeated simulations are carried out for each sea state in order to obtain a sample of sufficient size. A Gumbel distribution function is subsequently fitted to each sample of extreme values. Response surfaces are fitted for each of the two parameters (i.e., scale and shape) of these distributions as functions of the significant wave height and peak period. As an example, the response surface for the scale parameter of the tension at the top end is shown in Fig. 3.13. As observed, there is a strong dependency on the peak period with a maximum in the interval between 5 and 10 seconds.

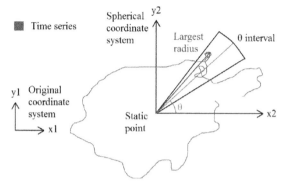

Figure 3.14. Construction of the convex hull.

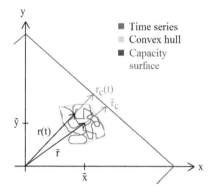

Figure 3.15. Capacity surface versus convex hull.

Considering next the joint statistics of a pair of response processes, a coordinate system which is centered at the mean value is referred to as (x_2, y_2). The extreme value of the radius vector within a certain angular interval in this coordinate system can then be determined based on a sample function as shown in Fig. 3.14. By repeating the simulation, a sample of extreme values is obtained and a corresponding Gumbel distribution function can be fitted. Generalization to three components can be achieved by introducing a second angle which corresponds to the out-of-plane response component.

The capacity surface for the riser cross section can also be introduced in the response space as illustrated for two response components in Fig. 3.15. The response process along the horizontal axis may, e.g., correspond to the pipe tension and the process along the vertical axis to one of the curvature components. A simplified capacity surface which is based on linear interpolation between the tensile capacity and the critical curvature is shown. The vector which represents the static response (i.e., the mean value vector) is denoted by $\bar{\mathbf{r}}$ in the figure.

In three dimensions, the normalized failure surface is accordingly expressed as follows:

$$\left|\frac{F}{F_{Critical}}\right| + \left|\frac{c_y}{c_{y,cr}}\right| + \left|\frac{c_z}{c_{z,cr}}\right| \leq 1 \quad , F \geq 0 \; (tension) \tag{3.48}$$

where the three denominators are the critical value along the three respective response axes for the tension and the two curvature components.

For each point in time, a scale factor can be computed which represents the utilization, i.e., the ratio of the dynamic radius vector along a specific direction to the distance from the origin to the capacity surface for the same direction, $\mathbf{r}_{Capacitysurface}$:

$$SF(t)$$

$$= Scale\; factor(t) = \frac{\sqrt{F(t)^2 + c_y(t)^2 + c_z(t)^2}}{\mathbf{r}_{Capacitysurface}} \tag{3.49}$$

A similar scale factor can also be defined for the static component as shown in Fig. 3.14.

Based on the probability distribution for the extreme radius vector in each of the angular sectors for a particular sea state, the corresponding extreme value distributions corresponding to return periods of 1 year, 10 years and 100 years can be established. By subsequently performing a weighting across all the sea states in the scatter diagram, the resulting long-term extreme value distributions for the same return periods are computed. The result is illustrated in Fig. 3.16, where the yellow color corresponds to the 100-year return period, the green color corresponds to the 10-year return period and the red color to the 1-year return period. It is seen that the axial force has a low value relative to the critical one while the curvatures exceed the capacity surface. This is due to the "naïve" model of the flexible riser which presently is applied for the upper end section. The presence of a bending

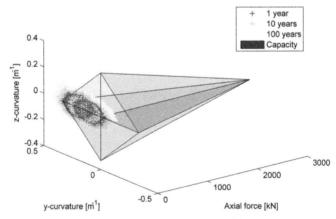

Figure 3.16. Long-term convex hull versus capacity surface for the cross-section at the upper end of the riser. Beneficial effect of bending stiffener is not accounted for.

stiffener in this region is completely neglected in the present numerical model. Such a bending stiffener will reduce the magnitudes of the curvature components significantly and can hence be designed such that there is sufficient reserve margin, i.e., the scale factor will have an acceptable value.

The number of sea states which are required in order that the long-term extreme value will converge is next considered. The grid in Fig. 3.10 corresponds to a 10 x 10 grid, and by comparison with results for grids of 6 × 6 and 8 × 8 sea states it is found that this is sufficiently accurate (i.e., with an error of less than 0.5 percent for the extreme value of the axial force for a 1-year return period). Already the 8 × 8 grid provides good results.

The extreme response values can next be computed by application of the response contour tubes. The "cross-section" of the contour response tube is represented by the three-dimensional convex hull for each sea state along the contour. As observed, the three-dimensional convex hull for the present case is a essentially two-dimensional since the variation of the normalized axial force is quite modest.

The highest value of the scaling factor SF(t) for the convex hulls (i.e., the tube cross-section) along the contour can then be determined. Some of the sea-states which are located along the contours are given in Table 3.1 below.

Subsequently, a comparison between this value to the corresponding value which is obtained from the long-term analysis can be made. The results for the individual response components are given in Table 3.2. It is seen that for the dominant responses, application of the 90% fractile for the short term distribution along the contour will give quite accurate results (i.e., with only a slight over prediction).

Table 3.1. Some of the sea states along the 1-year, 10-year and 100-year environmental contour lines.

Sea state no.	1-year		10-years		100 years	
	T_p [sec]	H_s [m]	T_p [sec]	H_s [m]	T_p [sec]	H_s [m]
1	6.2	3.9	6.0	4.2	5.8	4.5
5	7.8	5.4	7.9	6.1	8.0	6.7
10	9.7	7.3	10.0	8.4	10.2	9.4
15	11.3	9.0	11.7	10.5	12.1	11.8
20	12.7	10.3	13.2	12.0	13.7	13.7
25	13.8	11.0	14.6	12.9	15.2	14.7
30	14.7	11.1	15.7	13.0	16.5	14.8
35	15.4	10.0	16.5	12.3	17.4	14.0
40	15.8	9.3	16.9	10.8	17.9	12.3
45	16.1	7.7	17.1	8.8	18.1	9.9
50	16.6	5.8	17.6	6.5	18.4	7.2
55	17.6	3.9	19.1	4.2	20.5	4.5

Table 3.2. Comparison of results from full long-term and response contour-tube analysis.

Response type	Long-term response (dynamic)	Contour line	
		Sea state no.	X_{90}/X_{LT}
Axial force	4.44 kN	10	1.15
y-curvature	0.27 m^{-1}	18	1.03
z-curvature	0.15 m^{-1}	18	1.07

X_{90}/X_{LT} designates the ratio between the response obtained by application of the 90% fractile of the extreme value distribution for the critical sea state along the contour to the response obtained by application of the long-term analysis.

3.2.7.6 Concluding Remarks

The joint dynamic response of tension and curvature at the cross section of a flexible riser configuration was considered. Comparison was made between extreme response levels which were computed from a full long-term analysis versus a contour tube analysis. A good agreement was observed if the 90% fractile for the extreme sea-state along the contour was applied. Mainly, a slight over prediction was observed by applying the contour tube analysis versus the full long-term analysis for this particular choice of fractile.

As a next step, corresponding results which are obtained by application of more realistic Finite Element models of the upper part of the riser is of interest. This is in order to verify that the long-term analysis and the contour line approach will give comparative results also for that case.

A comparison between different ways of constructing the "tube cross-sections" is highly relevant. These cross-sections are presently established by application of time domain response analysis and convex hulls. The accuracy of less time-consuming methods, e.g., based on response contours with different time scales for the different response quantities can then be investigated.

One of the benefits of the present geometric approach is that the dominant response processes relative to a given capacity surface can be easily identified. The effect of applying capacity surfaces of different types can also be studied in a straightforward manner.

3.2.8 *Structural Reliability Analysis*

3.2.8.1 Introduction

Failure of a structure generally designates the event that the structure does not satisfy a specific set of functional requirements. Hence, it is a fairly wide concept which comprises such diversified phenomena as loss of stability, excessive response levels in terms of displacements, velocities or accelerations, as well as plastic deformations or fracture, e.g., due to overload or fatigue.

The consequences of different types of failure also vary significantly. Collapse of a single sub-component does not necessarily imply that the structure as a system immediately loses the ability to carry the applied loads. At the other extreme, a sudden loss of global stability is frequently accompanied by a complete and catastrophic collapse of the structure. Failure can also consist of a complex sequence of unfortunate events, possibly due to a juxtaposition of low-probability external or man-made actions and internal defects.

In engineering design, distinction is typically made between different categories of design criteria. These are frequently also referred to as limit states. The three most common categories are the Serviceability Limit State (SLS), the Ultimate Limit State (ULS) and the Fatigue Limit State (FLS). Many design documents also introduce the so-called Accidental Limit State in order to take care of unlikely structural conditions that may imply high losses.

Engineering design rules are generally classified as Level I reliability methods. These design procedures apply point values for the various design parameters and also introduce specific codified safety factors (also referred to as partial coefficients) which are intended to reflect the inherent statistical scatter associated with the parameters.

At the next level, second-order statistical information (i.e., information on variances and correlation properties in addition to mean values) can be applied if such is available. The resulting reliability measure and analysis method are then in general referred to as a Level II reliability method. At Level III, it is assumed that a complete set of probabilistic information is at hand (i.e., in terms of joint probability density and distribution functions).

3.2.8.2 Failure Function and Probability of Failure

The common basis for the different levels of reliability methods is the introduction of a so-called failure function (or limit state function, or g-function) which gives a mathematical definition of the failure event in mechanical terms. In order to be able to estimate the failure probability, it is necessary to know the difference between the maximum load a structure is able to withstand, R (often referred to as *resistance*), the loads it will be exposed to, Q, and the associated *load effects S*. The latter are typically obtained by means of (more or less) conventional structural analysis methods. For this "generic" case, the "g-function" is then expressed as:

$$g(R,S) = R - S \tag{3.50}$$

For positive values of this function (i.e., for $R > S$), the structure is in a safe state. Hence, the associated parameter region is referred to as the *safe domain*. For negative values (i.e., $R < S$), the structure is in a failed condition. The associated parameter region is accordingly referred to as the *failure domain*. The boundary between these two regions is the failure surface (i.e., $R = S$). The reason for application of these generalized terms is that the scalar quantities R and S in most cases are functions of a number of more basic design parameters. This implies that the simplistic two-dimensional formulation in reality involves a much larger number of such parameters corresponding to a reliability formulation of (typically) high dimension.

Here, a brief introduction is given to the basis for Level III structural reliability methods. Further details of these methods are found, e.g., in Refs. Madsen et al. (1986) and Melchers (1999). When concerned with waves, wind and dynamic structural response, it is common to assume that the statistical parameters are constant over a time period with a duration of (at least) 1 hour. This is frequently referred to as a *short-term statistical analysis*. A further assumption is typically that the stochastic dynamic excitation processes (i.e., the surface elevation or the wind turbulence velocity) are of the Gaussian type.

If the joint probability density function (or distribution function) of the strength and the load effect, i.e., $f_{RS}(r,s)$ is known, the probability of failure can generally be expressed as

$$p_f = P(Z = R - S \leq 0) = \iint\limits_{R \leq S} f_{R,S}(r,s) dr ds \qquad (3.51)$$

where the integration is to be performed over the failure domain, i.e., the region where the strength is smaller than or equal to the load effect.

This is illustrated in Fig. 3.17a, where both the joint density function and the two marginal density functions $f_R(r)$ and $f_S(s)$ are shown (the latter are obtained by a one-dimensional integration of the joint density function with respect to each of the variables from minus to plus infinity). The joint density function can then be split into two separate pieces as shown in parts (b) and (c) of the same figure. The failure probability can now be interpreted in a geometric sense as the volume of the joint density function which is located in the failure domain, i.e., the part of the plane to the right of the line $R = S$ (i.e., the region for which $S > R$). This corresponds to the slice of the volume of the joint pdf which is shown in Fig. 3.17c.

For the case of independent variables, the joint density function is just expressed as the product of the two marginal density functions. The resulting expression for the failure probability then becomes:

$$p_f = P(Z = R - S \leq 0) = \iint\limits_{R \leq S} f_R(r) \cdot f_S(s) dr ds \qquad (3.52)$$

where it is assumed that R and S are independent. By performing the integration with respect to the resistance variable, this can also be expressed as

$$p_f = P(Z = R - S \leq 0) = \int_{-\infty}^{+\infty} F_R(s) f_s(s) ds \qquad (3.53)$$

where

$$F_R(s) = P(R \leq s) = \int_{-\infty}^{s} f_R(r) dr \qquad (3.54)$$

Pdf of r and s

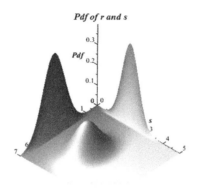

(a) Joint Pdf of *r* and *s*

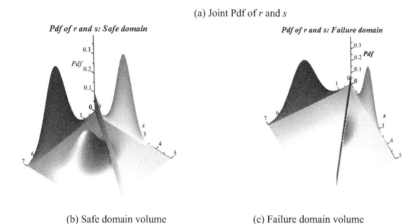

(b) Safe domain volume   (c) Failure domain volume

Figure 3.17. Interpretation of failure probability as a volume.

This situation is illustrated in Fig. 3.18 where the two marginal density functions now are shown in the same plane. The interval which contributes most to the failure probability is where both of the density functions have non-vanishing values (i.e., in the range between 1 and 3.5 for this particular example).

The integral in Eq. (3.53) is known as a convolution integral, where $F_R(r)$ denotes the cumulative distribution function of the mechanical resistance variable R. Closed-form expressions for this integral can be obtained for certain distributions, such as the Gauss distribution as will be discussed below. The resistance probability density function in Eq. (3.59), $f_R(r)$, is in many cases represented as a Gaussian or Lognormal variable. The density function of the load-effect, $f_S(s)$, typically corresponds to extreme environmental conditions (such as wind and waves) and is frequently assumed to be described by a Gumbel distribution, see, e.g., Gumbel (1958).

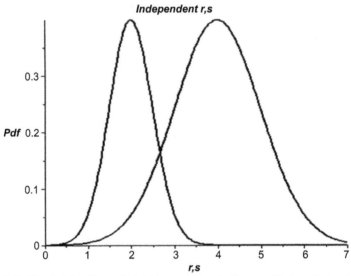

Figure 3.18. Marginal densities projected into same plane for the case of independent variables.

3.2.8.3 Time Variation of Load and Response

As the loads on marine structures are mainly due to wave-, wind and current, the statistical properties will fluctuate with time. The resistance will also in general be a function of time, e.g., due to deterioration processes such as corrosion (this cán clearly be counteracted by repair or other types of strength upgrading). A prominent example of the influence of corrosion is the time-varying reliability of pipeline systems, see, e.g., Mohd et al. (2014) and Leira et al. (2016) for probabilistic representations of such effects.

For ageing structures, a number of other deterioration processes than corrosion are generally also present. Some examples of phenomena that need to be considered are, e.g., fatigue cracks, deficiencies, misalignments and reported non-conformances, stress concentrations due to change of geometry (e.g., caused by corrosion).

Furthermore, a typical situation is that the extreme load effects increase with the duration of the time interval (i.e., the 20 year extreme value is higher than the 10 year extreme value, and the 3-hour extreme load-effect during a storm is higher than the 1-hour extreme load-effect).

This situation is illustrated by Fig. 3.19 for a relatively long time horizon. Here, t denotes time, and $t_0 = 0$ is the start time. The second "time slice" is at 10 years, and the third slice is at 20 years. The corresponding probability density functions of the mechanical resistance and the load effect are also shown for each of the three slices.

The figure illustrates that the structure will fail if (at any time during the considered time interval)

Pdf Time variation

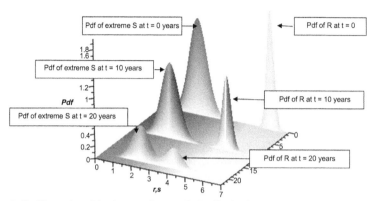

Figure 3.19. Illustration of the time-varying marginal Pdfs of resistance, $r(t)$, and extreme load-effect $s(t)$.

$$Z(t) = r(t) - s(t) < 0, \tag{3.55}$$

where $Z(t)$ is referred to as the safety margin (which varies with time). The probability that the event described by Eq. (3.55) will take place can be evaluated from the amount of overlap by the two probability density functions $f_R(r)$ and $f_S(s)$ at each time step, as shown in Fig. 3.19. At $t = 0$ and 10 years, these two functions barely touch each other, while at $t = 20$ years, they have a significant amount of overlap. The latter case represents a corresponding increase of the failure probability.

If it is chosen to use time-independent values of either R or S (or both), the minimum value of (2.5) during the interval $[0,T]$ must be used, where T denotes the design life time or the duration of a specific operation under consideration. In relation to the maximum load effect, an extreme value distribution, such as the Gumbel distribution (also referred to as the type I asymptotic form), as mentioned above. The Gumbel distribution may be applied in cases where the initial distribution has an exponentially decaying tail which is the case, e.g., for stochastic processes of the Gaussian type. Similarly, the probability density and distribution function of the minimum value is relevant. For durations of the order of a few days or less, simplifications can typically be introduced, since decrease of the strength properties on such limited time scales can usually be neglected.

The variation of the density functions will furthermore be different for the different types of limit states. For the fatigue limit state, the "resistance" can, e.g., correspond to the permissible cumulative damage (i.e., given by a Miner-Palmgren sum equal to 1.0). This is a time-independent quantity which may still be represented by a (time-invariant) random variable. The "load-effect" will now correspond to the (random) cumulative damage which is obtained from the probability distribution of the stress range cycles. If there are other deterioration processes present (such as corrosion), the "resistance" will clearly decrease with time also for this type of limit state.

3.2.8.4 Simplified Case

As a special (and simplified) case, we next consider the situation when both R and S are Gaussian random variables, with mean values μ_R and μ_S and variances σ^2_R and σ^2_S. Furthermore, the two variables are assumed to be uncorrelated. The quantity $Z = R - S$ will then also be Gaussian, with the mean value and variance being given by

$$\mu_Z = \mu_R - \mu_S$$

and $$\sigma^2_Z = \sigma^2_R + \sigma^2_S$$

(3.56)

The probability of failure may then be written as

$$p_f = P\,(Z = R - S \leq 0) = \Phi\left(\frac{0 - \mu_Z}{\sigma_Z}\right) = \Phi\left(-\frac{\mu_Z}{\sigma_Z}\right)$$

(3.57)

where $\Phi(.)$ is the standard normal distribution function (corresponding to a Gaussian variable with mean value 0 and standard deviation of 1.0). By inserting Eq.(3.56) into Eq. (3.57), we get

$$p_f = \Phi\left(-\frac{\{\mu_R - \mu_S\}}{\sqrt{\sigma^2_R + \sigma^2_S}}\right) = \Phi(-\beta)$$

(3.58)

where

$$\beta = \frac{\{\mu_R - \mu_S\}}{\sqrt{\sigma^2_R + \sigma^2_S}}$$

(3.59)

is defined as the safety index, see Cornell (1969). By defining an acceptable failure probability (i.e., $p_f = p_A$) on the left-hand side of this equation, one can find the corresponding value of β, i.e., β_A, that represents an acceptable lower bound on β (since decreasing β results in a higher failure probability). This value can be used to determine in a probabilistic sense whether the resistance R is within an acceptable range as compared to the load effect, S.

It is emphasized that these expressions for the failure probability are overly simplistic. Firstly, the effect of correlation between the variables is not included. Secondly, only two random variables are involved and thirdly, both variables are assumed to be Gaussian. However, the main idea here has been to introduce the concepts of reliability index and probability of failure and to illustrate how the statistical properties of the variables will influence the result. A summary is given in the next section of more accurate and refined methods for calculation of the failure probability.

3.2.9 Calculation of Failure Probability for Non-Gaussian Random Variables

The index above can also be extended to handle reliability formulations which involve random vectors of arbitrary dimensions. Typically, both the load effect term, S, and the resistance term, R, are expressed as functions of a number of basic parameters of a random nature. Assembling these in the respective vectors $\mathbf{X}_S$ and $\mathbf{X}_R$, the failure function becomes a function of the vector $\mathbf{X} = [\mathbf{X}^T_S, \mathbf{X}_R]^T$. The failure surface will accordingly be defined by the equation $g(\mathbf{X}) = 0$.

The safety index can readily be extended to comprise correlated as well as non-Gaussian variables. For general types of probability distributions, the failure probability as expressed by the integral in Eqs. (3.57) and (3.58) can be computed, e.g., by numerical integration. However, there also exist efficient approximate methods based on transformation into uncorrelated and standardized Gaussian variables.

In the case of non-Gaussian and uncorrelated variables this transformation is based on the following expressions (which is identical to the transformation according to Eq. (3.45) above if independence is assumed):

$$\Phi\left(u_1\left(t\right)\right) = F_{X_1}\left(x_1\left(t\right)\right)$$

$$..$$
$$\qquad\qquad\qquad\qquad\qquad\qquad\qquad\qquad\qquad\qquad (3.60)$$
$$..$$

$$\Phi\left(u_n\left(t\right)\right) = F_{X_n}\left(x_n\left(t\right)\right)$$

The simplified $g(R,S) = (R - S)$ reliability formulation may serve to illustrate how this transformation works for two different cases. As a first reference case, the two basic variables are taken to be uncorrelated Gaussian variables with mean values ($\mu_R = 3.0$, $\mu_s = 1.0$) and standard deviations ($\sigma_R = 0.1, \sigma_S = 0.2$). The relationship between the original basic variables and the transformed standardized Gaussian variables are then simply expressed as $R = 0.1U_1 + 3$ and $S = 0.2U_2 + 1$. The corresponding failure function in the transformed and normalized (U_1, U_2)-plane is shown as the upper surface in Fig. 3.20.

As a second case, the basic random variables R and S are both taken to be uncorrelated lognormal variables with the same mean values and standard deviations as before. The two corresponding transformations based on Eq. (3.60) are then expressed as:

$$\ln(r) = \sigma_{z1}u_1 + \mu_{Z1} \quad \ln(s) = \sigma_{z2}u_2 + \mu_{Z2} \qquad\qquad (3.61)$$

where

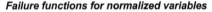

Failure functions for normalized variables

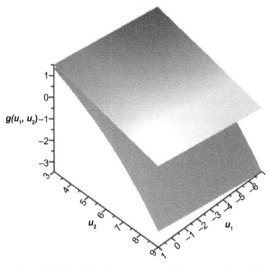

Figure 3.20. Comparison of failure functions expressed in terms of transformed (normalized) Gaussian variables u_1 and u_2. Upper: Gaussian basic random variables. Lower: Lognormal basic random variables.

$$\left(\sigma_{z1}^2 = \ln\left(1+\left(\frac{\sigma_R^2}{\mu_R}\right)\right) = 0.0011, \sigma_{z2}^2 = \ln\left(1+\left(\frac{\sigma_S^2}{\mu_S}\right)\right) = 0.039\right)$$

and   (3.62)

$$\left(\mu_{z1} = \ln(\mu_R) - 0.5\sigma_{z1}^2 = 1.098, \mu_{z2} = \ln(\mu_S) - 0.5\sigma_{z2}^2 = -0.0196\right)$$

The corresponding failure function $g(u_1, u_2) = \exp(\sigma_{z1}u_1 + \mu_{z1}) - \exp(\sigma_{z2}u_2 + \mu_{z2})$ is shown in the lower part of Fig. 3. 19.

Transformation of *correlated* non-Gaussian variables requires that both marginal and conditional distribution functions are applied. For a number of n random variables the expressions become (see also Eq. (3.40)):

$$\Phi\left(u_1(t)\right) = F_{X_1}\left(x_1(t)\right)$$

..

..   (3.63)

$$\Phi\left(u_n(t)\right) = F_{X_n|X_1,X_2,\ldots\ldots,X_{n-1}}\left(x_n(t)\,|\,x_1(t), x_2(t), \ldots..x_{(n-1)}(t)\right)$$

where the conditional cumulative distribution functions of increasing order are required.

Computation of the failure probability is frequently based on approximating the failure surface by its tangent plane at a proper point, or a second order surface at the same point. This point is identified by means of numerical iteration and is the point which is closest to the origin in the transformed space of standardized Gaussian variables (i.e., all of which have mean value zero and unit standard deviation).

For a more detailed description of these procedures (i.e., related to transformation of the variables and searching for the design point), reference is, e.g., made to Madsen et al. (1986), Melchers (1999), Hasofer and Lind (1974), Ditlevsen (1981), Hohenbichler and Rackwitz (1983), Rosenblatt (1952), Breitung (1984), Tvedt (1989) and Der Kiureghian and Li (1986).

3.3 Current Rules and Industry Practices

3.3.1 General

There are a large number of design codes and standards that apply for marine structures, and most of these address the topic of stochastic dynamic response analysis in some form. Design codes are developed both at the national and international level. In addition, a number of Classification societies have issued their own set of rules, which are becoming more and more harmonized based on collaboration within their joint organization IACS.

There are typically different rules for bottom-fixed versus floating (including compliant) systems as exemplified by the NORSOK and DNV set of rules which are applied for the Norwegian Continental Shelf, see, e.g., NORSOK (1997; 2007; 2012; 2013; 2015) and DNVGL (1992; 1992a; 2010; 2015; 2015a; 2015b; 2015c; 2015d; 2015e; 2016; 2016a; 2016b; 2017; 2017a; 2017b; 2017c). Similar differences between bottom-fixed and floating systems are also found in the ISO and API set of rules, see, e.g., ISO (2004; 2006; 2006a; 2006b; 2009; 2010; 2012; 2013; 2013a; 2014; 2014a; 2015; 2016; 2016a; 2016b; 2017) and API (2007; 2009; 2011; 2011a; 2014; 2014a; 2015; 2015a; 2015b; 2015c).

3.3.2 Limit States and Design Formats

Although the characteristics of these sets of rules are significantly different, there are still some common features with respect to which categories of limit states that are applied. As already mentioned these correspond to the Ultimate Limit State (ULS), Fatigue Limit State (FLS), Serviceability Limit State (SLS) and the Accidental Limit State (ALS).

The most common types of mechanical failure modes to be prevented by a proper design are generally yielding, buckling and fracture with the latter two being most critical. Interaction between these modes also relevant, such as initial yielding followed by full plastification of the cross-section and subsequent buckling instability leading to fracture. Conversely, initial buckling failure can lead to plastic deformations and subsequent fracture.

Regarding design formats, there are two main types. These are typically referred to as the Working Stress Design (WSD) format versus the Partial Load and Resistance Factors Design (LRFD) format. In both cases, values of the characteristic loads and the characteristic material strength are first obtained. Based on these values, the dimensioning load effect and the dimensioning resistance of the structure (or structural component) are subsequently calculated. For the WSD, the dimensioning check is simply performed by verifying that the ratio between the characteristic resistance and the characteristic load effect is higher or equal to a specified safety factor. For the LRFD, separate partial safety factors are applied for the loads and for the material strength. For different types of loads (e.g., functional loads versus environmental loads) the values of the partial safety factors are generally also different. By application of these safety factors, the design load effect and the design resistance are computed. The corresponding design check is then performed by verifying that the latter is higher than or equal to the former. Historically, the WSD was very much applied until it was (and still is) more and more replaced by the more refined LRFD. While the WSD is very easy to apply, the LRFD gives more refined design solutions and also provides a more uniform safety level for different types of structures and structural components.

3.3.3 Types of Structural Components

It also has to be remembered that the structural components under consideration are clearly different for the different categories of marine structures. For floating structures two-dimensional plate and shell type of components are highly relevant. This also includes stiffened panels which are composed of plate (shell) and beam members. For the main load-carrying system of bottom-fixed structures, beams (and rods) are most relevant. The connecting joints are also very important as they represent critical structural components. Cable-type members are much applied for mooring and position-keeping purposes.

Methods for computation of loads and load effects are typically also described in the design codes as discussed in Section 3.2.3 above. In some cases, these models are updated as new editions of the codes are issued. This may both reflect advancements associated with load and resistance models as well as calculation methods. In principle, these updates should also be reflected in a modified set of partial coefficients associated with the loads and load effects to be applied for design. However, this is not always the case if the modifications of the design codes are of a modest nature.

3.3.4 Structural Integrity Management, Requalification and Life Extension

Design standards and guidelines related to re-assessment, requalification and life extension of marine structures are increasingly focused upon. As, for example, the average age of North Sea offshore structures are increasing as these documents come more into focus.

Codes and guidelines covering structural integrity management typically introduce a classification of structures and structural components. Such a classification makes it possible to establish the basis for risk screening and determination of what types of inspection and what methods to use.

The classification of structural components will typically consider the following issues: (i) likelihood of corrosion, (ii) likelihood of overload, (iii) likelihood of fatigue cracks, (iv) consequence of component failure (i.e., the ability of the remaining structure to withstand external actions, e.g., by means of damage tolerance or redundancy), (v) possibility of progressive failure, (vi) deficiencies, misalignments and non-conformances reported in the DFI résumé, (vii) inspection history, (viii) presence of stress concentrations and critical load transfer connections, (ix) access and preparations to facilitate access for inspection, maintenance and repair, (x) extent of monitoring for the structural component and (xi) suitability of inspection method (i.e., deployment, reliability, accuracy).

A comprehensive evaluation of likelihood and consequences of failure are frequently required. This implies that transition from a classification based on type of structural category to a fully risk based approach may be required. In many cases it may be of benefit to apply probabilistic design methods based on first principles. The available information about the structure and its environment can then be reflected in an optimal way such that the various sources of uncertainty and statistical scatter are properly taken into account. This applies to structural members and components which are primarily used for fabrication, transport and installation phases, see, e.g., NORSOK N-005 (1997) and N-006 (2015).

3.4 Recommendations for Advanced Engineering Practice

Design of marine structures involves different design stages, starting with preliminary design (concept studies), followed by more detailed design and subsequently final design verification. For the early studies, fast and efficient calculation methods are in demand which may allow comparison of a relatively large number of alternative solutions. Frequency domain analysis methods offer a useful tool for such purposes. The contour line approach can also facilitate estimation of extreme load effects at this design stage. Relatively few sea states can also be applied for estimation of fatigue damage in order to minimize the computational effort.

During detailed design increasingly accurate calculation models and more advanced numerical simulations are generally required. Time domain analysis is highly relevant, possibly starting with regular wave analysis followed by non-linear stochastic time domain analysis. In this connection, accurate extreme value estimation becomes increasingly important. This requires that several samples of the environmental load processes and corresponding response time series are generated. This allows the scatter of simulated extreme values to be quantified and kept at an acceptable level. For estimation of fatigue damage, convergence studies in relation to the applied number of sea states in the scatter diagram should also be carried out.

For the purpose of design verification, specialized studies may be required in relation to critical design issues. Direct design methods based on first principles are highly relevant, involving detailed numerical models of the environment, load mechanisms and structural details. The computational power which is available in for such calculations is steadily increasing. Application of Computational Fluid Dynamics (CFD) is becoming more and more common, sometimes fully integrated with the structural Finite Element Models (FEM).

As already mentioned, refined and increasingly advanced calculation methods are becoming available for the purpose of life extension and reassessment of structures are also becoming available. In this connection it is also important that the effects of applied structural repair methods and other types of upgrading are accurately quantified. This also includes their influence related to the stochastic load effects which are present in different parts of the structure.

Fully probabilistic design for the purpose of a final verification of the structural reliability level is highly relevant. Quantification of model uncertainties which are associated with applied design formulas according to the code. Furthermore, by application of probabilistic formulations, the proper choice of updated design safety factors can be identified by calibration against reliability levels in existing codes. As improvements of the predictive power and the accuracy associated with refined formulations takes place, this should also be reflected in the applied design formats at large.

3.5 Concluding Remarks and Further Studies

Increasingly, refined models and methods are being applied for stochastic dynamic analysis of marine structures. Although powerful computers are now available, fully nonlinear stochastic time domain analysis and probability domain analysis may in general require significant efforts. This applies not the least to post-processing of response time series and estimation of the corresponding statistical properties. It is accordingly important that skilled designers can assess the results of such complex calculations in an adequate manner.

A future challenge (which has only to a limited extent been addressed in the research literature) is incorporation of control forces for highly dynamic systems such as floating wind turbines and vessels equipped with DP-systems. The analysis of such systems requires cooperation between several engineering disciplines which will be necessitated in order to obtain more and more optimized marine structures of tomorrow.

An interesting trend is the application of environmental data records in connection with definition of structural loads and load combinations. Full-scale measurements and structural monitoring systems are increasingly being utilized which serves to assess the structural integrity level. At the same time this allows quantification of the inherent model uncertainties associated with design of marine structures. This development holds a great promise for increasingly more reliable marine structural concepts and operations in the future.

Nomenclature

Scalars

A_{kl}	Amplitude corresponding to frequency number k and direction number l
$a(x,t)$	Drift coefficient in diffusion equation for state x at time t
$\tilde{a}, m$	Constants defining SN-fatigue curve
$b(x,t)$	Diffusion coefficient in diffusion equation for state x at time t
$c_x, c_{x,crit}$	Curvature of flexible riser pipe in the x-direction and its corresponding critical value
$c_z, c_{z,crit}$	Curvature of flexible riser pipe in the z-direction and its corresponding critical value
$E[x_{E,T}]$	Expected largest value of stochastic process x(t) for duration T
$E[D(T)]$	Expected fatigue damage during time period T
F, F_{crit}	Axial force in flexible riser and the corresponding critical value
$F_X(x), F_{\mathbf{X}}(\mathbf{x})$	Probability distribution function of scalar X and vector **X**, respectively
$f_X(x), f_{\mathbf{X}}(\mathbf{x})$	Probability density function of scalar X and vector **X**, respectively
$F_{R,L}(r)$	Long-term cumulative distribution of response maxima
$F_R(r\|\mathbf{x})$	Short-term distribution of response maxima for given values of environmental parameters which are contained in the vector **x**
$F_{L,E}(r_E)$	Long-term cumulative distribution of extreme response for an arbitrary sea state
$F_{S,E}(r_E\|\mathbf{x})$	Short-term distribution of extreme response for a sea state with given values of environmental parameters which are contained in the vector **x**
$g(\mathbf{x})$	Failure function expressed in terms of variables contained in the vector **x**
H_s	Significant wave height
$K_j(\mathbf{x})$	Intensity coefficient, i.e., limit for the ratio $\mu_j(\mathbf{x'})/\Delta t$ when Δt goes to zero
$p(\mathbf{x},t)$	Probability density function of state vector **x** at time t
$p(\mathbf{x},t\|\mathbf{x'},t')$	Transition probability density from state **x** at time t, to **x'** at time t'
$p(H_s,T_p)$	Joint density function of significant wave height and peak period

$p(T_p|H_s)$	Conditional probability density function of peak period for given value of significant wave height
$p(H_s)$	Marginal probability density function of significant wave height
R	Random variable representing the capacity (strength)
S	Random variable representing the load effect (e.g., stress or strain)
$S_{\zeta\zeta}(\omega)$	Wave spectral density function as a function of frequency
T_p	Peak period
$Z(t) = r(t) - s(t)$	Time dependent safety margin, i.e., difference between time dependent resistance r(t) and time dependent load effect s(t)

Vectors and matrices

$\mathbf{B}(\omega)$	Frequency dependent transfer matrix of total structural and hydrodynamic system
$\mathbf{C}(\mathbf{r},\dot{\mathbf{r}},t-\tau)$	Displacement and velocity dependent damping retardation matrix
$\mathbf{F}(\omega)$	Frequency dependent hydrodynamic transfer function matrix
$\mathbf{H}(\omega)$	Virtual frequency response function
$\mathbf{K}(\mathbf{r})$	Displacement dependent stiffness matrix
$\mathbf{M}(t-\tau)$	Mass retardation matrix
$\mathbf{Q}(t)$	Time varying load vector
$\mathbf{Q}(\omega,\theta)$	Frequency and direction dependent load vector
$\mathbf{r}_{Capacitysurface}$	Distance from the origin to the capacity surface for a given vector direction
$\mathbf{S}_x(\omega)$	Spectral density matrix of vector process $\mathbf{X}(t)$ also including cross-spectral densities

Greek letters

α, u	Parameters of Gumbel distribution
β	Reliability index
ε	Bandwidth parameter
$\gamma = 0.5772$	Euler constant
ω, ω_p	Frequency (in radians/second) and peak frequency
$\Delta\omega$	Increment of frequency

θ	Direction angle. This letter also designating the mean value of $\ln(H_s)$
$\Delta\theta$	Increment of direction angle
$\Delta\sigma$	Stress cycle
Δt	Increment of time
κ^2	Variance of $\ln(H_s)$
$\sigma_x^2, \dot{\sigma}_x^2, \ddot{\sigma}_x^2$	Variance of response process, its velocity and acceleration processes
$\sigma^2(H_s)$	Variance of $\ln(T_p)$ as a function of significant wave height
σ_R	Standard deviation of resistance
σ_S	Standard deviation of load effect
$\mu_j(\mathbf{x'})$	The j'th moment of the vector transition from state $\mathbf{x}$ to state $\mathbf{x'}$
$\mu(H_s)$	Mean value of $\ln(T_p)$ as a function of significant wave height
μ_R	Mean value of resistance
μ_S	Mean value of load effect
$v_x^+(s)$	Up-crossing rate for the level s
$v_x^+(0)$	Zero-crossing rate
$v_{x,max}^+$	Maximum possible value of the up-crossing rate
ζ, β	Weibull parameters of probability distribution of significant wave height
$\eta(\mathbf{x},t)$	Sea surface elevation at position $\mathbf{x}$ at time t
ω_k	Discrete frequency number k
$\Psi(\theta)$	Directional spreading function
$\Phi(x)$	Standard Normal cumulative distribution function
φ_{kl}	Phase angle corresponding to frequency number k and direction number l
θ_l	Discrete direction number k

Mathematical symbols

$\dfrac{\partial}{\partial}$	Partial derivative with respect to variable t (e.g., time)
$J_{ij}(x_i) = \dfrac{\partial u_i}{\partial x_j}$	Element (i,j) of Jacobian matrix for Rosenblatt transformation

References

Almar-Naess, A. 1999. Fatigue Handbook: Offshore Steel Structure, 3rd revision, Tapir Forlag, Trondheim.

API RP 2FPS (2011) – Recommended Practice for Planning, Designing, and Constructing Floating Production Systems, Second Edition.

API RP 2GEO (2011a) – Geotechnical and Foundation Design Considerations.

API RP 2MET (2014) – Petroleum and natural gas industries - Specific requirements for offshore structures – Part 1: Metocean design and operating considerations, First Edition.

API RP 2MOP (2015)-Marine Operations, Petroleum and natural gas industries-Specific requirements for offshore structures-Part 6: Marine Operations, First Edition, Includes Errata (2015), 07/01/2010.

API RP 2RD (2009)-Design of Risers for Floating Production Systems (FPSs) and Tension-Leg Platforms (TLPs) standard by American Petroleum Institute.

API RP 2SIM (2014a) – Structural Integrity Management of Fixed Offshore Structures, First Edition Standard by American Petroleum Institute.

API RP 2SK (2015a) – Design and Analysis of Stationkeeping Systems for Floating Structures, Third Edition (Includes 2008 Addendum).

API RP 2SM (2007) – Recommended Practice for Design, Manufacture, Installation, and Maintenance of Synthetic Fiber Ropes for Offshore Mooring (Includes 2007 Addendum) standard by American Petroleum Institute.

API RP 2T (2015b)-Recommended Practice for Planning, Designing and Constructing Tension Leg Platforms, Third Edition standard by American Petroleum Institute.

API Spec 2F (2015c) – Mooring Chain, Sixth Edition.

Athanassoulis, G.A., Skarsoulis, E.K. and Belibassakis, K.A. 1994. Bivariate distributions with given marginals with an application to wave climate description. Applied Ocean Research 16: 1–17.

Baarholm, G. Sagli and Moan, T. 2001. Application of contour line method to estimate extreme ship hull loads considering operational restrictions. Journal of Ship Research 45(3): 227–239.

Bathe, K.J. 1996. Finite Element Procedures, Prentice Hall, Englewood Cliffs, NJ.

Beck, A.T. and Melchers, R.E. 2004. On the ensemble crossing rate approach to time variant reliability analysis of uncertain structures. Probabilistic Engineering Mechanics 19: 9–19.

Belytschko, T. and Schoeberle, D.F. 1975.On the unconditional stability of an implicit algorithm for nonlinear structural dynamics. J. Applied Mechs 97: 865–869.

Bitner-Gregersen, E. and Guedes Soares, C. 1997. Overview of probabilistic models of the wave environment for reliability assessment of offshore structures. *In*: Guedes Soares, C. (ed.). Advances in Safety and Reliability. Pergamon 2: 1445–1456.

Bitner-Gregersen, E. and Guedes Soares, C. 2007.Uncertainty of average steepness prediction from global wave databases. Proc. MARSTRUCT, Glasgow, UK, pp. 3–10.

Borgman, L.E. 1969. Ocean wave simulation for engineering design. J. Waterways and Harbours Division, ASCE 95(WW4): 556–583.

Breitung, K. 1984. Asymptotic approximations for multinormal integrals. ASCE, J. Eng. Mech. Div. 110: 357–366.

Cartwright, D.E. and Longuet-Higgins, M.S. 1956. On the statistical distribution of the maxima of a random function. Proc. of the Royal Society of London, Vol. A237, pp. 1706–1711.

Chai, W., Naess, A. and Leira, B.J. 2015. Stochastic dynamic analysis and reliability of a vessel rolling in random beam seas. Journal of Ship Research 59(2): 113–131.

Chai, W., Naess, A. and Leira, B.J. 2016. Stochastic nonlinear ship rolling in random beam seas by the path integration method. Probabilistic Engineering Mechanics.

Clough, R. and Penzien, J. 1975. Dynamics of Structures, McGraw-Hill.

Cornell, C.A. 1969. A probability-based structural code. Journal of the American Concrete Institute 60(12): 974–985.

Der Kiureghian, A. and Liu, P.L. 1986. Structural reliability under incomplete probability information. ASCE, Journal of Engineering Mechanics, 112 (1): 85–104.

Der Kiureghian, A. and Liu, P.L. 1986. Structural reliability under incomplete probability information. ASCE, J. Eng. Mech. Div. 112(1): 85–104.

Ditlevsen, O. 1981. Principle of normal tail approximation. ASCE, J. Eng. Mech. Div. 107: 1191–1208.

DNV (1992): Classification Note 30.4 Foundations.

DNV (1992a): Classification Note 30.6 Structural Reliability Analysis of Marine Structures.

DNV (2010): RP-C201 Buckling strength of plated structures.

DNVG (2016a): RP-C203 Fatigue strength analysis of offshore steel structures.

DNVGL (2015): CG-0128 Buckling.

DNVGL (2015a): CG-0129 Fatigue assessments of ship structures.

DNVGL (2015c): OS-C103 Structural design of column-stabilised units - LRFD method.

DNVGL (2015d): OS-C105 Structural design of TLP - LRFD method.

DNVGL (2015e): OS-C106 Structural design of deep draught floating units - LRFD method.

DNVGL (2016) OS-C101 Edition (April 2016) Design of offshore steel structures, general – LRFD method.

DNVGL (2016b): RP-C208 Determination of structural capacity by Non-linear FE analysis methods.

DNVGL (2017): RP-C202 Buckling Strength of Shells.

DNVGL (2017a): RP-C204 Design against Accidental Loads.

DNVGL (2017b): RP-C205 Environmental Conditions and Environmental Loads.

DNVGL (2017c): OS-C104 Structural design of self-elevating units - LRFD method.

DNVGL(2015b): OS-C102 Structural design of offshore ships.

Faltinsen, O.M. 1990. Sea Loads on Ships and Offshore Structures, Cambridge University Press.

Fergestad, D., Leira, B.J. and Hoen, C. 1994. Troll-GBS: Comparison between Measurements and Numerical Calculations, Proc BOSS, Boston.

Ferreira, J.A. and Guedes Soares, C. 2001. Modelling Bivariate Distributions of Significant Wave Height and Mean Wave Period 24: 31–45.

FerryBorges, J .and Castanheta, M. 1971. Structural Safety, course 101, 2nd edn. Laboratorio National de Engenharia Civil., Lisbon.

Fossen, T.I. 2002. Marine Control Systems. Guidance, Navigation and Control of Ships, Rigs and Underwater Vehicles, Marine Cybernetics, Trondheim, Norway.

Fouques, S., Myrhaug, D. and Nielsen, F.G. 2004. Seasonal modelling of multivariate distributions of metocean parameters with application to marine operations. Journal of Offshore Mechanics and Arctic Engineering 126: 202–212.

Fu, P., Leira, B.J. and Myrhaug, D. 2016. Parametric Study Related to the Collision Between Two Risers, Paper No. 54637, Proc. OMAE 2016, Busan, South Korea.

Gumbel, E.J. 1958. Statistics of Extremes, Columbia University Press, New York, US.

Hagen, Ø. and Tvedt, L. 1991. Vector process out-crossings as parallel system sensitivity measure. Journal of Engineering Mechanics ASCE 117(10): 2201–2220.

Hammersley, F. and Handscomb, P. 1964. Monte Carlo Methods, Methuen, London.

Hasofer, A.M. and Lind, N.C. 1974. Exact and invariant second moment code format. ASCE, J. Eng. Mech. Div. 100: 111–121.

Haver, S. 1985. Wave climate off Northern norway. Applied Ocean Research 7(2): 85–92.

Haver, S., Sagli, G. and Gran, T.M. 1998. Long-term response analysis of fixed and floating structures. Proceedings, Wave'98-Ocean Wave Kinematics. Dynamics and Loads on Structures, International OTRC Symposium, Houston.

Hohenbichler, M. and Rackwitz, R. 1981. Non-normal dependent vectors in structural safety. ASCE, J. Eng. Mech. Div. 107: 1227–1258.

Hohenbichler, M. and Rackwitz, R. 1983. First-order concepts in system reliability. Structural Safety 1: 177–188.

Hughes, T.J.R. 1976. Stability, convergence and decay of energy of the average acceleration method in nonlinear structural dynamics. Computers and Structures 6: 313–324.

Hughes, T.J.R. 1977. Note on the stability of newmarks algorithm in nonlinear structural dynamics. Int. J. Num. Meth. Eng. 11: 383–386.

ISO 19900: (2013): Petroleum and natural gas industries - General requirements for offshore structures.

ISO 19901-1: (2015): Petroleum and natural gas industries - Specific requirements for offshore structures – Part 1: Metocean design and operating considerations.

ISO 19901-2: (2004): Petroleum and natural gas industries - Specific requirements for offshore structures – Part 2: Seismic design procedures and criteria

ISO 19901-3: (2014): Petroleum and natural gas industries - Specific requirements for offshore structures – Part 3: Topsides structure.

ISO 19901-4:(2016): Petroleum and natural gas industries – Specific requirements for offshore structures – Part 4: Geotechnical and foundation design considerations.

ISO 19901-5:(2016a): Petroleum and natural gas industries – Specific requirements for offshore structures – Part 5: Weight control during engineering and construction.

ISO 19901-6:(2009): Petroleum and natural gas industries – Specific requirements for offshore structures – Part 6: Marine operations.

ISO 19901-7:(2013a): Petroleum and natural gas industries – Specific requirements for offshore structures – Part 7: Stationkeeping systems for floating offshore structures and mobile offshore units.

ISO 19901-8:(2014a): Petroleum and natural gas industries – Specific requirements for offshore structures – Part 8: Marine soil investigations.

ISO 19903: (2006) Petroleum and natural gas industries – Fixed concrete offshore structures.

ISO 19903:(2006a): Petroleum and natural gas industries – Fixed concrete offshore structures.

ISO 19904-1:(2006b): Petroleum and natural gas industries – Floating offshore structures – Part 1: Monohulls, semi-submersibles and spars.

ISO 19905-1:(2016b): Petroleum and natural gas industries – Site-specific assessment of mobile offshore units – Part 1: Jack-ups.

ISO 19905-3: (2017): Petroleum and natural gas industries – Site-specific assessment of mobile offshore units – Part 3: Floating unit.

ISO 19906 (2010): Petroleum and natural gas industries – Arctic offshore structures.

ISO/TR 19905-2:(2012): Petroleum and natural gas industries – Site-specific assessment of mobile offshore units – Part 2: Jack-ups commentary and detailed sample calculation.

Jia, J. 2014 .Essentials of Applied Dynamic Analysis, Springer.

Johannesen, K., Meling, T.S. and Haver, S. 2001. Joint distribution for wind and waves in the Northern North Sea. Proc. Int. Offshore and Polar Engineering Conference, ISOPE, Stavanger, Norway.

Johnson, N.L. and Kotz, S. 1972. Distributions in Statistics: Continuous Multivariate Distributions, John Wiley & Sons, New York.

Jonathan, P., Flynn, J. and Ewans, K. 2010. Joint Modelling of Wave Spectral Parameters for Extreme Sea States. Ocean Engineering. doi:10.1016/j.oceaneng.2010.04.004.

Krenk, S. 2008. Extended state-space time integration with high-frequency energy dissipation. International Journal for Numerical Methods in Engineering 73: 1767–1787.

Kumar, P. and Narayanan, S. 2006. Solution of Fokker-Planck Equation by Finite Element and Finite Difference Methods for Nonlinear Systems, Vol. 31, Part 4, pp. 445–461.

Leira B.J., Karunakaran, D. and Hoen, C. 1994. Nonlinear behaviour and extreme dynamic response of the troll gravity platform. Proc. OMAE, Houston.

Leira, B.J., Næss, A. and Næss, O.E. 2016. Reliability analysis of corroding pipelines by enhanced Monte Carlo simulation. International Journal of Pressure Vessels and Piping 144: 11–17.

Lindstad, H.B. 2013. Contour Methods for Estimation of Multi-dimensional Extreme Riser Response, Master thesis. Department of Marine Technology, NTNU, Trondheim, Norway.

Longuet-Higgins, M.S. 1952. On the statistical distribution of the heights of sea waves. Journ. Maritime Research 11(.3).

Madsen, H., Krenk, S. and Lind, N.C. 1986. Methods of Structural Safety. Prentice-Hall, Englewood Cliffs, New Jersey.

Masud, A. and Bergman, L.A. 2005. Solution of the four-dimensional fokker-planck equation: Still a challenge. Proc. ICOSSAR 2005, Millpress, Rotterdam.

Mathiesen, J. and Bitner-Gregersen, E. 1990. Joint distributions for significant wave height and wave zero-upcrossing period. Applied Ocean Research 12(2): 93–103.

Melchers, R.E. 1999. Structural Reliability Analysis and Prediction. John Wiley & Sons, Chichester, England.

Mohd, M.H., Kim, D.K., Kim, D.W. and Paik, J.K. 2014. A time-variant corrosion wastage model for subsea gas pipelines. Ships and Offshore Structures 9:2: 161–176.

N-001 Integrity of offshore structures (Rev. 8, September 2012).

N-003 Actions and action effects (Edition 2, September 2007).

N-004 Design of steel structures (Rev. 3, February 2013).

N-005 Condition monitoring of loadbearing structures (Rev. 1, Dec. 1997).

N-006 Assessment of structural integrity for existing offshore load-bearing structures (Edition 2, April 2015).

Næss, A. and Leira, B.J. 1999. Load Effect Combination for Snow and Wind Action, ICASP'99, Sydney, Australia.

Nataf, A. 1962. Determination des distributions dont les marges sont donnees. Comptes Rendus de l'Academie des Sciences, Paris, 225: 42–43.

Nerzic, R. and Prevosto, M. 2000. Modelling of wind and wave joint occurence probability and persistence duration from satellite observation data. Proceedings of the 10th International Offshore and Polar Engineering Conference, Seattle, USA, pp. 154–158.

Newland, D.E 1993. An Introduction to Random Vibrations (Third Edn.) Longman Scientific & Technical.

Newmark, N.M. 1959. A Method of computation for structural dynamics. J. Eng. Mech. Div., ASCE 85(EM3): 67–94.

Ochi, M.K. 1978. Wave statistics for the design of ships and ocean structures. Trans. Soc. Naval. Architects and Marine Engrs. 60: 47–76.

Ottesen, T. and Aarstein, J. 2006. The statistical boundary polygon of a two-parameter stochastic process. Proc. OMAE 2006, Hamburg, Germany.

Prince-Wright, R. 1995. Maximum likelihood models of joint environmental data for TLP design. pp. 535–445. *In*: Guedes Soares, C. et al. (eds.). Proceedings of the 14th International Conference on Offshore Mechanics and Arctic Engineering. ASME, New York, Vol. II.

Rice, S.O. 1944. Mathematical Analysis of Random Noise. Bell System Technical Journal, Vol. 23, pp. 282–332 and Vol. 24, pp. 46–156.

Risken, H. 1989. The Fokker-Planck Equation (2nd edn). Springer, Berlin.

Rosenblatt, M. 1952. Remarks on a multivariate transformation. Ann. Math. Stat. 23: 470–472.

Shinozuka, M. 1972. Monte carlo solution of structural dynamics. Computers and Structures 2: 855.

SINTEF. 2012. RIFLEX Theory Manual v4, Marintek Report, Trondheim, Norway.

Soares, C.S. and Guedes Soares, C. 2007. Comparison of bivariate models of the distribution of significant wave height and peak wave period. Proc. OMAE 2007, San Diego, USA.

Turkstra, C.J. 1970. Theory of Structural Design Decisions, Study No. 2, Solid Mechanics Division, University of Waterloo, Waterloo, Canada.

Tvedt, L. 1989. Second order reliability by an exact integral. Lecture Notes in Engineering, Vol. 48, Springer Verlag, pp. 377–384.

Wen, Y.K. and Chen, H.C. 1989. System reliability under time varying loads, Part I and II. ASCE Journal of Engineering Mechanics 115(4): 808–839.

Winterstein, S.R., Ude, T.C., Cornell, C.A., Bjerager, P. and Haver, S. 1993. Environmental parameters for extreme response: Inverse FORM with omission factors. pp. 77–84. *In*: Schueller, G.I., Shinozuka, M. and Yao, J.T.P. (eds.). Proc. of ICOSSAR '93 (Innsbruck), Balkema, Rotterdam.

4

Mitigation of Stresses and Vibrations in Deep-water Steel Riser Design by Introducing a Lazy Wave Geometry

Felisita, Airindy, Gudmestad, Ove T.* and Karunakaran, Daniel

4.1 Introduction

Deep-water fields are generally developed using a combination of floating production platforms and subsea production systems. Another development strategy is to use subsea tiebacks to existing facilities offshore or onshore. Floating platforms come in many different forms; however, there are four generic field-proven deep-water floating platform types: Floating Production Storage and Offloading Platforms, Semisubmersibles, Spars or Tension Leg Platforms. In addition, there are also modifications of these generic platforms. Ronalds (2002) mentioned that the selection process for the floater type generally depends on eight key drivers: well pattern, export methodology, service life and geographical region, gas to oil ratio, topside weight, well count and water depth.

Regardless of the selected type of floating platform, there is always a need for risers. The next section firstly addresses the challenges that the risers have to face and, secondly, discusses the state of the art of various deep-water riser systems.

University of Stavanger, Ullandhaug, 4036 Norway.
Emails: airindy@gmail.com; daniel.karunakaran@subsea7.com
* Corresponding authr: ove.t.gudmestad@uis.no

4.2 Riser Systems

4.2.1 Challenges for Riser Systems in Deep Water with Harsh Environments

Some of the challenges for risers in harsh-weather deep-water areas are:

Water depth: The most apparent influence of increased water depth is the increase in the riser's length, thus increasing the force exerted on the platform from the riser. The force on the riser can be lowered by reducing the riser's top angle. However, this will also reduce the riser's curvature at the touchdown area, which in turn may increase the riser's fatigue loading. Increased water depth is also an important influence factor in respect of planning the field layout, as the riser's spread increases with the water depth. Catenary risers typically have a radial spread between 1 to 1.5 times the water depth. Hence, in a water depth of 3000 m, the total spread between diametrically opposed risers would be in the range of 6000 to 9000 m (Howells and Hatton, 1997).

Riser sizing: In shallower water, the internal pressure often governs the riser's wall thickness. However, in deeper water, one of the important criteria for riser's wall thickness is the high external hydrostatic pressure. For installation conditions, since pipes are generally laid in an empty state, the pipe's wall thickness alone must be able to resist collapse from the external pressure. This may result in excessive riser wall thickness. Some recommendations from Howells and Hatton (1997) for example are installing the line flooded or varying the riser's wall thickness with depth, instead of using uniform wall thickness for the entire depth.

Interaction between riser and the floating platform: The design process for the riser is highly dependent on the motion responses and the drift offsets of the host facility. Deeper water may reduce the direct effect of wave loading on the riser's lower sections. However, a harsh environment leads to severe dynamic motions, which will increase the cyclic stress loadings at the critical touchdown area of the riser.

Fatigue due to Vortex Induced Vibration (VIV): Deep-water risers have higher levels of exposure to VIV, due to increased length and the strong current that is often seen in deep-water locations. This means that the fatigue life of a deep-water riser is often dominated by VIV. To reduce the riser's VIV responses, VIV suppression devices can be used; however, these devices will increase the overall costs of the riser. In addition, there are still uncertainties in predicting the VIV responses of deep-water risers, for example on the staggered buoyancy part of a steel lazy wave riser (SLWR) configuration.

Challenges related to installation: As mentioned above, the increase in length for deep-water risers causes challenges not only to the host facility but also to the installation vessels. The top tension often exceeds the capability of many existing installation vessels. This kind of offshore operation calls for the use of state-of-the-art installation vessels that are equipped with higher capability. In addition, the installation window is often quite limited for areas with harsh environments. This is because the installation can only be executed during the calmer weather in the summer weather window.

Detailed planning and scheduling is important in order to ensure an efficient process during the complete installation campaign. One of the trends for deep-water installation is to wet store the riser, hence separating the riser's campaign from the floater's delivery. Furthermore, careful thought should also be put into the installation method in order to minimize fatigue damage that may be acquired during the installation campaign.

4.2.2 Various Deep-Water Riser Systems

The motions of a floater have a significant influence on a riser's long-life performance. A similar influence is also seen at the floater-riser interphase: the riser's presence has static and dynamic effects on the floater's responses. The floater, the risers and the mooring systems create a global system with complex responses to environmental loading. This complex interaction effect is called the *coupling effect.*

In general, risers can be grouped into two classes based on the coupling process:

- A riser system that is directly connected to the floater—hence, the riser's performance is strongly correlated with the floater's motions—is called a coupled riser system, i.e., flexible risers, top-tensioned risers, steel catenary risers (SCRs) or steel lazy wave risers (SLWRs). The buoyancy section of the SLWR configuration acts as a damper and separates the floater's motion from the critical touchdown area. This generally causes better performance from the SLWR than the SCR.

- Members of the other group are called un-coupled riser systems, since the 'main' riser is not connected to the floater directly but through a series of flexible jumpers and various types of submerged buoys. The flexible jumper and buoy significantly reduce the amount of dynamic motion that the 'main' riser (usually in the form of a steel riser or bundle) is exposed to, hence improving the riser's performance. This group of risers is also often known as hybrid riser systems as they employ a combination of flexible and steel risers.

Flexible risers are pipes with low bending stiffness and which are designed to sustain repeated deflections. The main characteristic of this type of pipes is that several layers of steel and flexible materials form its wall. Flexible risers are often preferred to steel pipes since these risers are less sensitive to fatigue, as they are more compliant than steel pipes. Flexible pipes are also easier to transport or to install since they can be fabricated in long lengths and stored on reels. However, for deep-water applications, flexible risers face big challenges due to the limitations in their diameter and pressure rating capabilities. At the time of writing, the record for the deepest flexible risers is held by a 9" ID pipe from Technip that can go down to 3000 m (Vidigal da Silva and Damiens, 2016).

In recent years, steel catenary risers (SCRs) have been enjoying widespread acceptability, as deep-water risers often demand large bore risers that can withstand high pressure, high temperature and aggressive fluids. The SCRs are mainly installed in regions with benign environments, such as the West of Africa, Gulf of Mexico and Brazil. This popularity is mostly due to the simplicity and the low cost offered by SCRs. However, SCRs are quite challenging, when applied in regions with harsher climates, since they are very sensitive to the dynamic motions of the floating host structure. In addition, VIV due to current is also another concern to be considered. A combination of floater's dynamic motions and VIV leads to fatigue problems at the two most critical areas: the hang-off zone and the touchdown area. As a result, fatigue damage of the SCR dominates all aspects of the riser's integrity, governing the choice of material for the riser's structure and driving high quality welding requirements for the fabrication process.

With the limited applicability of flexible risers and the high costs of uncoupled risers, consequently there is a need to improve SCR performance, in particular for applications in deep-water locations with harsh environments. One method for improving SCR performance is to introduce buoyancy modules to the riser configuration in order to achieve the so-called steel lazy wave riser (SLWR) configuration. An early reference to the SLWR configuration was obtained from the study of Karunakaran et al. (1996). However, the first SLWR configuration was not installed until 2008 in the BC10 field in Brazil. Since then, SLWRs have gained popularity for deep-water applications and have been applied in the Caesar Tonga field (2012) and the Stones field (scheduled to be finished in 2016) in the Gulf of Mexico.

4.3 SLWR Configurations

This section provides information correlated with the analysis model, such as the pipe's properties, buoyancy modules' properties and the environmental conditions. These properties are considered typical for harsh environmental conditions.

4.3.1 Model Overview—General

Figure 4.1 shows a schematic drawing of a typical SLWR model. The L1 symbol in the picture represents the length of the riser from the topside end to the beginning of the buoyancy section; the L2 symbol represents the length of the riser's buoyancy section, while the L3 symbol represents the riser's length from the end of the buoyancy section to the end of the riser. The term 'riser's buoyancy section' refers to the riser's sections that are covered with buoyancy modules. The term 'riser's wave height' in the figure refers to the vertical distance between the lowest point at the sagbend and the highest point at the hogbend of the riser.

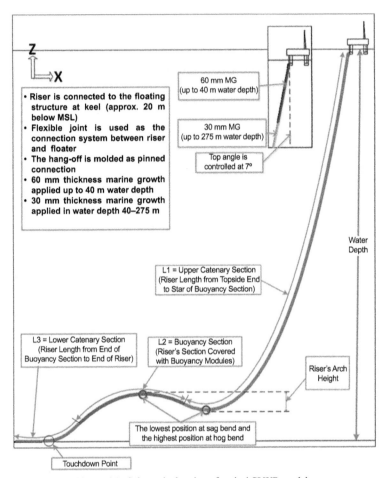

Figure 4.1. Schematic drawing of typical SLWR model.

The SLWR is normally to be applied with a semi-submersible or an FPSO as the host. For either type of host, the riser's top end is positioned at the keel of the floater, approximately 20 m below sea level. The top angle is selected at 7°. This angle is considered as common practice for risers installed in water depth ranges between 1000 and 2000 m.

A flexible joint is used as the connection point between the steel riser and the floating structure. In most analysis, the flex joint is modeled as a pinned connection. For extreme loading, the flex joint's stiffness will not influence the response. However, the rotational stiffness of the flex joint will influence the fatigue responses of the riser sections adjacent to it. When considering the design of the hang-off connection, fatigue at the riser's top section must be carried out separately in a detailed analysis. The issue of fatigue near the flex joint is normally addressed using a 5 to 10 m long tapered section (Karunakaran et al., 2005).

Marine growths of different thicknesses are to be applied onto the riser from the top section up to a water depth of 275 m. Typical details of marine growth application are provided in Table 4.3.

For operational conditions, three different floater offsets are normally considered; these are the *Nominal, Far and Near* offsets. The offsets are set in the direction of the global-X axis. The far and near offsets are assumed to have ± 8% WD (water depth) distance from the nominal position. For the accidental condition, two additional floater offsets are introduced: *Far Accidental and Near Accidental* offsets, where each of these accidental offsets has ± 10% WD (water depth) distance from the nominal position. These distances are considered generic for deep-water application.

4.3.2 SLWR Variations

Various lazy wave outlines for the SLWR configuration have to be considered to determine the optimum geometry. The base case (in this work referred as riser configuration C_0) would typically be selected by positioning the hog and sag bends as close to the seabed as possible, for example, at approximately 160 m above the seabed. Modification against the base case riser can be carried out by several methods, i.e. varying the length of the buoyancy section (L2 in Fig. 4.1), applying different dimensions of buoyancy modules (as shown in Table 4.2), changing the water depth, changing the type of hydrocarbon contents (oil or gas) or by applying different pitch distances between the buoyancy modules. However, this work only considers one variation of pitch distance between the buoyancy modules as detailed in the following.

4.3.3 Pipe's and Buoyancy Modules' Properties

A 10" ID steel riser pipe is used to represent atypical SLWR. The properties of the typical SLWR pipe are presented in Table 4.1, while Table 4.2 provides data on the buoyancy modules. The distance between the buoyancy modules is held at 12 m, since it is assumed that this spacing is the most practical for installation on steel pipes offshore.

4.3.4 Environmental Conditions

4.3.4.1 Marine Growth

Marine growth of different thicknesses should be applied. Typical values for marine growth are given in Table 4.3. This work applies marine growth up to a water depth of approximately 275 m.

Table 4.1 SLWR typical pipe properties.

Properties	Value	Unit
Inner Diameter (10" pipe)	254	mm
Wall thickness	26	mm
Coating thickness	76	mm
Total outer diameter	458	mm
Pipe material	Carbon steel, grade X65	
Young's modulus (E)	207000	MPa
Specified Minimum Yield Stress (SMYS)	448	MPa
Pipe material density	7850	kg/m³
Coating density	700	kg/m³
Pipe weight including coating (in air)	244	kg/m
Design pressure	5000	psi
Internal fluid density	Oil = 800, Gas = 180	kg/m³

Table 4.2. Typical buoyancy modules' dimensions and properties.

Type	Outside Diameter (OD) [m]	Length [m]
1	1.81	1.8
2	1.87	1.8
3	1.92	1.8
4	1.96	1.8
5	1.76	1.8
6	1.80	1.3
7	1.875	1.3
8	1.935	1.3
9	1.99	1.3
10	1.74	1.3
11	1.57	1.8
12	1.684	1.684
Buoyancy module material density	500 kg/m³	
Assumed weight of inner clamps	50 kg	
Distance between modules	12 m	

Table 4.3. Marine growth thicknesses.

Water depth [m]	Marine growth thickness [mm]	Ref.
Above +2	0	StandardsNorway (2007)
+2 to –40	60	
Under –40	30	

4.3.4.2 Waves and Current

ULS and ALS limit state conditions are generally represented using 100-year and 10,000-year sea states, respectively. Typical design sea states for the deep-water area of the Norwegian continental shelf (NCS) are given in Table 4.4, while current profiles used for the limit state analysis are given in Table 4.5 for the deepwater on the NCS.

For wave-induced fatigue analysis, 12 wave directions with 18 sea-state blocks should be sufficient. The fatigue is to be calculated for the whole of the riser's length. A stress concentration factor (SCF) of 1.251 and S-N curve D (DNV 2011) are recommended in the fatigue analysis. VIV-fatigue analysis must be carried out using a typical deep-water NCS current. The directional wave probability distribution and sea-state blocks for wave-induced fatigue calculation are given in Table 4.6 and Fig. 4.2, and typical current profiles and probability distribution for VIV-induced fatigue calculation are provided in Fig. 4.3 and Table 4.7.

Table 4.4. Typical design sea states for deepwater areas within the NCS.

1-year seastate	Significant wave height (H_s)	12 m
	Corresponding wave period (T_p)	16.1 s
10-year sea state	Significant wave height (H_s)	14.6 m
	Corresponding wave period (T_p)	17.5 s
100-year sea state	Significant wave height (H_s)	17 m
	Corresponding wave period (T_p)	18.9 s
10000-year sea state	Significant wave height (H_s)	21.5 m
	Corresponding wave period (T_p)	21.3 s

Table 4.5. Typical current profile for deepwater area within NCS.

Water depth [m]	1-Year current velocity [m/s]	10-Year current velocity [m/s]	100-Year current velocity [m/s]
−10	1.44	1.65	1.85
−50	1.11	1.26	1.40
−100	1.09	1.25	1.40
−200	0.97	1.09	1.20
−300	0.75	0.83	0.90
−400	0.67	0.74	0.80
−500	0.66	0.73	0.80
−600	0.54	0.60	0.65
−800	0.54	0.60	0.65
−1000	0.50	0.55	0.60
−1200	0.50	0.55	0.60
−3 m above sea bottom	0.41	0.46	0.5

Table 4.6. Directional wave probability distribution for fatigue analysis.

Sector No.	Direction	Sector probability (%)
1	0	11.89
2	30	10.5
3	60	2.72
4	90	1.16
5	120	1.41
6	150	2.64
7	180	4.61
8	210	14
9	240	19.98
10	270	12.61
11	300	8.68
12	330	9.8

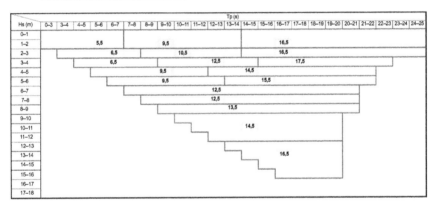

Figure 4.2. Sea-state blocks used in fatigue analysis.

4.3.4.3 Long-Term Wave Statistics

The long-term environmental condition could be estimated using the environmental contour lines approach. A typical contour lines diagram for deep-water area at the Norwegian Continental Shelf is shown in Fig. 4.4.

4.4 Long-Term and Limit Stress Analysis Results

Let us consider a case where the base case SLWR configuration was set to have the minimum number of buoyancy modules and positioned relatively close to the seabed (sag bend is approximately 160 m above the seabed). The low positioning means that the buoys are relatively small and therefore cost efficient.

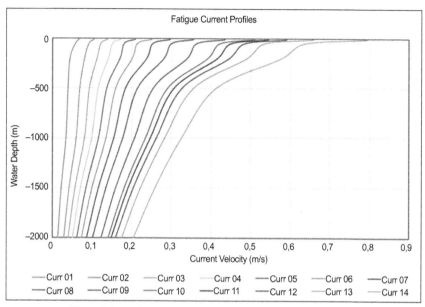

Figure 4.3. Typical current velocity profiles for VIV-fatigue analysis.

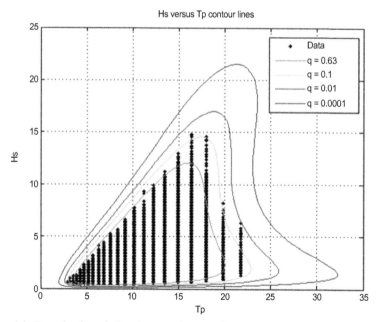

Figure 4.4. Example of a typical environmental contour line plot for a site at deepwater areas of the Norwegian Continental Shelf, q = annual probability of exceedance.

Table 4.7. Typical current velocity profiles and probability distribution for VIV-fatigue analysis.

Water depth	Current velocity profile (m/s)													
	1	2	3	4	5	6	7	8	9	10	11	12	13	14
−10	0,067	0,105	0,139	0,174	0,21	0,251	0,297	0,355	0,435	0,51	0,545	0,591	0,658	0,793
−50	0,058	0,092	0,122	0,152	0,182	0,215	0,253	0,3	0,363	0,422	0,449	0,484	0,535	0,636
−200	0,044	0,081	0,112	0,141	0,171	0,202	0,238	0,281	0,339	0,391	0,415	0,445	0,49	0,577
−500	0,039	0,069	0,093	0,115	0,137	0,16	0,185	0,215	0,254	0,29	0,306	0,326	0,355	0,412
−1000	0,034	0,06	0,081	0,099	0,116	0,135	0,155	0,178	0,208	0,235	0,247	0,262	0,283	0,325
−1500	0,022	0,042	0,057	0,071	0,085	0,1	0,116	0,135	0,16	0,182	0,192	0,205	0,223	0,259
−2000	0,014	0,029	0,04	0,051	0,062	0,074	0,087	0,102	0,123	0,141	0,149	0,16	0,176	0,206
Probability	**0,1**	**0,1**	**0,1**	**0,1**	**0,1**	**0,1**	**0,1**	**0,1**	**0,1**	**0,02**	**0,02**	**0,02**	**0,02**	**0,02**

This section will present the analysis results for the base case SLWR configuration and variations. The long-term response analysis is carried out, followed by the limit stress analysis and fatigue analysis. The analyses were carried out using computer program OrcaFlex version 9.8e (Orcina, 2016).

4.4.1 Long-Term Response Analysis Results

The long-term response analysis is the first analysis stage to be carried out in order to verify the acceptability of the riser configuration. This is because a marine structure should be designed such that it can withstand all expected load events with a sufficient safety margin (Haver, 2007).

An SLWR can be considered as a complex problem, whose responses are influenced by many aspects; hence, the environmental contour line approach may be selected as the most convenient method in order to obtain a reasonable estimate for the q-probability value (i.e., the response value corresponding to an annual exceedance probability of q). The following steps recommended by the study of Haver (2007) should be used when applying the environmental contour line approach:

- Establish the q-probability contour or surface for the involved metocean characteristics, e.g., significant wave height and spectral peak period. A typical environmental contour line for a deep-water area on the Norwegian continental shelf is given in Fig. 4.4.
- Identify the most unfavorable metocean condition along the q-probability contour/surface, with respect to extremes of the response problem under consideration. The most unfavorable metocean condition was selected from various sea states along the peak of the $q = 10^{-2}$ contour line, as shown in Fig. 4.5.
- Establish the distribution function for a 3-hour maximum response for the most unfavorable metocean condition.
- The estimate for the q-probability response value is obtained by the α-percentile of the established extreme value distribution. As per the recommendation from Norsok Standards N-003 (StandardsNorway, 2007), the α value is taken as the 90% percentile and 95% percentile for ULS and ALS conditions, respectively.

For each of the sea states selected for identifying the most unfavorable metocean conditions, several (at least five) observations (using varied seed numbers) were allocated.

From the results above, the most unfavorable metocean condition was identified as $H_s = 17.0$ m and $T_p = 19.5$ s. A distribution function should then be developed based on the combination of metocean conditions. In SLWR studies by Felisita et al. (2015), it was found that the variation in effective tension is lower compared to the variation of the bending moment. Effective tension has approximately a 4%

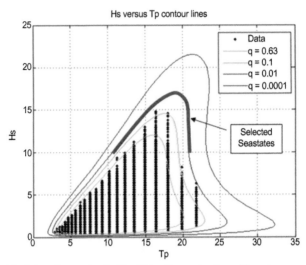

Figure 4.5. Selected sea states for identifying the most unfavorable metocean condition.

spread at the hang-off (top tension) and a 9–10% spread at the other parts of the riser. On the other hand, the bending moment has approximately 21–25% spread at the sag bend, hog bend and touchdown.

4.4.2 Limit State Analysis Results

4.4.2.1 Effective Tension

Figure 4.6 shows typical trends of effective tension along the riser for a SLWR configuration. The terms 'maximum' and 'minimum' refer to the highest and lowest tension loads in the riser during the dynamic analysis phase.

The highest effective tension along the riser is found at the topside end, i.e., the riser's end connected to the floating structure. Top tension is one of the important design criteria when designing a riser, in particular for any riser system that involves a disconnectable hang-off system. In a SLWR configuration, the top tension is a function of the length of the riser's upper catenary section (L1 in Fig. 4.1) from the hang-off to the lowest position in the sag bend. On the other hand, the lowest point of the sag bend is influenced by the amount of buoyancy. Hence, the highest top tension will occur at a riser, which has the longest upper catenary section and the lowest buoyancy force, as shown in Fig. 4.7.

The sag bend area of an SLWR is designed to bear most of the downward motions induced by the vessel's motions and the heave motions of the upper catenary section. This makes the sag bend prone to compression, particularly at the sag bend area prior to the buoyancy section, as seen in Fig. 4.6. Even though the compression force generally occurs instantaneously—only when the

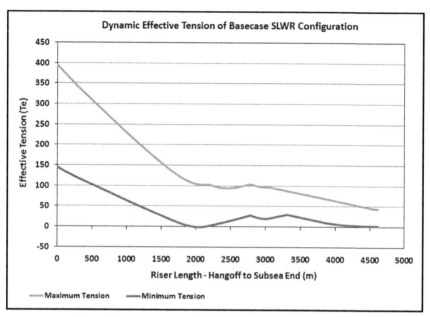

Figure 4.6. Trends in effective tension load along the riser length.

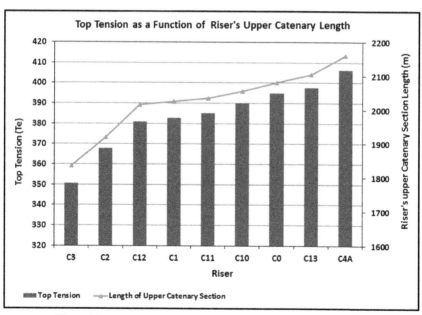

Figure 4.7. A comparison of top tension from various riser configurations.

floater experiences maximum downward velocity at the hang-off position—the compression is particularly important as it implies the possibility of global buckling of the pipe as a bar.

The dimension of the lazy wave's arch (i.e., riser's arch height, see Fig. 4.1), has a strong influence on the amount of compression experienced by the riser. Risers with shorter arches experience higher negative tension compared to risers with higher arches. This is thought to happen due to the longer upper catenary section (L1) of the risers with lower arches. The risers with lower arches are generally positioned with the buoyancy section close to the seabed, and a longer catenary section gives a higher downward force towards the sag bend. In addition, it is also observed that lazy wave risers with shorter arches have higher riser motions than risers with higher arches, which causes larger compression.

Even though general riser design standards such as those from DNV-GL (i.e., DNV OS-F201 (DNV, 2010)) and British Standards (i.e., PD 8010-2: 2004 (BSI, 2004)) accept some amount of compression for a steel riser, it is recommended that the riser does not experience any compression. As the arch shape of the lazy wave riser represents a significant role in handling the compressive forces from the vessel's motion, the configuration of the lazy wave riser needs to be designed in order to minimize or to avoid the compression at the sag bend region.

4.4.2.2 Utilization Criteria

The pipe's utilization is normally calculated using LRFD principle as per DNV OS-F201 (DNV, 2010), using the combination of effective tension, bending moment and pressure exposure, as given in Eq. 4.1.

$$\{\gamma_{SC} \cdot \gamma_m\} \left\{ \left[\left(\frac{|M_d|}{M_k} \cdot \sqrt{1 - \left(\frac{p_{ld} - p_e}{p_b(t)} \right)^2} \right) + \left(\frac{T_{ed}}{T_k} \right)^2 \right] + \left(\frac{p_{ld} - p_e}{p_b(t)} \right)^2 \right\} \leq 1 \qquad (4.1)$$

Where:

γ_{SC} = safety class resistance factor, which is dependent on riser's safety class. This study uses the value of γ_{SC} = 1.26, which corresponds to safety class high (Table 5.3 in DNV OS-F201 (DNV, 2010)).

γ_M = material resistance factor, to take into account the uncertainties in the material quality. This study uses the value of γ_M = 1.15 (Table 5.4 in DNV OS-F201 (DNV, 2010)).

M_d = design bending moment

T_{ed} = design effective tension

P_{ld} = local internal design pressure

P_e = local external pressure

M_k = plastic bending moment resistance

T_k = plastic axial force resistance

$P_b(t)$ = burst resistance

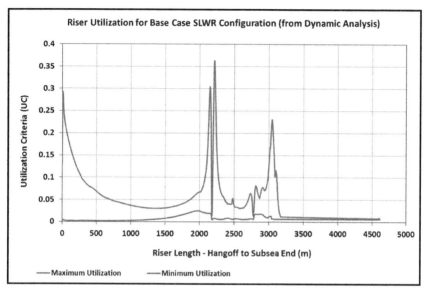

Figure 4.8. Typical trend of riser utilization along its length based on DNV OS-F201.

Figure 4.8 shows the typical trend of a riser's utilization along its length. The terms 'maximum' and 'minimum' refer to the highest and lowest utilization in the riser during the dynamic analysis phase. From Eq. (4.1), a riser's utilization is calculated using its tension, bending and pressure capacity. The riser generally has sufficient capacity with regard to the tension and pressure containment. However, the bending capacity may become critical, particularly at areas with high curvature such as the sag bend and touchdown. This explains why the highest utilization values are generally found in these areas.

Since the utilization is correlated with the bending stress, a riser with a higher arch may have higher utilization than that of a riser with a low arch height. Steel lazy wave risers with higher arches often have a high curvature that leads to high stresses at their sag and hog bend areas. This may cause the sag or hog bends, rather than the touchdown, to represent the governing values in terms of the riser's utilization.

4.5 Wave Induced Fatigue Analysis Results

Figure 4.9 shows the calculated fatigue life along the base case SLWR configuration. The critical areas are typically the top section, the sag bend and the touchdown. The top section is critical against fatigue due to the combination of the floater's motions and the connection stiffness. The rotational stiffness of the flexible joint influences the fatigue responses of the riser's sections within the vicinity of the flexible joint. It was observed that fatigue life is very low between the hang-off point and the first 10 m of the riser. However, this low fatigue life near the flexible

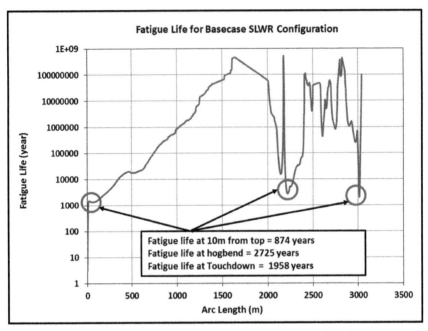

Figure 4.9. Typical trend of fatigue life along the SLWR configuration.

joint is normally controlled using a 5 to 10- m long tapered section (Karunakaran et al., 2005).

The other critical area for fatigue damage is the touchdown area, since this area is less compliant due to the seabed stiffness. Fatigue lives at sag and hog bends may be lower than at other parts of the riser due to the higher curvature levels, but fatigue lives there are generally higher than at the touchdown.

One of the important observations is the influence of the lazy wave geometry (i.e., riser's arch) on the fatigue life at the touchdown. It was observed that SLWR configurations with larger arch heights have longer calculated fatigue lives at the touchdown compared to configurations with lower arches. Fatigue lives at the sag and hog bends also show an improvement with the increase in the lazy wave arch height, although the improvement is not as significant as that at the touchdown. Further discussion on the correlation between the risers' fatigue lives and their arch dimensions are given in the study of Felisita et al. (2016a).

4.6 Fatigue Due to Vortex Induced Vibrations (VIV)

Felisita et al. (2016b) presented a study on the VIV responses of steel lazy wave risers. Based on the results reported in this paper, the following conclusions were drawn:

- *The trend of VIV fatigue follows the curvature trend of the riser.*
  The riser's curvature has a significant influence on the trend of the VIV-induced fatigue along the riser's length. Hence, parts of the riser with higher levels of curvature (i.e., sag and hog bends and touchdown) experience higher fatigue damage compared to other parts of the riser. High levels of curvature mean high levels of bending stresses, which will lead to high levels of fatigue damage. Maximum VIV-induced fatigue damage is found at the touchdown, since the pipe's movement at the touchdown is restricted due to the soil.

- *Water depth*
  Despite the shape of the lazy wave, risers at the same water depth tend to have a similar degree (or very little variation) of fatigue damage. Furthermore, risers in shallower water have lower fatigue lives compared to risers in deeper water. This is due to the fact that, in shallower water, the risers (particularly for parts of the riser which have high levels of curvature) are exposed to larger current velocities.

  In addition, for an equivalent type of configuration, risers in shallow water have greater system stiffness than risers in deeper water. This causes the risers in shallow water to experience a higher response frequency for the same excited mode number, compared to risers in deeper water. Furthermore, the greater system stiffness also leads to risers in shallower water experiencing higher stress amplitude for the same excited mode number, even though the response amplitudes are almost the same. The combination of higher current exposure, higher stress and higher response frequencies (caused by the greater system stiffness) leads to the risers in shallower water having lower fatigue lives.

  The VIV-induced fatigue life of a riser in shallower water can be improved by creating a lazy wave configuration with as little curvature as possible (i.e., with a short arch shape). However, it should be noticed that this type of lazy wave configuration only has limited capability to absorb dynamic motions; hence, the riser may have a low wave-induced fatigue life.

- *The arch of the lazy wave configuration*
  Contrary to the observation from wave-induced fatigue, the arch's dimension (i.e., arch height/length) of the lazy wave shape only exerts little to modest influence on the VIV-induced fatigue life of a riser.

- *The hydrocarbon content (self-weight)*
  The self-weight of the riser mainly affects the VIV-induced fatigue life at the upper catenary part of the riser. Even though the VIV-induced fatigue life at this area of a riser is generally quite high, risers whose content has lower density have lower fatigue lives at this part compared to risers whose content has higher density.

4.7 Conclusion

Mitigation of stresses and vibrations in deep-water steel risers can be achieved by introducing lazy wave geometry. The geometry will have to be optimized with respect to the location of buoyancy modules and top tension.

In order to obtain a cost-efficient riser configuration, it is better to position the arch (the sag and hog bends) quite low towards the seabed. However, it is important for the arch to have an adequate arch dimension in order to reduce the dynamic motions of the riser such that sufficient fatigue life is obtained, in particular near the touchdown location.

On the other hand, the stress levels at the sag and hog bends of the lazy wave riser should also be optimized, since these areas may experience high compression and bending stresses in cases where the bending radius becomes small. The minimum bending radius will become a critical design criterion for lazy wave risers with high arches.

VIV-mitigation tools must be considered in order to ensure sufficient VIV-fatigue lives for the risers located in deep water.

Acknowledgements

The first author wishes to thank DEA E&P Norge for financial support during the thesis work. The last author would like to thank SubSea7 for permission to publish the paper.

References

BSI. 2004. PD 8010-2:2004 Code of Practice for Pipelines – Part 2: Subsea Pipelines. Standards. British Standards, London, UK.

DNV-GL. 2010. DNV-OS-F201 Dynamic Risers.Standards. Det Norske Veritas, Høvik, Oslo, Norway.

DNV-GL. 2011. DNV-RP-C203 Fatigue Design of Offshore Steel Structures. Standards. Det Norske Veritas, Høvik, Oslo, Norway.

Felisita, A., Gudmestad, O.T., Karunakaran, D. and Martinsen, L.O. 2015. Review of steel lazy wave riser concepts for North Sea. Paper No: OMAE2015-41182. Proceedings of the 34th International Conference on Ocean, Offshore and Arctic Engineering (OMAE). ASME, St. John's, Newfoundland, Canada.

Felisita, A., Gudmestad, O.T., Karunakaran, D. and Martinsen, L.O. 2016a. Review of Steel Lazy Wave Riser Concepts for the North Sea. Paper accepted for publication in Journal of Offshore Mechanics and Arctic Engineering. ASME.

Felisita, A., Gudmestad, O.T., Karunakaran, D. and Martinsen, L.O. 2016b. A review of VIV responses of steel lazy wave riser. Paper No: OMAE2016-54321. Proceedings of the 35th International Conference on Ocean, Offshore and Arctic Engineering (OMAE). ASME, Busan, South Korea.

Felisita, A. 2017. On the Application of Steel Lazy Wave Riser for Deepwater Locations with Harsh Environment. Doctoral Thesis, University of Stavanger, Stavanger, Norway.

Haver, S. 2007. A discussion of long term response versus mean maximum response of the selected design sea state. Paper No: OMAE2007-29552. Proceedings of the 26th International Conference on Offshore Mechanics and Arctic Engineering (OMAE). ASME, San Diego, California, USA.

Howells, H. and Hatton, S. 1997. Challenges for Ultra-Deep Water Riser System. Conference paper presented at Floating Production Systems. IIR, London, UK. Available from: http://2hoffshore.com/technical-papers/challenges-for-ultra-deepwater-riser-systems/.

Karunakaran, D., Nordsve, N.T. and Olufsen, A. 1996. An efficient metal riser configuration for ship and semi based production systems. Paper No: I-96-106. Proceedings of the Sixth International Offshore and Polar Engineering Conference (ISOPE). The International Society of Offshore and Polar Engineers, Los Angeles, USA.

Karunakaran, D., Meling, T.S., Kristoffersen, S. and Lund, K.M. 2005. Weight-optimized SCRs for deepwater harsh environments. Paper No: OTC-17224-MS. Proceedings of Offshore Technology Conference. OTC, Houston, Texas, USA.

Orcina. 2016. OrcaFlex version 9.8e. Computer program. Orcina, Cumbria, UK.

Ronalds, B.F. 2002. Deepwater facility selection. Paper No: OTC-14259-MS. Proceedings of Offshore Technology Conference. OTC, Houston, Texas, USA.

StandardsNorway. 2007. NORSOK Standard N-003 Actions and Action Effects. Standards. Standards Norway, Lysaker, Norway.

Vidigal da Silva, J. and Damiens, A. 2016. 3000 m water depth flexible pipe configuration portfolio. Paper No: OTC-26933-MS. Proceedings of Offshore Technology Conference. OTC, Houston, Texas, USA.

5

Wind-Induced Dynamic Response Calculations

Einar N. Strømmen

5.1 Introduction

It is in the following taken for granted that the cause of the dynamic loads stems from a combination of wind turbulence and motion induced forces, i.e., the presentation covers the so-called buffeting theory. The effects of vortex shedding are separately handled elsewhere in this book, while special effects such as rain-wind induced vibrations have not been included. Thus, the dynamic load comprise distributed drag, lift and pitching moment components q_D, q_L, q_M, while in structural axis the corresponding components are q_y, q_z and q_θ (please note that in wind engineering it is customary to define the pitching moment positive in the opposite direction of that which is shown in Fig. 5.1 (see also Fig. 5.2). The subject is given a more comprehensive presentation by E. Strømmen in Refs. (Strømmen, 2010; Strømmen, 2014). The basic assumptions behind the buffeting theory are that the instantaneous velocity pressure may be converted into loads based on load coefficients that have been obtained from static tests, and that linearization of any fluctuating parts will render results with sufficient accuracy. It is a requirement for linearization that structural displacements and cross sectional rotations are small and that the turbulence components u, v and w are small as compared to the mean wind velocity V. It is taken for granted that the local element x axis is either horizontal or vertical (i.e., parallel to either y_f or z_f as illustrated in Fig. 5.1). The situation for a horizontal element is show in Fig. 5.2. (The situation for a vertical element is

Department of Structural Engineering, Norwegian University of Science and Technology.

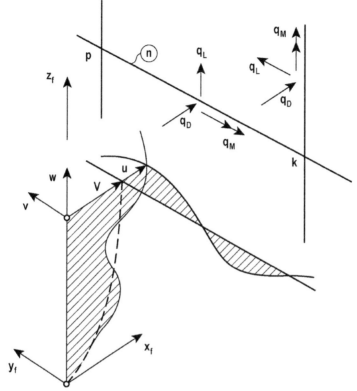

Figure 5.1. Line-like structure in a turbulent wind field.

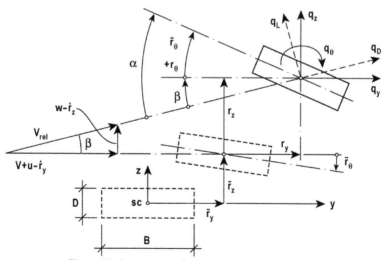

Figure 5.2. Instantaneous flow and displacement quantities.

identical, except that w should be replaced by v.) The cross section at an arbitrary position x is first given the time invariant (mean, static) displacements $\bar{r}_y(x)$, $\bar{r}_z(x)$ and $\bar{r}_\theta(x)$. In this position the wind velocity vector is $V + u(x, t)$ in the along wind horizontal direction and $w(x, t)$ in the vertical across wind direction. It is about this position that the structure oscillates.

The principle of d'Alambert is adopted and the cross section is given an arbitrary dynamic displacement $r_y(x, t)$, $r_z(x, t)$ and $r_\theta(x, t)$. The instantaneous cross sectional drag, lift and moment forces in flow axes are then given by

$$
\begin{bmatrix} q_D(x,t) \\ q_L(x,t) \\ q_M(x,t) \end{bmatrix} = \frac{1}{2}\rho V_{rel}^2 \cdot \begin{bmatrix} D \cdot C_D(\alpha) \\ B \cdot C_L(\alpha) \\ B^2 \cdot C_M(\alpha) \end{bmatrix} \tag{5.1}
$$

where C_D, C_L and C_M are cross sectional characteristic load coefficients from static tests (see Fig. 5.3), V_{rel} is the instantaneous relative wind velocity and α is the angle of flow incidence. Transformation into structural axis is obtained by

$$
\mathbf{q}_{tot}(x,t) = \begin{bmatrix} q_y \\ q_z \\ q_\theta \end{bmatrix}_{tot} = \begin{bmatrix} \cos\beta & -\sin\beta & 0 \\ \sin\beta & \cos\beta & 0 \\ 0 & 0 & 1 \end{bmatrix} \cdot \begin{bmatrix} q_D \\ q_L \\ q_M \end{bmatrix} \quad \text{where } \beta = \arctan\left(\frac{w - \dot{r}_z}{V + u - \dot{r}_y}\right) \tag{5.2}
$$

The first linearizationn involves the assumption that the fluctuating flow components $u(x, t)$ and $w(x, t)$ are small as compared to V, and that structural displacements are small. Then $\cos\beta \approx 1$ and $\sin\beta \approx \tan\beta \approx \beta \approx (w - \dot{r}_z)/(V + u - \dot{r}_y) \approx (w - \dot{r}_z)/V$, and thus: $V_{rel}^2 = (V + u - \dot{r}_y)^2 + (w - \dot{r}_z)^2 \approx V^2 + 2Vu - 2V\dot{r}_y$ and $\alpha = \bar{r}_\theta + r_\theta + \beta \approx \bar{r}_\theta + r_\theta + w/V - \dot{r}_z/V$. The second linearization involves the flow incidence

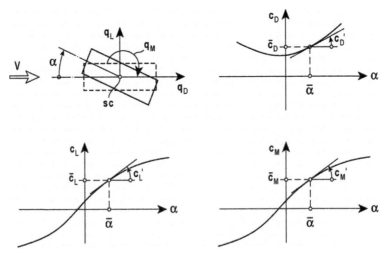

Figure 5.3. Load coefficients obtained from static tests.

dependent load coefficients. As illustrated in Fig. 5.3, the nonlinear variation of the load coefficient curves is replaced by the following linear approximation $\begin{bmatrix} C_D & C_L & C_M \end{bmatrix}^T = \begin{bmatrix} \bar{C}_D & \bar{C}_L & \bar{C}_M \end{bmatrix}^T + \alpha_f \cdot \begin{bmatrix} C'_D & C'_L & C'_M \end{bmatrix}^T$ where $\bar{\alpha}$ and α_f are the mean value and the fluctuating part of the angle of flow incidence, C'_D, C'_L and C'_M are the slopes of the load coefficient curves at $\bar{\alpha}$, and where $\bar{\alpha} = \bar{r}_\theta$ and $\alpha_f = r_\theta + w/V - \dot{r}_z/V$. Combining Eqs. (5.1–5.3) and discarding higher order terms then the following is obtained

$$\mathbf{q}_{tot}\left(x,t\right) = \bar{\mathbf{q}} + \mathbf{b}_q \cdot \mathbf{v} + \mathbf{c}_{q_{ae}} \cdot \dot{\mathbf{r}} + \mathbf{k}_{q_{ae}} \cdot \mathbf{r} \text{ where } \mathbf{v}\left(x,t\right) = \begin{bmatrix} u \\ w \end{bmatrix} \text{ and } \mathbf{r}\left(x,t\right) = \begin{bmatrix} r_y \\ r_z \\ r_\theta \end{bmatrix}$$

$$(5.3)$$

where $\bar{\mathbf{q}}(x) = \begin{bmatrix} \bar{q}_y \\ \bar{q}_z \\ \bar{q}_\theta \end{bmatrix} = \dfrac{\rho V^2 B}{2} \begin{bmatrix} (D/B)\bar{C}_D \\ \bar{C}_L \\ B\bar{C}_M \end{bmatrix}$ and

$$\frac{\mathbf{b}_q\left(x\right)}{0.5\rho VB} = \begin{bmatrix} 2(D/B)\bar{C}_D & \left((D/B)C'_D - \bar{C}_L\right) \\ 2\bar{C}_L & \left(C'_L + (D/B)\bar{C}_D\right) \\ 2B\bar{C}_M & BC'_M \end{bmatrix}$$

$$\frac{\mathbf{c}_{q_{ae}}\left(x\right)}{0.5\rho VB} = -\begin{bmatrix} 2(D/B)\bar{C}_D & \left((D/B)C'_D - \bar{C}_L\right) & 0 \\ 2\bar{C}_L & \left(C'_L + (D/B)\bar{C}_D\right) & 0 \\ 2B\bar{C}_M & BC'_M & 0 \end{bmatrix} \text{ and}$$

$$\frac{\mathbf{k}_{q_{ae}}\left(x\right)}{0.5\rho V^2 B} = \begin{bmatrix} 0 & 0 & (D/B)C'_D \\ 0 & 0 & C'_L \\ 0 & 0 & BC'_M \end{bmatrix}$$

It is seen that the total load vector comprises a static $\bar{\mathbf{q}}(x) = \begin{bmatrix} \bar{q}_y & \bar{q}_z & \bar{q}_\theta \end{bmatrix}^T$ and a dynamic part $\mathbf{q}(x,t) = \begin{bmatrix} q_y & q_z & q_\theta \end{bmatrix}^T = \mathbf{b}_q \cdot \mathbf{v} + \mathbf{c}_{q_{ae}} \cdot \dot{\mathbf{r}} + \mathbf{k}_{q_{ae}} \cdot \mathbf{r}$ where $\mathbf{b}_q \cdot \mathbf{v}$ is the dynamic loading associated with turbulence (u and w) in the oncoming flow, while $\mathbf{c}_{q_{ae}} \cdot \dot{\mathbf{r}}$ and $\mathbf{k}_{q_{ae}} \cdot \mathbf{r}$ are motion induced loads associated with structural velocity and displacement. (For an element that is vertical in the flow, the local axis system is maintained and thus, the necessary load equations above may simply be obtained by replacing w by v.) This theory is applicable in original degrees of freedom and it may readily be expanded into modal coordinates. It may be used in time domain as well as in frequency domain. While the theory is well accepted for load effects at mean wind velocities well below any stability limit, it is well known that the theory is inadequate for significant parts of the wind and motion induced load effects at higher mean wind velocities, i.e., it does not predict many of the most important wind induced stability problems that are observed in experiments or in full scale structural behaviour. The main problem with the buffeting theory arises from the basic hypothesis shown in Eq. (6.1), i.e., from the assumption that fluctuating motion induced loads are converted into dynamic load effects by the combination of an interpretation of the relative instantaneous wind velocity pressure and static load coefficients (d'Alambert and Bernoulli).

Thus, if a solution in original coordinates (i.e., in the element degrees of freedom) is pursued, then the theory needs modifications. The modified hypothesis which is usually adopted (see, e.g., Salvatori and Borri (Salvatori and Borri, 2007)) is that

- a proportional relationship between load and structural velocity and displacements are still applicable, but only at an incremental level (i.e., within a small time step $\Delta\tau$), and
- that the motion induced load coefficients depend on the history of the motion.

Thus,

$$\lim_{\Delta\tau\to 0}\frac{\Delta\mathbf{q}_{ae}(x,t)}{\Delta\tau}=\frac{d\mathbf{q}_{ae}(x,t)}{dt}=\mathbf{c}_{ae}(s)\cdot\frac{d}{d\tau}\begin{bmatrix}\dot{r}_y(x,\tau)\\ \dot{r}_z(x,\tau)\\ \dot{r}_\theta(x,\tau)\end{bmatrix}+\mathbf{k}_{ae}(s)\cdot\frac{d}{d\tau}\begin{bmatrix}r_y(x,\tau)\\ r_z(x,\tau)\\ r_\theta(x,\tau)\end{bmatrix}$$

$$(5.4)$$

$$\text{where } \mathbf{c}_{ae}(s)=-\frac{\rho VB}{2}\begin{bmatrix}2\dfrac{D}{B}\bar{C}_D\Phi_{D\dot{y}} & \left(\dfrac{D}{B}C_D'-\bar{C}_L\right)\Phi_{D\dot{z}} & DC_D'\Phi_{D\dot{\theta}}\\[2ex] 2\bar{C}_L\Phi_{L\dot{y}} & \left(C_L'+\dfrac{D}{B}\bar{C}_D\right)\Phi_{D\dot{z}} & BC_L'\Phi_{L\dot{\theta}}\\[2ex] 2B\bar{C}_M\Phi_{M\dot{y}} & BC_M'\Phi_{D\dot{z}} & B^2C_M'\Phi_{M\dot{\theta}}\end{bmatrix}$$

$$\text{and } \mathbf{k}_{ae}(s)=\frac{\rho V^2}{2}\begin{bmatrix}\bar{C}_D\Phi_{Dy} & \left(\dfrac{D}{B}C_D'-\bar{C}_L\right)\Phi_{Dz} & -DC_D'\Phi_{D\theta}\\[2ex] \bar{C}_L\Phi_{Ly} & \left(C_L'+\dfrac{D}{B}\bar{C}_D\right)\Phi_{Dz} & -BC_L'\Phi_{L\theta}\\[2ex] B\bar{C}_M\Phi_{My} & BC_M'\Phi_{Dz} & B^2C_M'\Phi_{M\theta}\end{bmatrix}$$

where τ is a time history variable, $s=t-\tau$ and $\Phi_{ij}(s)\begin{cases}i=D,L \text{ or } M\\ j=\dot{y},\dot{z} \text{ or } \theta\end{cases}$ is a set of altogether eighteen indicial functions, each associated with interaction between drag, lift or moment forces and velocity or displacement of motion in y, z or θ directions. These functions describe how an incremental structural motion is giving rise to a corresponding change of motion induced loads. Thus, $\mathbf{q}_{ae}(x, t)$ may be obtained from history integration:

$$\mathbf{q}_{ae}(x,t) = \int_0^t \left\{ \mathbf{c}_{ae}(s) \cdot \frac{d}{d\tau} \begin{bmatrix} \dot{r}_y(x,\tau) \\ \dot{r}_z(x,\tau) \\ \dot{r}_\theta(x,\tau) \end{bmatrix} + \mathbf{k}_{ae}(s) \cdot \frac{d}{d\tau} \begin{bmatrix} r_y(x,\tau) \\ r_z(x,\tau) \\ r_\theta(x,\tau) \end{bmatrix} \right\} d\tau \tag{5.5}$$

As time goes towards infinity it is a physical requirement that loads obtained from indicial functions will asymptotically approach quasi-static loads, i.e., $\Phi_{ij}(s) \xrightarrow[s \to \infty]{} 1$. It is also a physical requirement that if the structural motion is expressed in modal coordinates then loads obtained from indicial functions must be identical to those obtained from aerodynamic derivatives (see below), a requirement which may be used to determine indicial functions from known aerodynamic derivatives.

Alternatively, if a solution in modal coordinates is pursued, then aerodynamic loads obtained from the quasi-static buffeting theory (given in Eq. 6.3) may be replaced by aerodynamic derivatives (as first suggested by Scanlan and Tomko (Scanlan and Tomko, 1971)), i.e.,

$$\left. \begin{aligned} \mathbf{c}_{q_{ae}} &= \frac{\rho B^2}{2} \omega_i \hat{\mathbf{c}}_{q_{ae}} \\ \mathbf{k}_{q_{ae}} &= \frac{\rho B^2}{2} \omega_i^2 \hat{\mathbf{k}}_{q_{ae}} \end{aligned} \right\} \text{where } \hat{\mathbf{c}}_{q_{ae}} = \begin{bmatrix} P_1^* & P_5^* & BP_2^* \\ H_5^* & H_1^* & BH_2^* \\ BA_5^* & BA_1^* & B^2 A_2^* \end{bmatrix} \text{ and}$$

$$\hat{\mathbf{k}}_{q_{ae}} = \begin{bmatrix} P_4^* & P_6^* & BP_3^* \\ H_6^* & H_4^* & BH_3^* \\ BA_6^* & BA_4^* & B^2 A_3^* \end{bmatrix} \tag{5.6}$$

and where the non-dimensional coefficients P_k^*, H_k^*, A_k^*, $k = 1-6$ contained in $\hat{\mathbf{c}}_{q_{ae}}$ and $\hat{\mathbf{k}}_{q_{ae}}$ are the so-called aerodynamic derivatives, B is the cross sectional width and ω_i is the resonance frequency associated with the relevant mode shapes of the system. The reason why these quantities can only be used in modal coordinates is that they have been determined from aero-elastic section model test, an experimental set up which is a model of two dominant mode shapes of the system. Thus, the model laws (the similarity requirements) will require that an extension of the test results from such model test can only be applied to the modal equations of the full scale structure. (Further explained in Sections 5.3 and 5.4.)

5.2 Theory

Basic definitions of displacements and forces are presented in Figs. 5.4 and 5.5. The overall problem of a structural system of line like members in a turbulent wind field, defined by the wind velocity components $V(z_f) + u(y_f, z_f, t)$, $v(y_f, z_f, t)$ and $w(y_f, z_f, t)$ is illustrated in Fig. 5.1. At an arbitrary position on an element n (between nodes p and k) the wind field and the interaction between flow and structural motion will generate three load components, one in the direction of drag, one in the across wind direction (vertical or horizontal depending on the orientation of the element in relation to the flow) and one torsion moment. Adopting a system of six degrees of freedom in each node, then there is load and displacement vectors $[R_1 \ R_2 \ R_3 \ R_4 \ R_5 \ R_6]^T$ and $[r_1 \ r_2 \ r_3 \ r_4 \ r_5 \ r_6]^T$ in each node (see Fig. 5.6). It is taken for granted that the global axis X, Y and Z coincide with the flow axis $-y_f$, x_f and z_f, and that the structural system is two-dimensional and perpendicular to the main wind direction. Unfortunately, a two-dimensional system is at the moment a necessary restriction, due to lack of experimental evidence. Also, experimental support of wind load on an element at an arbitrary attitude in the flow is insufficient, and thus the theory below is only supported by experimental data if the elements of the system are either horizontal or vertical, i.e., that the roll angle γ (see Fig. 5.8) is either 0 or 90°.

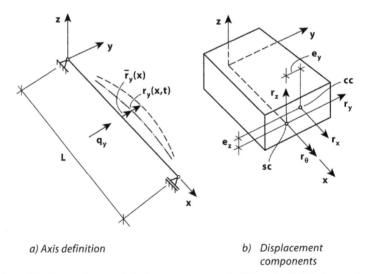

a) Axis definition

b) Displacement components

Figure 5.4. Structural axes and displacement components (CC: centroid, SC: shear centre).

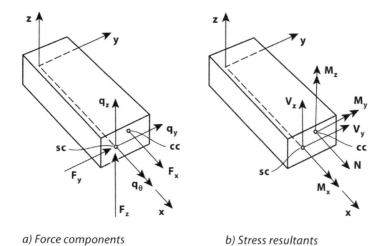

a) Force components   b) Stress resultants

Figure 5.5. Basic axis and vector definitions (CC: centroid, SC: shear centre).

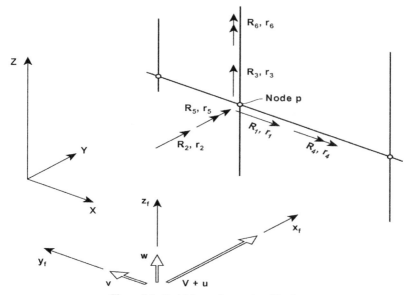

Figure 5.6. Nodal forces from external loads.

It should be noted that experimental data from structural aerodynamics usually complies with the force and displacement definitions where pitching moment and cross sectional rotation are defined by windward edge up. It is in the following assumed that this definition applies to all the aerodynamic data (e.g.,

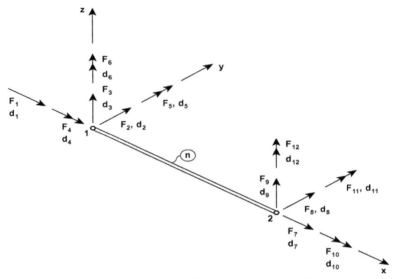

Figure 5.7. Element degrees of freedom and element end forces.

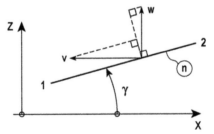

Figure 5.8. Definition of roll angle γ.

load coefficients, aerodynamic derivatives, etc.), that are adopted for numerical calculations. However, in the finite element theory it has been chosen to strictly comply with the usual convention that all external and internal forces and displacement degrees of freedom are vectors in global as well as local coordinates (see Figs. 5.4 and 5.5). A free body diagram of an arbitrary beam (line-like) type element n, with local axis x, y and z is illustrated in Fig. 5.7. For simplicity, it is in the following taken for granted that all displacements as well as forces are limited to stationary fluctuating (dynamic) contributions, i.e., the time-invariant mean value (static) part is disregarded. Then, at an arbitrary spanwise position x the element is subject to a distributed dynamic load containing buffeting and motion induced contributions:

$$\mathbf{q}_{tot}(x,t) = \begin{bmatrix} q_x & q_y & q_z & q_\theta \end{bmatrix}_{tot}^T = \mathbf{q}(x,t) + \mathbf{q}_{ae}(x,t) \tag{5.7}$$

where the buffeting load $\mathbf{q}(x,t)$ is defined by $\mathbf{q}(x,t) = \mathbf{B}_q(V)\boldsymbol{\beta}_0 \hat{\mathbf{v}}_0(x,t)$ where

$$
\mathbf{B}_q = \frac{\rho V^2 B}{2}
\begin{bmatrix}
0 & 0 & 0 \\
2\dfrac{D}{B}\bar{C}_D I_u & \left(\dfrac{D}{B}C'_D - \bar{C}_L\right)I_v & \left(\dfrac{D}{B}C'_D - \bar{C}_L\right)I_w \\
2\bar{C}_L I_u & \left(C'_L + \dfrac{D}{B}\bar{C}_D\right)I_v & \left(C'_L + \dfrac{D}{B}\bar{C}_D\right)I_w \\
-2B\bar{C}_M I_u & -BC'_M I_v & -BC'_M I_w
\end{bmatrix}
\tag{5.8}
$$

where I_u, I_v and I_w are the turbulence intensities, $\boldsymbol{\beta}_0$ is a three by three matrix taking account of the element orientation (roll angle $\gamma = 0$ or $90°$) in the flow (see Fig. 5.8) and where $\hat{\mathbf{v}}_0$ is the reduced turbulence velocity vector, i.e.,

$$
\boldsymbol{\beta}_0 =
\begin{bmatrix}
1 & 0 & 0 \\
0 & \sin\gamma & 0 \\
0 & 0 & \cos\gamma
\end{bmatrix}
\quad \text{and} \quad
\hat{\mathbf{V}}_0 =
\begin{bmatrix}
\dfrac{u(x,t)}{\sigma_u} & \dfrac{v(x,t)}{\sigma_v} & \dfrac{w(x,t)}{\sigma_w}
\end{bmatrix}^T
\tag{5.9}
$$

The motion induced part $\mathbf{q}_{ae}(x,t)$ is defined by $\mathbf{q}_{ae}(x,t) = \mathbf{c}_{ae0}\dot{\mathbf{r}}_{el}(x,t) + \mathbf{k}_{ae0}\mathbf{r}_{el}(x,t)$ where $\mathbf{r}_{el}(x,t) = \begin{bmatrix} r_x & r_y & r_z & r_\theta \end{bmatrix}_{el}^T$ is the cross sectional displacements and rotation (torsion) at an arbitrary position x along the element span, and where $\mathbf{c}_{ae0}$ and $\mathbf{k}_{ae0}$ contain the load coefficients associated with motion induced forces at cross sectional level, i.e., they are four by four matrices containing zeros on first row and first column and elsewhere filled with $\mathbf{c}_{ae}$ and $\mathbf{k}_{ae}$, whichever solution strategy may be chosen. At ends 1 and 2 the element have nodal forces and displacements

$$
\mathbf{F}(t) =
\begin{bmatrix}
\mathbf{F}_1 \\
\mathbf{F}_2
\end{bmatrix}
\quad \text{where} \quad
\begin{cases}
\mathbf{F}_1 = \begin{bmatrix} F_1 & F_2 & F_3 & F_4 & F_5 & F_6 \end{bmatrix}^T \\
\mathbf{F}_2 = \begin{bmatrix} F_7 & F_8 & F_9 & F_{10} & F_{11} & F_{12} \end{bmatrix}^T
\end{cases}
$$

$$
\mathbf{d}(t) =
\begin{bmatrix}
\mathbf{d}_1 \\
\mathbf{d}_2
\end{bmatrix}
\quad \text{where} \quad
\begin{cases}
\mathbf{d}_1 = \begin{bmatrix} d_1 & d_2 & d_3 & d_4 & d_5 & d_6 \end{bmatrix}^T \\
\mathbf{d}_2 = \begin{bmatrix} d_7 & d_8 & d_9 & d_{10} & d_{11} & d_{12} \end{bmatrix}^T
\end{cases}
\tag{5.10}
$$

It is assumed that the cross sectional displacement vector $\mathbf{r}_{el}(x,t)$ may with sufficient accuracy be described by the product of a shape function matrix $\boldsymbol{\psi}(x)$ and the nodal displacement vector $\mathbf{d}(t)$, i.e., that $\mathbf{r}_{el}(x,t) = \boldsymbol{\psi}(x) \cdot \mathbf{d}(t)$, where

$$
\boldsymbol{\psi}(x) =
\begin{bmatrix}
\psi_1 & 0 & 0 & 0 & 0 & 0 & \psi_7 & 0 & 0 & 0 & 0 & 0 \\
0 & \psi_2 & 0 & 0 & 0 & \psi_6 & 0 & \psi_8 & 0 & 0 & 0 & \psi_{12} \\
0 & 0 & \psi_3 & 0 & \psi_5 & 0 & 0 & 0 & \psi_9 & 0 & \psi_{11} & 0 \\
0 & 0 & 0 & \psi_4 & 0 & 0 & 0 & 0 & 0 & \psi_{10} & 0 & 0
\end{bmatrix}
\tag{5.11}
$$

Table 5.1. Shape functions ($\hat{x} = x/L$).

$\psi_1 = \psi_4$	$\psi_2 = \psi_3$	$\psi_5 = \psi_6$	$\psi_7 = \psi_{10}$	$\psi_8 = \psi_9$	$\psi_{11} = -\psi_{12}$
$1 - \hat{x}$	$1 - 3\hat{x}^2 + 2\hat{x}^3$	$-\hat{x}(1 - 2\hat{x} + \hat{x}^2)$	$\hat{x}$	$3\hat{x}^2 - 2\hat{x}^3$	$\hat{x}^2(1 - \hat{x})$

where ψ_i, $i = 1 - 12$ are the shape functions commonly used elsewhere in structural mechanics. (It should be noted that if the element lengths are beyond a certain limit, then the usual polynomial shape functions should be replaced by the harmonic and exponential functions which are required from an equivalent solution of the differential equation.) Applying d'Alambert's principle at a position of equilibrium defined by $\mathbf{r}_{el}(x,t)$, and letting the system be subject to an incremental virtual displacement $\delta\mathbf{r}_{el} = \begin{bmatrix} \delta r_x & \delta r_y & \delta r_z & \delta r_\theta \end{bmatrix}^T$ compatible with $\delta\mathbf{d} = \begin{bmatrix} \delta\mathbf{d}_1 \\ \delta\mathbf{d}_2 \end{bmatrix}$ where $\begin{cases} \delta\mathbf{d}_1 = \begin{bmatrix} \delta d_1 & \delta d_2 & \delta d_3 & \delta d_4 & \delta d_5 & \delta d_6 \end{bmatrix}^T \\ \delta\mathbf{d}_2 = \begin{bmatrix} \delta d_7 & \delta d_8 & \delta d_9 & \delta d_{10} & \delta d_{11} & \delta d_{12} \end{bmatrix}^T \end{cases}$ such that $\delta\mathbf{r}_{el} = \boldsymbol{\psi}(x)\cdot\delta\mathbf{d}$, then external and internal works performed during this motion are given by

$$\left.\begin{aligned} W_{ext} &= \int_L \delta\mathbf{r}_{el}^T \mathbf{q}dx + \int_L \delta\mathbf{r}_{el}^T \mathbf{c}_{aeo}\dot{\mathbf{r}}_{el}dx + \int_L \delta\mathbf{r}_{el}^T \mathbf{k}_{aeo}\mathbf{r}_{el}dx + \int_L \delta\mathbf{r}_{el}^T \cdot(-\mathbf{m}_0\ddot{\mathbf{r}}_{el})dx \\ W_{int} &= \int_L \delta\mathbf{r}^T \mathbf{c}_0\dot{\mathbf{r}}_{el}dx + \int_L \delta\mathbf{r}_{el}''^T \mathbf{k}_{bo}\mathbf{r}_{el}''dx + \int_L \delta\mathbf{r}_{el}'^T \mathbf{k}_{ao}\mathbf{r}_{el}'dx + \int_L \delta\mathbf{r}_{el}'^T \mathbf{k}_{to}\mathbf{r}_{el}'dx + \int_L \delta\mathbf{r}_{el}'^T \bar{\mathbf{n}}_0\mathbf{r}_{el}'dx \end{aligned}\right\}$$

$$(5.12)$$

where $\mathbf{m}_0 = \begin{bmatrix} m_x & 0 & 0 & 0 \\ & m_y & 0 & -e_z m_y \\ & & m_z & e_y m_z \\ sym. & & & m_\theta \end{bmatrix}$ and $\begin{cases} \mathbf{k}_{ao} = diag\begin{bmatrix} EA & 0 & 0 & 0 \end{bmatrix} \\ \mathbf{k}_{bo} = diag\begin{bmatrix} 0 & EI_z & EI_y & 0 \end{bmatrix} \\ \mathbf{k}_{to} = diag\begin{bmatrix} 0 & 0 & 0 & GI_t \end{bmatrix} \\ \bar{\mathbf{n}} = diag\begin{bmatrix} 0 & \bar{N} & \bar{N} & e_0^2\bar{N} \end{bmatrix} \end{cases}$

where primes indicate derivation with respect to x, EI_z and EI_y are cross sectional bending stiffness about y and z axis, GI_t is torsion stiffness, $\bar{N}$ is the time invariant (mean) axial force in the element and $e_0^2 = e_y^2 + e_z^2 + (I_y + I_z)/A$. The damping matrix, containing structural damping properties at a cross sectional level, is assumed diagonal, i.e., $\mathbf{c}_0 = diag\begin{bmatrix} c_x & c_y & c_z & c_\theta \end{bmatrix}$. (It should be noted that its content is non-existing, as only modal damping is known for overall structural behaviour.) Setting $W_{ext} = W_{int}$ it is seen that the equilibrium condition at element level is given by

$$\mathbf{m}\ddot{\mathbf{d}} + (\mathbf{c} - \mathbf{c}_{ae})\dot{\mathbf{d}} + (\mathbf{k} + \mathbf{k}_g - \mathbf{k}_{ae})\mathbf{d} = \mathbf{F}_q \qquad (5.13)$$

where

$$\mathbf{F}_q = \int_L \boldsymbol{\psi}^T \mathbf{q}dx \qquad\qquad \mathbf{m} = \int_L \boldsymbol{\psi}^T \mathbf{m}_0\boldsymbol{\psi}dx, \quad \mathbf{c} = \int_L \boldsymbol{\psi}^T \mathbf{c}_0\boldsymbol{\psi}dx$$

$$\mathbf{c}_{ae} = \int_L \boldsymbol{\psi}^T \mathbf{c}_{aeo}\boldsymbol{\psi}dx \qquad \mathbf{k} = \int_L \boldsymbol{\psi}''^T \mathbf{k}_{bo}\boldsymbol{\psi}''^T dx + \int_L \boldsymbol{\psi}'^T \mathbf{k}_{ao}\boldsymbol{\psi}'^T dx + \int_L \boldsymbol{\psi}'^T \mathbf{k}_{to}\boldsymbol{\psi}'^T dx$$

$$\mathbf{k}_{ae} = \int_L \boldsymbol{\psi}^T \mathbf{k}_{aeo}\boldsymbol{\psi}dx \qquad \mathbf{k}_g = \int_L \boldsymbol{\psi}'^T \mathbf{n}_0\boldsymbol{\psi}'^T dx$$

Let $\mathbf{d}_n$ be the degrees of freedom associated with an arbitrary element n and $\mathbf{A}_n$ the connectivity matrix describing the relationship between $\mathbf{d}_n$ and global degrees of freedom $\mathbf{r}$:

$$\mathbf{d}_n = \mathbf{A}_n \cdot \mathbf{r} \tag{5.14}$$

Applying a set of virtual displacements $\delta \mathbf{r}$ to the discretized global structural system, then $\delta \mathbf{d}_n = \mathbf{A}_n \cdot \delta \mathbf{r}$, and since the virtual work exerted by the external forces (at global as well as at element level) must be equal to the sum of the virtual work of the internal forces, then

$$\sum_{n=1}^{N} \delta \mathbf{d}_n^T \mathbf{F}_{qn} = \sum_{n=1}^{N} \delta \mathbf{d}_n^T \left[\mathbf{m}\ddot{\mathbf{d}}_n + \left(\mathbf{c} - \mathbf{c}_{ae} \right) \dot{\mathbf{d}}_n + \left(\mathbf{k} + \mathbf{k}_g - \mathbf{k}_{ae} \right) \mathbf{d}_n \right] \tag{5.15}$$

where N is the total number of elements in the system. Introducing Eq. (5.14) and it is seen that $\delta \mathbf{r}^T$ may be discarded and then the following is obtained

$$\mathbf{M}\ddot{\mathbf{r}}(t) + \left[\mathbf{C} - \mathbf{C}_{ae}(V) \right] \dot{\mathbf{r}}(t) + \left[\mathbf{K} + \mathbf{K}_g(\bar{N}) - \mathbf{K}_{ae}(V) \right] \mathbf{r}(t) = \mathbf{R}_q(V, t) \tag{5.16}$$

$$\text{where} \quad \begin{bmatrix} \mathbf{M} \\ \mathbf{C} \\ \mathbf{K} \end{bmatrix} = \sum_{n=1}^{N} \mathbf{A}_n^T \begin{bmatrix} \mathbf{m} \\ \mathbf{c} \\ \mathbf{k} \end{bmatrix}_n \mathbf{A}_n \quad \begin{bmatrix} \mathbf{K}_g \\ \mathbf{K}_{ae} \\ \mathbf{C}_{ae} \end{bmatrix} = \sum_{n=1}^{N} \mathbf{A}_n^T \begin{bmatrix} \mathbf{k}_g \\ \mathbf{k}_{ae} \\ \mathbf{c}_{ae} \end{bmatrix}_n \mathbf{A}_n \quad \mathbf{R}_q = \sum_{n=1}^{N} \mathbf{A}_n^T \mathbf{F}_{qn}$$

This equation represents the dynamic equilibrium condition in what is usually referred to as the original degrees of freedom. It may be used to calculate dynamic response in time domain as well as in frequency domain. In a time domain approach it will be necessary to perform a time domain simulation of the fluctuating flow components $u(X_k, Z_k, t)$, $v(X_k, Z_k, t)$ and $w(X_k, Z_k, t)$ at every node p. In a frequency domain approach it will be necessary to introduce the stochastic properties of the flow components and to perform spanwise averaging to obtain the corresponding stochastic properties of the relevant response quantities. The main problem is to fill $\mathbf{C}, \mathbf{C}_{ae}(V)$ and $\mathbf{K}_{ae}(V)$ with meaningful quantities. In general, a solution in original degrees of freedom will require the indicial functions Φ_{ij}. Particularly, response calculations in close vicinity to the stability limit may become unmanageable, not only because of the problems with the experimental determination of indicial functions, but also because the solution will rely heavily on the precision of numerically small but important differences between entries in $\mathbf{C}$ and $\mathbf{C}_{ae}(V)$.

5.2.1 Response Calculations in Modal Coordinates

The dynamic equilibrium condition in original degrees of freedom in Eq. (5.16) may readily be transformed into a modal format by choosing

$$\mathbf{r}(t) = \mathbf{\Phi} \cdot \mathbf{\eta}(t) \quad \text{where} \quad \left. \begin{aligned} \mathbf{\eta}(t) &= \begin{bmatrix} \eta_1 & \eta_2 & \cdots & \eta_i & \cdots & \eta_{N_{\text{mod}}} \end{bmatrix}^T \\ \mathbf{\Phi} &= \begin{bmatrix} \boldsymbol{\varphi}_1 & \boldsymbol{\varphi}_2 & \cdots & \boldsymbol{\varphi}_i & \cdots & \boldsymbol{\varphi}_{N_{\text{mod}}} \end{bmatrix} \end{aligned} \right\} \tag{5.17}$$

where the vector $\boldsymbol{\eta}(t)$ contains N_{mod} modal coordinates η_i and $\boldsymbol{\Phi}$ contains the mode shapes and $\boldsymbol{\varphi}_i$ ($i = 1,2,\ldots,N_{mod}$) contains the mode shape values, each associated with the corresponding global degree of freedom number, i.e., $\boldsymbol{\varphi}_i = \begin{bmatrix} \phi_1 & \phi_2 & \cdots & \phi_k & \cdots & \phi_{N_r} \end{bmatrix}^T$. It should be noted that because $\mathbf{K}_{ae}$ depends of the mean wind velocity, so will the total stiffness of the system, and thus, the resonance frequencies and the associated mode shapes will change with increasing mean wind velocities. These changes are not negligible with respect to the overall behaviour of the system in the close vicinity of a stability limit, where it will usually be necessary to update system quantities due to these effects (resonance frequencies in particular). Sufficiently far from an instability limit these effects are usually small such that the modal solution strategy of most wind engineering problems may be based on eigenfrequencies ω_i and corresponding mode shapes $\boldsymbol{\varphi}_i$ as determined from the eigenvalue problem in still air, i.e., from $\left[\mathbf{K}+\mathbf{K}_g\left(\bar{N}\right)-\omega_i^2\mathbf{M}\right]\boldsymbol{\varphi}_i = \mathbf{0}$. Introducing Eq. (5.17) into Eq. (5.16), and pre-multiplying the entire equation by $\boldsymbol{\Phi}^T$, then the following dynamic equilibrium condition in modal coordinates is obtained

$$\tilde{\mathbf{M}}\ddot{\boldsymbol{\eta}}(t)+\left[\tilde{\mathbf{C}}-\tilde{\mathbf{C}}_{ae}(V)\right]\dot{\boldsymbol{\eta}}(t)+\left[\tilde{\mathbf{K}}-\tilde{\mathbf{K}}_{ae}(V)\right]\boldsymbol{\eta}(t) = \tilde{\mathbf{R}}_q(V,t) \tag{5.18}$$

where
$$\begin{cases} \tilde{\mathbf{M}} = \boldsymbol{\Phi}^T\mathbf{M}\boldsymbol{\Phi} = diag\left[\tilde{M}_i\right] = diag\left[\boldsymbol{\varphi}_i^T\mathbf{M}\boldsymbol{\varphi}_i\right] \\ \tilde{\mathbf{C}} = \boldsymbol{\Phi}^T\mathbf{C}\boldsymbol{\Phi} = diag\left[\tilde{C}_i\right] = diag\left[\boldsymbol{\varphi}_i^T\mathbf{C}\boldsymbol{\varphi}_i\right] \\ \tilde{\mathbf{K}} = \boldsymbol{\Phi}^T\mathbf{M}\boldsymbol{\Phi} = diag\left[\tilde{K}_i\right] = diag\left[\boldsymbol{\varphi}_i^T\mathbf{K}\boldsymbol{\varphi}_i\right] \end{cases}$$ and

$$\begin{cases} \tilde{\mathbf{C}}_{ae}(V) = \boldsymbol{\Phi}^T\mathbf{C}_{ae}(V)\boldsymbol{\Phi} \\ \tilde{\mathbf{K}}_{ae}(V) = \boldsymbol{\Phi}^T\mathbf{K}_{ae}(V)\boldsymbol{\Phi} \\ \tilde{\mathbf{R}}_q(V,t) = \boldsymbol{\Phi}^T\mathbf{R}_q(V,t) \end{cases}$$

　　Due to the orthogonal properties of the mode shapes, all the off diagonal terms in $\tilde{\mathbf{M}}$, $\tilde{\mathbf{C}}$ and $\tilde{\mathbf{K}}$ are zeros. (The content of $\tilde{\mathbf{K}}$ may be determined by pre-multiplying the eigenvalue problem $\left[\mathbf{K}-\omega_i^2\mathbf{M}\right]\boldsymbol{\varphi}_i = \mathbf{0}$ by $\boldsymbol{\varphi}_i^T$, from which it is readily seen that $\tilde{K}_i = \omega_i^2\tilde{M}_i$). It is common practice to introduce N_{mod} modal damping ratios ζ_i, each associated with the corresponding modal critical damping $2\tilde{M}_i\omega_i$, and thus the content of $\tilde{\mathbf{C}}$ is given by $\tilde{C}_i = 2\tilde{M}_i\omega_i\zeta_i$. In modal coordinates the aerodynamic derivatives are applicable. Since they have been determined from two-degrees of freedom aero elastic section model, they are strictly spoken only applicable to the two eigenmodes that was included in the tests. However, it may be argued that in a modal format the motion induced forces are primarily system properties, e.g., like the modal damping ratio, and because they have been determined as modal properties for the combination of the two most critical of modes, they are likely to be conservative if applied to any other combination of modes. If this hypothesis may be accepted, then the contents of

$$\tilde{\mathbf{C}}_{ae} = \begin{bmatrix} \ddots & & \\ & \tilde{C}_{ae_{ij}} & \\ & & \ddots \end{bmatrix} \quad \text{and} \quad \tilde{\mathbf{K}}_{ae} = \begin{bmatrix} \ddots & & \\ & \tilde{K}_{ae_{ij}} & \\ & & \ddots \end{bmatrix} \tag{5.19}$$

are given by $\begin{cases} \tilde{C}_{ae_{ij}}\left(\hat{V}_i,\hat{V}_j\right) = \boldsymbol{\varphi}_i^T \cdot \int_L \boldsymbol{\psi}^T(x) \mathbf{c}_{ae_0}\left(\hat{V}_i,\hat{V}_j\right) \boldsymbol{\psi}(x) dx \cdot \boldsymbol{\varphi}_j \\[2mm] \tilde{K}_{ae_{ij}}\left(\hat{V}_i,\hat{V}_j\right) = \boldsymbol{\varphi}_i^T \cdot \int_L \boldsymbol{\psi}^T(x) \mathbf{k}_{ae_0}\left(\hat{V}_i,\hat{V}_j\right) \boldsymbol{\psi}(x) dx \cdot \boldsymbol{\varphi}_j \end{cases} \tag{5.20}$

where the non-dimensional aerodynamic derivatives P_k^*, H_k^*, A_k^*, $k=1-6$, are functions of the reduced mean wind velocity $\hat{V}_i = V/(B\omega_i)$ or $\hat{V}_i = V/(B\omega_i)$, depending on which displacement component they belong to, and where ω_i and ω_j are the in-wind resonance frequencies associated with modes i and j, which are themselves affected by stiffness contributions from $\mathbf{K}_{ae}$. Hence, iterations may be required. It should be noted that in the theory above, any motion induced coupling between horizontal and vertical or torsion motion can only be accounted for if the section model tests have allowed for the detection of such effects. Otherwise, it will only cover the coupling between vertical motion and torsion, i.e., P_k^* must be taken from the quasi-static theory. A solution to Eq. (5.18) may be pursued in time domain as well as in frequency domain.

5.2.2 The Solution in Modal Coordinates in Frequency Domain

Taking the Fourier transform throughout Eq. (5.18), i.e., setting $\boldsymbol{\eta}(t) = \text{Re} \sum_\omega \mathbf{a}_\eta(\omega) \cdot e^{i\omega t}$ and $\tilde{\mathbf{R}}_q(t) = \text{Re} \sum_\omega \mathbf{a}_{\tilde{R}_q}(\omega) \cdot e^{i\omega t}$, where $\mathbf{a}_\eta(\omega)$, $\mathbf{a}_{\tilde{R}}(\omega)$ and $\mathbf{a}_{\tilde{R}_q}(\omega)$ are N_{mod} by 1 vectors containing the Fourier coefficients of the modal coordinates and the modal loads, and pre-multiplying by $\tilde{\mathbf{K}}^{-1}$, then the modal dynamic equilibrium equation is satisfied for each ω-setting if

$$\left\{ \mathbf{I} - \tilde{\mathbf{K}}^{-1}\tilde{\mathbf{K}}_{ae}(V) - \tilde{\mathbf{K}}^{-1}\tilde{\mathbf{M}}\omega^2 + \left[\tilde{\mathbf{K}}^{-1}\tilde{\mathbf{C}} - \tilde{\mathbf{K}}^{-1}\tilde{\mathbf{C}}_{ae}(V) \right] i\omega \right\} \cdot \mathbf{a}_\eta(\omega) = \tilde{\mathbf{K}}^{-1}\mathbf{a}_{\tilde{R}_q}(V,\omega) \tag{5.21}$$

Recalling that $\tilde{\mathbf{M}} = diag\left[\tilde{M}_i \right]$, $diag\left[K \right]$ and $\tilde{\mathbf{C}} = diag\left[\tilde{C}_i \right]$ where $\tilde{M}_i = \boldsymbol{\varphi}_i^T \mathbf{M} \boldsymbol{\varphi}_i$, $\tilde{K}_i = \omega_i^2 \tilde{M}_i$ and $\tilde{C}_i = 2\tilde{M}_i \omega_i \zeta_i$, then

$$\mathbf{a}_\eta(\omega) = \hat{\mathbf{H}}_\eta(\omega) \cdot \tilde{\mathbf{K}}^{-1}\mathbf{a}_{\tilde{R}_q}(V,\omega) \tag{5.22}$$

where the frequency response matrix is defined by

$$\hat{\mathbf{H}}_\eta(V,\omega) = \left\{ \mathbf{I} - \boldsymbol{\kappa}_{ae}(V) - \left(\omega \cdot diag\left[\frac{1}{\omega_i} \right] \right)^2 + 2i\omega \cdot diag\left[\frac{1}{\omega_i} \right] \cdot \left[\boldsymbol{\zeta} - \boldsymbol{\zeta}_{ae}(V) \right] \right\}^{-1} \tag{5.23}$$

where $\mathbf{I}$ is the N_{mod} by N_{mod} identity matrix, $\boldsymbol{\zeta} = diag\left[\zeta_i\right]$ and

$$\mathbf{K}_{ae} = \begin{bmatrix} \ddots & & \ddots \\ & \kappa_{ae_{ij}} & \\ \ddots & & \ddots \end{bmatrix} = \tilde{\mathbf{K}}^{-1}\tilde{\mathbf{K}}_{ae} \quad \text{and}$$

(5.24)

$$\boldsymbol{\zeta}_{ae} = \begin{bmatrix} \ddots & & \ddots \\ & \zeta_{ae_{ij}} & \\ \ddots & & \ddots \end{bmatrix} = \frac{1}{2} diag\left[\omega_i\right]\left(\tilde{\mathbf{K}}^{-1}\tilde{\mathbf{C}}_{ae}\right)$$

while the cross spectral density matrix of the modal coordinates is defined by

$$\mathbf{S}_{\eta\eta}\left(\omega\right) = \lim_{T\to\infty}\frac{\mathbf{a}_\eta^*\mathbf{a}_\eta^T}{\pi T} = \begin{bmatrix} S_{\eta_1\eta_1} & \cdots & S_{\eta_1\eta_m} & \cdots & S_{\eta_1\eta_{N_{\text{mod}}}} \\ \vdots & \ddots & \vdots & \ddots & \vdots \\ S_{\eta_n\eta_1} & \cdots & S_{\eta_n\eta_m} & \cdots & S_{\eta_n\eta_{N_{\text{mod}}}} \\ \vdots & \ddots & \vdots & \ddots & \vdots \\ S_{\eta_{N_{\text{mod}}}\eta_1} & \cdots & S_{\eta_{N_{\text{mod}}}\eta_m} & \cdots & S_{\eta_{N_{\text{mod}}}\eta_{N_{\text{mod}}}} \end{bmatrix} =$$

$$\lim_{T\to\infty}\frac{1}{\pi T}\left\{\left[\hat{\mathbf{H}}_\eta\left(\omega\right)\tilde{\mathbf{K}}^{-1}\mathbf{a}_{\tilde{R}_q}\right]^*\left[\hat{\mathbf{H}}_\eta\left(\omega\right)\tilde{\mathbf{K}}^{-1}\mathbf{a}_{\tilde{R}_q}\right]^T\right\} = \hat{\mathbf{H}}_\eta^*\tilde{\mathbf{K}}^{-1}\left\{\lim_{T\to\infty}\frac{1}{\pi T}\left(\mathbf{a}_{\tilde{R}_q}^*\mathbf{a}_{\tilde{R}_q}^T\right)\right\}\left(\tilde{\mathbf{K}}^{-1}\right)^T\hat{\mathbf{H}}_\eta^T$$

$$= \hat{\mathbf{H}}_\eta^*\left(V,\omega\right)\tilde{\mathbf{K}}^{-1}\mathbf{S}_{\tilde{R}_q\tilde{R}_q}\left(V,\omega\right)\left(\tilde{\mathbf{K}}^{-1}\right)^T\hat{\mathbf{H}}_\eta^T\left(V,\omega\right)$$

(5.25)

where $\mathbf{S}_{\tilde{R}\tilde{R}}\left(V,\omega\right) = \lim_{T\to\infty}\frac{1}{\pi T}\left(\mathbf{a}_{\tilde{R}_q}^*\cdot\mathbf{a}_{\tilde{R}_q}^T\right)$ is the cross spectral density matrix of the modal wind load. Since $\mathbf{r}\left(t\right) = \boldsymbol{\Phi}\boldsymbol{\eta}\left(t\right)$ then $\mathbf{a}_r = \boldsymbol{\Phi}\cdot\mathbf{a}_\eta$, and thus

$$\mathbf{S}_{rr}\left(V,\omega\right) = \lim_{T\to\infty}\frac{1}{\pi T}\left(\mathbf{a}_r^*\mathbf{a}_r^T\right) = \lim_{T\to\infty}\frac{1}{\pi T}\left[\left(\boldsymbol{\Phi}\mathbf{a}_\eta\right)^*\left(\boldsymbol{\Phi}\mathbf{a}_\eta\right)^T\right] = \boldsymbol{\Phi}\lim_{T\to\infty}\frac{1}{\pi T}\left(\mathbf{a}_\eta^*\mathbf{a}_\eta^T\right)\boldsymbol{\Phi}^T$$

$$= \boldsymbol{\Phi}\mathbf{S}_{\eta\eta}\left(V,\omega\right)\boldsymbol{\Phi}^T = \boldsymbol{\Phi}\hat{\mathbf{H}}_\eta^*\left(V,\omega\right)\tilde{\mathbf{K}}^{-1}\cdot\mathbf{S}_{\tilde{R}_q\tilde{R}_q}\left(V,\omega\right)\cdot\left(\tilde{\mathbf{K}}^{-1}\right)^T\hat{\mathbf{H}}_\eta^T\left(V,\omega\right)\boldsymbol{\Phi}^T$$

(5.26)

and since $\tilde{\mathbf{R}}_q\left(V,t\right) = \boldsymbol{\Phi}^T\mathbf{R}_q\left(V,t\right)$, then $\mathbf{a}_{\tilde{R}}\left(\omega\right) = \boldsymbol{\Phi}^T\cdot\mathbf{a}_R\left(\omega\right)$ and $\mathbf{a}_{\tilde{R}_q}\left(V,\omega\right) = \boldsymbol{\Phi}^T\cdot\mathbf{a}_{R_q}\left(V,\omega\right)$, and thus

$$\mathbf{S}_{\tilde{R}_q\tilde{R}_q}\left(V,\omega\right) = \lim_{T\to\infty}\frac{1}{\pi T}\left(\mathbf{a}_{\tilde{R}_q}^*\cdot\mathbf{a}_{\tilde{R}_q}^T\right) = \lim_{T\to\infty}\frac{1}{\pi T}\left[\left(\boldsymbol{\Phi}^T\cdot\mathbf{a}_{R_q}\right)^*\cdot\left(\boldsymbol{\Phi}^T\cdot\mathbf{a}_{R_q}\right)^T\right]$$

$$= \boldsymbol{\Phi}^T\cdot\lim_{T\to\infty}\frac{1}{\pi T}\left(\mathbf{a}_{R_q}^*\cdot\mathbf{a}_{R_q}^T\right)\cdot\boldsymbol{\Phi} = \boldsymbol{\Phi}^T\cdot\mathbf{S}_{R_qR_q}\left(\omega\right)\cdot\boldsymbol{\Phi}$$

(5.27)

The cross spectral response matrix is then given by:

$$\mathbf{S}_{rr}(V,\omega) = \mathbf{\Phi}\hat{\mathbf{H}}_\eta^*(V,\omega)\tilde{\mathbf{K}}^{-1}\left[\mathbf{\Phi}^T\mathbf{S}_{R_qR_q}(\omega)\mathbf{\Phi}\right]\left(\tilde{\mathbf{K}}^{-1}\right)^T\hat{\mathbf{H}}_\eta^T(V,\omega)\mathbf{\Phi}^T \qquad (5.28)$$

Recalling that $\mathbf{R}_q = \sum\limits_{n=1}^{N}\mathbf{A}_n^T\mathbf{F}_{q_n}$ and $\mathbf{F}_q = \int\limits_L\boldsymbol{\psi}^T\mathbf{q}dx$, where

$$\mathbf{q}(x,V,t) = \mathbf{B}_q(V)\boldsymbol{\beta}_0\hat{\mathbf{v}}_0(t), \text{ then}$$

$$\mathbf{S}_{R_qR_q}(V,\omega) = \lim_{T\to\infty}\frac{1}{\pi T}\mathbf{a}_{R_q}^*\mathbf{a}_{R_q}^T = \lim_{T\to\infty}\frac{1}{\pi T}\left(\sum_{n=1}^{N}\mathbf{A}_n^T\mathbf{a}_{F_{q_n}}\right)^*\left(\sum_{n=1}^{N}\mathbf{A}_n^T\mathbf{a}_{F_{q_n}}\right)^T$$

$$= \sum_{n=1}^{N}\sum_{m=1}^{N}\mathbf{A}_n^T\lim_{T\to\infty}\frac{1}{\pi T}\left(\mathbf{a}_{F_{q_n}}^*\mathbf{a}_{F_{q_m}}^T\right)\mathbf{A}_m = \sum_{n=1}^{N}\sum_{m=1}^{N}\mathbf{A}_n^T\mathbf{S}_{F_{q_n}F_{q_m}}(V,\omega)\mathbf{A}_m$$

$$\qquad (5.29)$$

where

$$\mathbf{S}_{F_{q_n}F_{q_m}}(V,\omega) = \lim_{T\to\infty}\frac{1}{\pi T}\left\{\left[\int_L\mathbf{a}_{\hat{v}_0}(x,\omega)dx\right]_n^*\left[\int_L\boldsymbol{\psi}^T\mathbf{B}_q(V)\boldsymbol{\beta}_0dx\right]_m^T\right\} =$$

$$\int_{L_n}\int_{L_m}\boldsymbol{\psi}^T(x_n)\left[\mathbf{B}_q(V)\boldsymbol{\beta}_0\right]_n\lim_{T\to\infty}\frac{1}{\pi T}\left[\mathbf{a}_{\hat{v}_0}^*(x_n,\omega)\mathbf{a}_{\hat{v}_0}^T(x_m,\omega)\right]\left[\mathbf{B}_q(V)\boldsymbol{\beta}_0\right]_m^T\boldsymbol{\psi}(x_m)dx_ndx_m$$

$$= \int_{L_n}\int_{L_m}\boldsymbol{\psi}^T(x_n)\left[\mathbf{B}_q(V)\boldsymbol{\beta}_0\right]_n\mathbf{S}_{\hat{v}_0\hat{v}_0}(\Delta x_{nm},\omega)\left[\mathbf{B}_q(V)\boldsymbol{\beta}_0\right]_m^T\boldsymbol{\psi}(x_m)dx_ndx_m$$

where

$$\mathbf{S}_{\hat{v}_0\hat{v}_0}(\Delta x_{nm},\omega) = \begin{bmatrix} S_{uu}(\Delta x_{nm},\omega)/\sigma_u^2 & 0 & 0 \\ 0 & S_{vv}(\Delta x_{nm},\omega)/\sigma_v^2 & 0 \\ 0 & 0 & S_{ww}(\Delta x_{nm},\omega)/\sigma_w^2 \end{bmatrix}$$

$$\qquad (5.30)$$

and $\Delta x_{nm} = |x_n - x_m|$. The response covariance matrix is then obtained by frequency domain integration of Eq. (5.28). Thus

$$\mathbf{Cov}_{rr}(V) = \begin{bmatrix} \sigma_1^2 & \cdots & Cov_{1i} & \cdots & Cov_{1j} & \cdots & Cov_{1N_r} \\ \vdots & \ddots & \vdots & & \vdots & & \vdots \\ Cov_{i1} & \cdots & \sigma_i^2 & \cdots & Cov_{ij} & & \vdots \\ \vdots & & \vdots & \ddots & \vdots & & \vdots \\ Cov_{j1} & \cdots & Cov_{ji} & \cdots & \sigma_j^2 & & \vdots \\ \vdots & & & & & \ddots & \vdots \\ Cov_{N_r1} & \cdots & \cdots & \cdots & \cdots & \cdots & \sigma_{N_r}^2 \end{bmatrix} = \int_0^\infty\mathbf{S}_{rr}(V,\omega)d\omega$$

$$\qquad (5.31)$$

where

$$\mathbf{S}_{rr}(V,\omega)=\boldsymbol{\Phi}\hat{\mathbf{H}}_{\eta}^{*}(V,\omega)\tilde{\mathbf{K}}^{-1}\left\{\boldsymbol{\Phi}^{T}\left[\sum_{n=1}^{N}\sum_{m=1}^{N}\mathbf{A}_{n}^{T}\mathbf{S}_{F_{q_{n}}F_{q_{m}}}(V,\omega)\mathbf{A}_{m}\right]\boldsymbol{\Phi}\right\}\left(\tilde{\mathbf{K}}^{-1}\right)^{T}\hat{\mathbf{H}}_{\eta}^{T}(V,\omega)\boldsymbol{\Phi}^{T}$$

(5.32)

5.2.3 Possible Simplifications

If the element lengths are sufficiently short then the allocation of wind load effects to element ends may be performed as illustrated in Fig. 5.9 above, i.e., the turbulence induced (buffeting) load vector $\mathbf{q}(x,t)=\begin{bmatrix} q_{1} & q_{2} & q_{3} & q_{4} & q_{5} & q_{6} \end{bmatrix}^{T}$ $=\begin{bmatrix} 0 & q_{y} & q_{z} & q_{\theta} & 0 & 0 \end{bmatrix}^{T}$ at element level is given by

$$\mathbf{R}_{n}(t)=\begin{bmatrix} \mathbf{R}_{1} \\ \mathbf{R}_{2} \end{bmatrix}_{n}$$

(5.33)

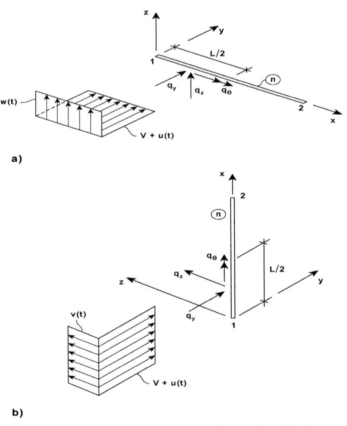

Figure 5.9. Wind load effects at horizontal and vertical positions of the element.

where $\mathbf{R}_{i_n}(t) = \left\{ \mathbf{B}_{q_n} \boldsymbol{\beta}_0 \hat{\mathbf{v}}_0 \right\}_n$,   where $i = 1$ or 2 refers to element ends, and where

$$\mathbf{B}_{q_n} = \begin{bmatrix} \mathbf{B}_{q0}(Z_1) & \mathbf{0} \\ \mathbf{0} & \mathbf{B}_{q0}(Z_2) \end{bmatrix}_n \quad \text{and} \quad \boldsymbol{\beta}_n = \begin{bmatrix} \boldsymbol{\beta}_0 & \mathbf{0} \\ \mathbf{0} & \boldsymbol{\beta}_0 \end{bmatrix}_n \quad \text{and} \quad \hat{\mathbf{v}}_n = \begin{bmatrix} \hat{\mathbf{v}}_0(Z_1) \\ \hat{\mathbf{v}}_0(Z_2) \end{bmatrix}$$

$$\boldsymbol{\beta}_0 = \begin{bmatrix} 1 & 0 & 0 \\ 0 & \sin\gamma & 0 \\ 0 & 0 & \cos\gamma \end{bmatrix} \qquad \hat{\mathbf{v}}_0 = \begin{bmatrix} \dfrac{u(x,t)}{\sigma_u} & \dfrac{v(x,t)}{\sigma_v} & \dfrac{w(x,t)}{\sigma_w} \end{bmatrix}^T$$

and

$$\mathbf{B}_{q0} = \frac{\rho \left[V(Z_i) \right]^2 L}{4} \begin{bmatrix} 0 & 0 & 0 \\ 2D\bar{C}_D I_u(Z_i) & \left(DC_D' - B\bar{C}_L \right) I_v(Z_i) & \left(DC_D' - B\bar{C}_L \right) I_w(Z_i) \\ 2B\bar{C}_L I_u(Z_i) & \left(BC_L' + D\bar{C}_D \right) I_v(Z_i) & \left(BC_L' + D\bar{C}_D \right) I_w(Z_i) \\ -2B^2 \bar{C}_M I_u(Z_i) & -B^2 C_M' I_v(Z_i) & -B^2 C_M' I_w(Z_i) \\ 0 & 0 & 0 \\ 0 & 0 & 0 \end{bmatrix}$$

Adopting a similar approach to the motion induced load effects

$$\mathbf{q}_{ae}(x,t) = \mathbf{c}_{ae0} \begin{bmatrix} 0 \\ \dot{r}_{yel} \\ \dot{r}_{zel} \\ \dot{r}_{\theta el} \\ 0 \\ 0 \end{bmatrix} + \mathbf{k}_{ae0} \begin{bmatrix} 0 \\ r_{yel} \\ r_{zel} \\ r_{\theta el} \\ 0 \\ 0 \end{bmatrix} \quad \text{then} \quad \begin{cases} \mathbf{c}_{ae_n} = \dfrac{L}{2} \begin{bmatrix} \mathbf{c}_{ae0}(Z_1) & \mathbf{0} \\ \mathbf{0} & \mathbf{c}_{ae0}(Z_2) \end{bmatrix}_n \\[4mm] \mathbf{k}_{ae_n} = \dfrac{L}{2} \begin{bmatrix} \mathbf{k}_{ae0}(Z_1) & \mathbf{0} \\ \mathbf{0} & \mathbf{k}_{ae0}(Z_2) \end{bmatrix}_n \end{cases}$$

$$(5.34)$$

If a quasi-static approach is adopted, then $\mathbf{c}_{ae0}$ and $\mathbf{k}_{ae0}$ are given by:

$$\mathbf{c}_{ae0} = \frac{\rho V}{2} \begin{bmatrix} 0 & 0 & 0 & 0 & 0 & 0 \\ 0 & -2D\bar{C}_D & -\left(DC_D' - B\bar{C}_L \right) & 0 & 0 & 0 \\ 0 & -2B\bar{C}_L & -\left(BC_L' + D\bar{C}_D \right) & 0 & 0 & 0 \\ 0 & 2B^2\bar{C}_M & B^2 C_M' & 0 & 0 & 0 \\ 0 & 0 & 0 & 0 & 0 & 0 \\ 0 & 0 & 0 & 0 & 0 & 0 \end{bmatrix} \quad \mathbf{k}_{ae0} = \frac{\rho V^2}{2} \begin{bmatrix} 0 & 0 & 0 & 0 & 0 & 0 \\ 0 & 0 & 0 & -DC_D' & 0 & 0 \\ 0 & 0 & 0 & -BC_L' & 0 & 0 \\ 0 & 0 & 0 & B^2 C_M' & 0 & 0 \\ 0 & 0 & 0 & 0 & 0 & 0 \\ 0 & 0 & 0 & 0 & 0 & 0 \end{bmatrix}$$

$$(5.35)$$

The content of $\mathbf{c}_{ae0}$ and $\mathbf{k}_{ae0}$ may be replaced by the aerodynamic derivatives if a modal approach is adopted. In any case, the modal load vector is defined by

$$\tilde{\mathbf{R}}(t) = \mathbf{\Phi}^T \mathbf{R}(t) = \mathbf{\Phi}^T \cdot \sum_{n=1}^N \mathbf{A}_n^T \cdot \mathbf{R}_n(t) = \mathbf{\Phi}^T \cdot \sum_{n=1}^N \mathbf{A}_n^T \cdot \left[\mathbf{B}_{q_n} \mathbf{\beta}_n \hat{\mathbf{v}}_n(t) \right] \tag{5.36}$$

Taking the Fourier transform renders $\mathbf{a}_{\tilde{R}}(\omega) = \mathbf{\Phi}^T \cdot \sum_{n=1}^N \mathbf{A}_n^T \cdot \left[\mathbf{B}_{q_n} \mathbf{\beta}_n \mathbf{a}_{\hat{v}_n}(\omega) \right]$ and thus, the cross spectral density matrix of the modal buffeting wind load is given by

$$\mathbf{S}_{\tilde{R}\tilde{R}}(\omega) = \lim_{T\to\infty} \frac{1}{\pi T} \left\{ \left[\mathbf{\Phi}^T \sum_{n=1}^N \mathbf{A}_n^T \left(\mathbf{B}_{q_n} \mathbf{\beta}_n \mathbf{a}_{\hat{v}_n}^* \right) \right] \cdot \left[\mathbf{\Phi}^T \sum_{n=1}^N \mathbf{A}_n^T \left(\mathbf{B}_{q_n} \mathbf{\beta}_n \mathbf{a}_{\hat{v}_n} \right) \right]^T \right\}$$

$$= \mathbf{\Phi}^T \left\{ \sum_{n=1}^N \sum_{m=1}^N \mathbf{A}_n^T \left[\mathbf{B}_{q_n} \mathbf{\beta}_n \cdot \lim_{T\to\infty} \frac{1}{\pi T} \left(\mathbf{a}_{\hat{v}_n}^* \cdot \mathbf{a}_{\hat{v}_m}^T \right) \cdot \mathbf{\beta}_m^T \mathbf{B}_{q_m}^T \right] \mathbf{A}_m \right\} \mathbf{\Phi} \tag{5.37}$$

$$= \mathbf{\Phi}^T \cdot \left\{ \sum_{n=1}^N \sum_{m=1}^N \mathbf{A}_n^T \cdot \left[\mathbf{B}_{q_n} \mathbf{\beta}_n \cdot \hat{\mathbf{S}}_{\hat{v}\hat{v}}(\omega) \cdot \mathbf{\beta}_m^T \mathbf{B}_{q_m}^T \right] \cdot \mathbf{A}_m \right\} \cdot \mathbf{\Phi}$$

where $\hat{\mathbf{S}}_{\hat{v}\hat{v}}(\omega) = \lim_{T\to\infty} \frac{1}{\pi T} \left(\mathbf{a}_{\hat{v}_n}^* \cdot \mathbf{a}_{\hat{v}_m}^T \right) = \sqrt{\hat{\mathbf{S}}_n(\omega)} \cdot \hat{\mathbf{Co}}_{nm}(\Delta s_{nm}, \omega) \cdot \sqrt{\hat{\mathbf{S}}_m(\omega)}$ and $\hat{\mathbf{S}}_j$, $j = n$ or m, is the diagonal cross spectral density matrix of the reduced turbulence velocity vector $\mathbf{a}_{\hat{v}_j}(\omega) = \left[a_{\hat{u}_1} \quad a_{\hat{v}_1} \quad a_{\hat{w}_1} \quad a_{\hat{u}_2} \quad a_{\hat{v}_2} \quad a_{\hat{w}_2} \right]_j^T$. Thus

$$\mathbf{S}_{\eta\eta}(\omega) = \tilde{\mathbf{H}}_\eta^*(\omega) \cdot \mathbf{S}_{\tilde{R}\tilde{R}}(\omega) \cdot \tilde{\mathbf{H}}_\eta^T(\omega) = \tilde{\mathbf{H}}_\eta^*(\omega) \cdot \left[\mathbf{\Phi}^T \cdot \mathbf{S}_{RR}(\omega) \cdot \mathbf{\Phi} \right] \cdot \tilde{\mathbf{H}}_\eta^T(\omega)$$

$$= \tilde{\mathbf{H}}_\eta^* \left(\mathbf{\Phi}^T \left\{ \sum_{n=1}^N \sum_{m=1}^N \mathbf{A}_n^T \left[\mathbf{B}_{q_n} \mathbf{\beta}_n \cdot \left(\hat{\mathbf{S}}_n^{1/2} \hat{\mathbf{Co}}_{nm} \hat{\mathbf{S}}_m^{1/2} \right) \cdot \mathbf{\beta}_m^T \mathbf{B}_{q_m}^T \right] \mathbf{A}_m \right\} \mathbf{\Phi} \right) \tilde{\mathbf{H}}_\eta^T \tag{5.38}$$

where
$$\hat{\mathbf{S}}_n(\omega) = diag\left[\hat{S}_{u1_n} \quad \hat{S}_{v1_n} \quad \hat{S}_{w1_n} \quad \hat{S}_{u2_n} \quad \hat{S}_{v2_n} \quad \hat{S}_{w2_n} \right]$$
$$\hat{\mathbf{S}}_m(\omega) = diag\left[\hat{S}_{u1_m} \quad \hat{S}_{v1_m} \quad \hat{S}_{w1_m} \quad \hat{S}_{u2_m} \quad \hat{S}_{v2_m} \quad \hat{S}_{w2_m} \right]$$

are the reduced auto spectral density matrices associated with elements n and m, and

$$\hat{\mathbf{Co}}_{nm}(\Delta s_{nm}, \omega) = \begin{bmatrix} \hat{\mathbf{Co}}_{1n1m} & \hat{\mathbf{Co}}_{1n2m} \\ \hat{\mathbf{Co}}_{1n2m} & \hat{\mathbf{Co}}_{2n2m} \end{bmatrix} \tag{5.39}$$

is the reduced covariance matrix between corresponding element ends, where

$$\hat{\mathbf{Co}}_{1n1m} = \begin{bmatrix} \hat{Co}_{uu}\left(\Delta s_{1_n 1_m}, \omega\right) & 0 & 0 \\ 0 & \hat{Co}_{vv}\left(\Delta s_{1_n 1_m}, \omega\right) & 0 \\ 0 & 0 & \hat{Co}_{ww}\left(\Delta s_{1_n 1_m}, \omega\right) \end{bmatrix}$$

$$\hat{\mathbf{Co}}_{2n2m} = \begin{bmatrix} \hat{Co}_{uu}\left(\Delta s_{2_n 2_m}, \omega\right) & 0 & 0 \\ 0 & \hat{Co}_{uu}\left(\Delta s_{2_n 2_m}, \omega\right) & 0 \\ 0 & 0 & \hat{Co}_{uu}\left(\Delta s_{2_n 2_m}, \omega\right) \end{bmatrix}$$

$$\hat{\mathbf{Co}}_{1n2m} = \begin{bmatrix} \hat{Co}_{uu}\left(\Delta s_{1_n 2_m}, \omega\right) & 0 & 0 \\ 0 & \hat{Co}_{vv}\left(\Delta s_{1_n 2_m}, \omega\right) & 0 \\ 0 & 0 & \hat{Co}_{ww}\left(\Delta s_{1_n 2_m}, \omega\right) \end{bmatrix}$$

$$\hat{\mathbf{Co}}_{2n1m} = \begin{bmatrix} \hat{Co}_{uu}\left(\Delta s_{2_n 1_m}, \omega\right) & 0 & 0 \\ 0 & \hat{Co}_{vv}\left(\Delta s_{2_n 1_m}, \omega\right) & 0 \\ 0 & 0 & \hat{Co}_{ww}\left(\Delta s_{2_n 1_m}, \omega\right) \end{bmatrix}$$

Adopting a Kaimal (Kaimal et al., 1972) type of auto spectrum and simple exponential co-spectrum decay, then

$$\left.\begin{aligned}
\hat{S}_{p_i}(\omega) &= \frac{S_{p_i}(\omega)}{\sigma_{p_i}^2} = \frac{A_p \cdot {}^{xf}L_p / V(Z_i)}{\left[1 + 1.5 A_p \omega {}^{xf} L_p / V(Z_i)\right]^{5/3}} \\[2mm]
\hat{S}_{p_j}(\omega) &= \frac{S_{p_j}(\omega)}{\sigma_{p_j}^2} = \frac{A_p \cdot {}^{xf}L_p / V(Z_j)}{\left[1 + 1.5 A_p \omega {}^{xf} L_p / V(Z_j)\right]^{5/3}} \\[2mm]
\hat{Co}_{pp}\left(\Delta s_{ij}, \omega\right) &= \exp\left\{-\omega\sqrt{\left[c_{px}\left(X_i - X_j\right)\right]^2 + \left[c_{pz}\left(Z_i - Z_j\right)\right]^2}\Big/\bar{V}_{ij}\right\} \\[2mm]
\bar{V}_{ij} &= \left[V(Z_i) + V(Z_j)\right]/2
\end{aligned}\right\} \quad \begin{aligned} p &= \begin{cases} u \\ v \\ w \end{cases} \\[3mm] i, j &= \begin{cases} 1_n \\ 2_n \\ 1_m \\ 2_m \end{cases} \end{aligned} \quad (5.40)$$

Finally, since $\mathbf{r}(t) = \boldsymbol{\Phi} \cdot \boldsymbol{\eta}(t)$, then $\mathbf{a}_r(\omega) = \boldsymbol{\Phi} \cdot \mathbf{a}_\eta(\omega)$, and thus, the cross spectral density matrix of the displacement response quantities is given by

$$\mathbf{S}_{rr}(\omega) = \lim_{T \to \infty} \frac{1}{\pi T}\left(\mathbf{a}_r^* \mathbf{a}_r^T\right) = \lim_{T \to \infty} \frac{1}{\pi T}\left[\left(\boldsymbol{\Phi}\mathbf{a}_\eta\right)^* \left(\boldsymbol{\Phi}\mathbf{a}_\eta\right)^T\right] = \boldsymbol{\Phi} \lim_{T \to \infty} \frac{1}{\pi T}\left(\mathbf{a}_\eta^* \mathbf{a}_\eta^T\right)\boldsymbol{\Phi}^T$$

$$= \boldsymbol{\Phi}\mathbf{S}_{\eta\eta}\boldsymbol{\Phi}^T = \boldsymbol{\Phi}\left[\hat{\mathbf{H}}_\eta^* \left(\boldsymbol{\Phi}^T \left\{\sum_{n=1}^{N}\sum_{m=1}^{N} \mathbf{A}_n^T \left[\mathbf{B}_{q_n}\boldsymbol{\beta}_n \left(\hat{\mathbf{S}}_n^{1/2}\hat{\mathbf{Co}}_{nm}\hat{\mathbf{S}}_m^{1/2}\right)\boldsymbol{\beta}_m^T\mathbf{B}_{q_m}^T\right]\mathbf{A}_m\right\}\boldsymbol{\Phi}\right)\hat{\mathbf{H}}_\eta^T\right]\boldsymbol{\Phi}^T$$

$$(5.41)$$

5.3 Current Practices

5.3.2 Simplified Mode by Mode Calculations of Dynamic Response

In many cases the most important modes contain only a single displacement component in horizontal, vertical or torsion direction (ϕ_y, ϕ_z, or ϕ_θ) and the corresponding eigenfrequencies (ω_y, ω_z, or ω_θ) are well separated, such that the response calculations may with sufficient accuracy be carried out mode by mode. Then it is convenient to express the calculations on a continuous format and the response of each component may be obtained by simple frequency domain summation, i.e.,

$$
\begin{bmatrix} S_{r_y}(\omega,x) \\ S_{r_z}(\omega,x) \\ S_{r_\theta}(\omega,x) \end{bmatrix} \approx \sum_{n=1}^{N_{mod}} \begin{bmatrix} S_{r_{yn}}(\omega,x) \\ S_{r_{zn}}(\omega,x) \\ S_{r_{\theta n}}(\omega,x) \end{bmatrix} \quad \Rightarrow \quad \begin{bmatrix} \sigma_{r_y}^2(x) \\ \sigma_{r_z}^2(x) \\ \sigma_{r_\theta}^2(x) \end{bmatrix} = \int_0^\infty \begin{bmatrix} S_{r_y}(\omega,x) \\ S_{r_z}(\omega,x) \\ S_{r_\theta}(\omega,x) \end{bmatrix} d\omega \tag{5.42}
$$

where N_{mod} is the number of eigenmodes deemed necessary to include, and

$$
S_{r_{yn}}(\omega,x) = \left[\phi_{yn}(x) \cdot \frac{\rho B^2 D}{\tilde{m}_{yn}} \cdot \left(\frac{V}{B\omega_{yn}} \right)^2 \cdot \bar{C}_D I_u \cdot \left| \hat{H}_{yn}(\omega) \right| \cdot \hat{J}_{yn}(\omega) \right]^2 \cdot \frac{S_u(\omega)}{\sigma_u^2}
$$

$$
S_{r_{zn}}(\omega,x) = \left[\phi_{zn}(x) \frac{\rho B^3}{2\tilde{m}_{zn}} \cdot \left(\frac{V}{B\omega_{zn}} \right)^2 \cdot C_L' I_w \cdot \left| \hat{H}_{zn}(\omega) \right| \cdot \hat{J}_{zn}(\omega) \right]^2 \cdot \frac{S_w(\omega)}{\sigma_w^2} \Bigg\}
$$

$$
S_{r_{\theta n}}(\omega,x) = \left[\phi_{\theta n}(x) \frac{\rho B^4}{2\tilde{m}_{\theta n}} \cdot \left(\frac{V}{B\omega_{\theta n}} \right)^2 \cdot C_M' I_w \cdot \left| \hat{H}_{\theta n}(\omega) \right| \cdot \hat{J}_{\theta n}(\omega) \right]^2 \cdot \frac{S_w(\omega)}{\sigma_w^2}
$$

$$
\tag{5.43}
$$

where ϕ_{yn}, ϕ_{zn} and $\phi_{\theta n}$ are the relevant mode shapes, ω_{yn}, ω_{zn} and $\omega_{\theta n}$ are the corresponding eigenfrequencies. The frequency-response functions are defined by:

$$
\hat{H}_{yn}(\omega) = \left[1 - \kappa_{ae_{yn}} - \left(\omega/\omega_{yn} \right)^2 + 2i\left(\zeta_{yn} - \zeta_{ae_{yn}} \right) \cdot \omega/\omega_{yn} \right]^{-1}
$$

$$
\hat{H}_{zn}(\omega) = \left[1 - \kappa_{ae_{zn}} - \left(\omega/\omega_{zn} \right)^2 + 2i\left(\zeta_{zn} - \zeta_{ae_{zn}} \right) \cdot \omega/\omega_{zn} \right]^{-1} \Bigg\} \tag{5.44}
$$

$$
\hat{H}_{\theta n}(\omega) = \left[1 - \kappa_{ae_{\theta n}} - \left(\omega/\omega_{\theta n} \right)^2 + 2i\left(\zeta_{\theta n} - \zeta_{ae_{\theta n}} \right) \cdot \omega/\omega_{\theta n} \right]^{-1}
$$

where ζ_{y_n}, ζ_{z_n} and ζ_{θ_n} are damping ratios associated with mode n. The joint acceptance functions are given by

$$
\left.
\begin{aligned}
\hat{J}_{y_n}(\omega) &= \left(\iint_{L_{\exp}} \phi_{y_n}(x_1) \cdot \phi_{y_n}(x_2) \cdot \hat{C}o_{uu}(\Delta x, \omega) dx_1 dx_2 \right)^{1/2} \Bigg/ \int_L \phi_{y_n}^2 dx \\
\hat{J}_{z_n}(\omega) &= \left(\iint_{L_{\exp}} \phi_{z_n}(x_1) \cdot \phi_{z_n}(x_2) \cdot \hat{C}o_{ww}(\Delta x, \omega) dx_1 dx_2 \right)^{1/2} \Bigg/ \int_L \phi_{z_n}^2 dx \\
\hat{J}_{\theta_n}(\omega) &= \left(\iint_{L_{\exp}} \phi_{\theta_n}(x_1) \cdot \phi_{\theta_n}(x_2) \cdot \hat{C}o_{ww}(\Delta x, \omega) dx_1 dx_2 \right)^{1/2} \Bigg/ \int_L \phi_{\theta_n}^2 dx
\end{aligned}
\right\}
\tag{5.45}
$$

while the aerodynamic (motion induced forces are given by:

$$
\begin{bmatrix} \kappa_{ae_{y_n}} \\ \zeta_{ae_{y_n}} \end{bmatrix}
= \begin{bmatrix} \dfrac{\tilde{K}_{ae_{y_n}}}{\omega_{y_n}^2 \tilde{M}_{y_n}} \\[2ex] \dfrac{\tilde{C}_{ae_{y_n}}}{2\omega_{y_n} \tilde{M}_{y_n}} \end{bmatrix}
= \begin{bmatrix} \dfrac{\dfrac{\rho B^2}{2} \omega_{y_n}^2 P_4^* \displaystyle\int_{L_{\exp}} \phi_{y_n}^2 dx}{\omega_{y_n}^2 \tilde{m}_{y_n} \displaystyle\int_L \phi_{y_n}^2 dx} \\[3ex] \dfrac{\dfrac{\rho B^2}{2} \omega_{y_n} P_1^* \displaystyle\int_{L_{\exp}} \phi_{y_n}^2 dx}{2\omega_{y_n} \tilde{m}_{y_n} \displaystyle\int_L \phi_y^2 dx} \end{bmatrix}
= \dfrac{\rho B^2}{\tilde{m}_{y_n}} \cdot \dfrac{\displaystyle\int_{L_{\exp}} \phi_{y_n}^2 dx}{\displaystyle\int_L \phi_{y_n}^2 dx} \cdot \begin{bmatrix} \dfrac{1}{2} P_4^* \\[2ex] \dfrac{1}{4} P_1^* \end{bmatrix}
\tag{5.46}
$$

$$
\begin{bmatrix} \kappa_{ae_{z_n}} \\ \zeta_{ae_{z_n}} \end{bmatrix}
= \begin{bmatrix} \dfrac{\tilde{K}_{ae_{z_n}}}{\omega_{z_n}^2 \tilde{M}_{z_n}} \\[2ex] \dfrac{\tilde{C}_{ae_{z_n}}}{2\omega_{z_n} \tilde{M}_{z_n}} \end{bmatrix}
= \begin{bmatrix} \dfrac{\dfrac{\rho B^2}{2} \omega_{z_n}^2 H_4^* \displaystyle\int_{L_{\exp}} \phi_{z_n}^2 dx}{\omega_{z_n}^2 \tilde{m}_{z_n} \displaystyle\int_L \phi_{z_n}^2 dx} \\[3ex] \dfrac{\dfrac{\rho B^2}{2} \omega_{z_n} H_1^* \displaystyle\int_{L_{\exp}} \phi_{z_n}^2 dx}{2\omega_{z_n} \tilde{m}_{z_n} \displaystyle\int_L \phi_{z_n}^2 dx} \end{bmatrix}
= \dfrac{\rho B^2}{\tilde{m}_{z_n}} \cdot \dfrac{\displaystyle\int_{L_{\exp}} \phi_{z_n}^2 dx}{\displaystyle\int_L \phi_{z_n}^2 dx} \cdot \begin{bmatrix} \dfrac{1}{2} H_4^* \\[2ex] \dfrac{1}{4} H_1^* \end{bmatrix}
\tag{5.47}
$$

$$
\begin{bmatrix} \kappa_{ae\theta_n} \\[2mm] \zeta_{ae\theta_n} \end{bmatrix} = \begin{bmatrix} \dfrac{\tilde{K}_{ae\theta_n}}{\omega_{\theta_n}^2 \tilde{M}_{\theta_n}} \\[4mm] \dfrac{\tilde{C}_{ae\theta_n}}{2\omega_{\theta_n}\tilde{M}_{\theta_n}} \end{bmatrix} = \begin{bmatrix} \dfrac{\dfrac{\rho B^2}{2}\omega_{\theta_n}^2 A_3^* \displaystyle\int_{L_{exp}} \phi_{\theta_n}^2 dx}{\omega_{\theta_n}^2 \tilde{m}_{\theta_n} \displaystyle\int_{L} \phi_{\theta_n}^2 dx} \\[8mm] \dfrac{\dfrac{\rho B^2}{2}\omega_{\theta_n} A_2^* \displaystyle\int_{L_{exp}} \phi_{\theta_n}^2 dx}{2\omega_{\theta_n}\tilde{m}_{\theta_n}\displaystyle\int_{L}\phi_{\theta_n}^2 dx} \end{bmatrix} = \dfrac{\rho B^4}{\tilde{m}_{\theta_n}} \dfrac{\displaystyle\int_{L_{exp}} \phi_{\theta_n}^2 dx}{\displaystyle\int_{L}\phi_{\theta_n}^2 dx} \cdot \begin{bmatrix} \dfrac{1}{2} A_3^* \\[3mm] \dfrac{1}{4} A_2^* \end{bmatrix}
\tag{5.48}
$$

The evenly distributed equivalent modal masses are defined by:

$$
\left.
\begin{aligned}
\tilde{m}_{y_n} &= \tilde{M}_{y_n} \Big/ \int_{L} \phi_{y_n}^2 dx = \int_{L} m_{y_n}\phi_{y_n}^2 dx \Big/ \int_{L}\phi_{y_n}^2 dx \\
\tilde{m}_{z_n} &= \tilde{M}_{z_n} \Big/ \int_{L} \phi_{z_n}^2 dx = \int_{L} m_{z_n}\phi_{z_n}^2 dx \Big/ \int_{L}\phi_{z_n}^2 dx \\
\tilde{m}_{\theta_n} &= \tilde{M}_{\theta_n} \Big/ \int_{L} \phi_{\theta_n}^2 dx = \int_{L} m_{\theta_n}\phi_{\theta_n}^2 dx \Big/ \int_{L}\phi_{\theta_n}^2 dx
\end{aligned}
\right\}
\tag{5.49}
$$

In some cases it will suffice to include only a single mode single component, and then the above further simplified into:

$$
\left.
\begin{aligned}
\sigma_{r_y}(x) &= |\phi_y(x)| \cdot \frac{\rho B^2 D}{\tilde{m}_y} \cdot \bar{C}_D I_u \cdot \left(\frac{V}{B\omega_y}\right)^2 \cdot \left[\int_0^\infty |\hat{H}_y(\omega)|^2 \cdot \frac{S_u(\omega)}{\sigma_u^2} \cdot \hat{J}_y^2(\omega)d\omega\right]^{1/2} \\
\sigma_{r_z}(x) &= |\phi_z(x)| \cdot \frac{\rho B^3}{2\tilde{m}_z} \cdot C_L' I_w \cdot \left(\frac{V}{B\omega_z}\right)^2 \cdot \left[\int_0^\infty |\hat{H}_z(\omega)|^2 \cdot \frac{S_w(\omega)}{\sigma_w^2} \cdot \hat{J}_z^2(\omega)d\omega\right]^{1/2} \\
\sigma_{r_\theta}(x) &= |\phi_\theta(x)| \cdot \frac{\rho B^4}{2\tilde{m}_\theta} \cdot C_M' I_w \cdot \left(\frac{V}{B\omega_\theta}\right)^2 \cdot \left[\int_0^\infty |\hat{H}_\theta(\omega)|^2 \cdot \frac{S_w(\omega)}{\sigma_w^2} \cdot \hat{J}_\theta^2(\omega)d\omega\right]^{1/2}
\end{aligned}
\right\}
\tag{5.50}
$$

In some cases it will also suffice to use the quasi-static versions of the aerodynamic derivatives, in which case:

$$
\left.
\begin{aligned}
& P_1^* = -2\bar{C}_D \frac{D}{B}\frac{V}{B\omega_{y_n}} && H_1^* = -\left(C_L' + \bar{C}_D \frac{D}{B}\right)\frac{V}{B\omega_{z_n}} && A_2^* = 0 \\
& P_4^* = 0 && H_4^* = 0 && A_3^* = C_M' \left(\frac{V}{B\omega_{\theta_n}}\right)^2
\end{aligned}
\right\}
\tag{5.51}
$$

5.3.2 Description of the Wind Field

The natural logarithmic profile for the height variation of the mean wind velocity (first shown by Millikan (Millkan, 1938)) averaged over a time window of ten minutes

$$\frac{V(z_f)}{V(10)} = \begin{cases} k_T \cdot \ln(z_f/z_0) \text{ when } z_f > z_{min} \\ k_T \cdot \ln(z_{min}/z_0) \text{ when } z_f \leq z_{min} \end{cases} \tag{5.52}$$

where k_T, z_0 and z_{min} are parameters characteristic to the terrain in question. The height z_{min} has been introduced because such a velocity profile has a limited validity close to the ground, where turbulence and directional effects prevail. z_0 is usually called the roughness length. Typical values of k_T and z_0 varies from about 0.15 and 0.01 for open sea and countryside without obstacles to about 0.25 and 1.0 for built up urban areas. Corresponding values of z_{min} varies between 2 and about 15 m. The turbulence intensities are defined by

$$I_n(z_f) = \sigma_n(z_f)/V(z_f) \text{ where } n = u, v, w \tag{5.53}$$

A typical variation of the turbulence intensity for the along wind u component is given by

$$I_u(z_f) \approx \begin{cases} 1/\ln(z_f/z_0) \text{ when } z_f > z_{min} \\ 1/\ln(z_{min}/z_0) \text{ when } z_f \leq z_{min} \end{cases} \tag{5.54}$$

Under isotropic conditions (e.g., high above the ground) $I_u \approx I_v \approx I_w$. In homogeneous terrain up to a height of about 200 m and not unduly close to the ground

$$\begin{bmatrix} I_v \\ I_w \end{bmatrix} \approx \begin{bmatrix} 3/4 \\ 1/2 \end{bmatrix} \cdot I_u \tag{5.55}$$

There is a number of different expressions for the auto spectral densities of the turbulence components. The following non-dimensional expression proposed by Kaimal et al. (Kaimal et al., 1972) is often encountered in the literature:

$$\frac{f \cdot S_n \{f\}}{\sigma_n^2} = \frac{A_n \cdot \hat{f}_n}{\left(1 + 1.5 \cdot A_n \cdot \hat{f}_n\right)^{5/3}} \quad \text{where} \quad n = u, v, w \tag{5.56}$$

where $\hat{f}_n = f \cdot {}^{xf}L_n/V$, and ${}^{xf}L_n$ is the integral length scale of the relevant turbulence component. In general, the determination of spatial properties of the turbulence components should be based on full scale recordings on the site in question. However, for a first approximation and under homogeneous conditions not unduly close to the ground, the following may be adopted

$$\begin{bmatrix} ^{yf}L_u \\ ^{zf}L_u \\ ^{xf}L_v \\ ^{yf}L_v \\ ^{zf}L_v \\ ^{xf}L_w \\ ^{yf}L_w \\ ^{zf}L_w \end{bmatrix} \approx \begin{bmatrix} 1/3 \\ 1/4 \\ 1/4 \\ 1/4 \\ 1/12 \\ 1/12 \\ 1/16 \\ 1/16 \end{bmatrix} \cdot {}^{xf}L_u \quad \text{where:} \quad \begin{cases} \dfrac{{}^{xf}L_u\left(z_f\right)}{{}^{xf}L_u\left(z_{f0}\right)} \approx \left(\dfrac{z_f}{z_{f0}}\right)^{0.3} \\[4mm] z_f \geq z_{f0} = 10m \\[2mm] {}^{xf}L_u\left(z_{f0}\right) = 100m \end{cases} \tag{5.57}$$

Since the wind field is usually assumed homogeneous and perpendicular to the span of the structure, phase spectra may be neglected. The normalized co-spectrum is defined by

$$\hat{Co}_{nn}\left(\Delta s, f\right) = \frac{\text{Re}\left[S_{nn}\left(\Delta s, f\right)\right]}{S_n\left(f\right)} \qquad \begin{cases} n = u, v, w \\ \Delta s = \Delta x_f, \Delta y_f, \Delta z_f \end{cases} \tag{5.58}$$

For a first approximation and under homogeneous conditions, the following may be adopted

$$\hat{Co}_{nn}\left(\Delta s, f\right) = \exp\left(-c_{ns} \cdot \frac{f \cdot \Delta s}{V\left(z_f\right)}\right) \qquad \text{where:} \quad c_{ns} = \begin{cases} c_{uy_f} = c_{uz_f} \approx 9 \\ c_{vy_f} = c_{vz_f} = c_{wy_f} \approx 6 \\ c_{wz_f} \approx 3 \end{cases} \tag{5.59}$$

Caution should be exercised as the variation in c_{ns} values is considerable (see Solari and Piccardo (Solari and Piccardo, 2001)). The simple expression in Eq. (5.59) has the obvious disadvantage that the normalised co–spectrum becomes unity at all Δs when $f = 0$, whereas a typical normalised co–spectrum will decay at any value of f. Under the assumption of isotropic conditions, an alternative version has been developed by Krenk (Krenk, 1995) applicable for the along–wind u component.

5.3.3 The Static Load Coefficients

The section model test set-up is illustrated in Fig. 5.10. It may be used in

- a static set up, or in
- a dynamic set up.

In a static set up all the helical springs (item 4 in Fig. 5.10) are held tight (or replaced by rods). In this set up the section model represents a short segment of the system. Force recordings in load cells 1 – 6 at a suitable variation of the angle of incidence (α) determines static cross sectional load coefficients, as shown in Fig. 5.3. For a line-like system close to that of a flat plate $C'_L \to 2\pi$ while $C'_M \to \pi/2$.

5.3.4 Aero-Elastic Tests for the Determination of Aerodynamic Derivatives

In an aero-elastic test set-up as shown in Fig. 5.10, the model

- is either suspended by helical springs and free to move at its own will, or the helical spring are removed, and the model
- is subject to actuators setting it into a prescribed motion.

The aero-elastic section model setup is used to:

- determine the stability limit,
- detect possible vortex shedding effects,
- provide indications of the buffeting (dynamic) response behavior, and
- determine aerodynamic derivatives.

The aero-elastic section model has its origin in the observation that a bridge system with a flat plate like cross section may exhibit a flutter stability limit, and that the detection of this limit is the most important issue for the design of the bridge.

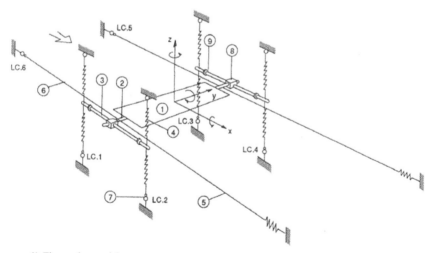

1) The section model
2) Horizontal shaft (axle) (often extended through the wind tunnel wall)
3) Extension arms (often outside of wind tunnel wall)
4) Helical springs
5) Drag wire on leeward side
6) Drag wire on windward side
7) Single component load cells LC.1–LC.7.
8) Centre piece
9) Adjustable additional mass

Figure 5.10. The aero-elastic section model.

It is well known that a flutter phenomenon comprise the combination of the lowest mode whose content is mainly torsion together with the shape-wise similar and lowest mode whose content is mainly vertical. Therefore, the aero elastic section model is designed to model the most onerous combination of a torsional and a vertical mode. After the choice of geometric model scale, the design of the setup starts with a numerical identification of these two modes (see Fig. 5.11). Their modal properties are given by

- $\phi_\theta, \tilde{M}_\theta, \omega_z \left(\Rightarrow \tilde{K}_\theta = \omega_z^2 \tilde{M}_\theta \right)$
- $\phi_z, \tilde{M}_z, \omega_\theta \left(\Rightarrow \tilde{K}_z = \omega_z^2 \tilde{M}_z \right)$

Thus, it is important to realize that the two modelled degrees of freedom in the aero elastic setup do not represent any physical degrees of freedom in the full scale structure itself. Rather, they represent the chosen modal degrees of freedom:

- one in torsion mode (usually the lowest symmetric mode), and
- the other represents the lowest vertical mode (shape-wise similar to the torsion mode).

Thus, all the quantities in section model set up are modal, and therefore, all test results from the dynamic aero-elastic set up are also modal quantities. The obvious reason for this comes from the basic idea of the model experiment; that the model represents a full scale system, and then the similarity requirements (model laws

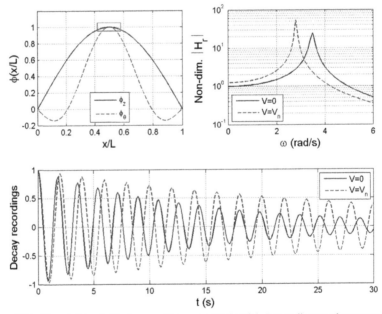

Figure 5.11. Upper left hand side diagram: Chosen modes ϕ_z and ϕ_θ. Lower diagram: decay recording. Upper right: corresponding frequency-response function.

according to the Buckingham π theorem) must be fulfilled in designing the model as well as in the application of results (i.e., in the extension of test results to full scale conditions). If the model set up is modal, then test results are modal, and therefore they can only be extended in to full scale modal conditions.

The motion induced forces manifest themselves by changes to the in-wind

- damping and
- stiffness properties

of the two degrees of freedom in the setup, as illustrated in Fig. 5.11 (from tests with a freely suspended model). It is these changes that are expressed by the aerodynamic derivatives. The experimental procedures for the extraction of these effects are described in Section 5.4 below.

5.4 Recommendations for Advanced Engineering Practice

From the section model the aerodynamic derivatives may be extracted

- from tests with a freely suspended section model, or
- by forced vibration tests.

5.4.1 Determination of AD's by Tests with a Freely Suspended Section Model

In a suspended section model setup, the equilibrium conditions are given by

$$
\begin{bmatrix} \tilde{M}_z & 0 \\ 0 & \tilde{M}_\theta \end{bmatrix} \begin{bmatrix} \ddot{r}_z(t) \\ \ddot{r}_\theta(t) \end{bmatrix} + \begin{bmatrix} \tilde{C}_z & 0 \\ 0 & \tilde{C}_\theta \end{bmatrix} \begin{bmatrix} \dot{r}_z(t) \\ \dot{r}_\theta(t) \end{bmatrix} + \begin{bmatrix} \tilde{K}_z & 0 \\ 0 & \tilde{K}_\theta \end{bmatrix} \begin{bmatrix} r_z(t) \\ r_\theta(t) \end{bmatrix} = \int_L \begin{bmatrix} q_z((V,t)) \\ q_\theta(V,t) \end{bmatrix} dx
$$

$$
+ \int_L \frac{\rho B^2}{2} \begin{bmatrix} \omega_z H_1^* & \omega_\theta BH_2^* \\ \omega_z BA_1^* & \omega_\theta B^2 A_2^* \end{bmatrix} dx \begin{bmatrix} \dot{r}_z(t) \\ \dot{r}_\theta(t) \end{bmatrix} + \int_L \frac{\rho B^2}{2} \begin{bmatrix} \omega_z^2 H_4^* & \omega_\theta^2 BH_3^* \\ \omega_z^2 BA_4^* & \omega_\theta^2 B^2 A_3^* \end{bmatrix} dx \begin{bmatrix} r_z(t) \\ r_\theta(t) \end{bmatrix}
$$

$$
(5.60)
$$

where all quantities are at model scale. It is taken for granted that the mass and stiffness matrices are known. The tests procedure starts with

- simple decay recordings without wind (i.e., at $V = 0$), from which the model eigen-frequencies and damping properties (see Fig. 5.11) as well as the cross sectional signature load may be determined.

Then equivalent decay recordings are performed at various mean wind velocities (i.e., at $V \neq 0$, as illustrated in the lower diagram in Fig. 5.11),

- first in a purely vertical motion,
- then in a purely torsional motion, and
- finally in a combined vertical and torsional motion,

from which the changes to damping properties and resonance (eigen-) frequencies are recorded and subsequently used to extract the aerodynamic derivatives from the equilibrium conditions given in Eq. 6.60. The most common strategy is to adopt ambient vibration tests at a suitable number of mean wind velocities V_n, and apply system identification routines on the response (or the response co-variances), e.g., as described by Jakobsen (Jakobsen, 1995a; Jakobsen, 1995b). A more direct strategy is to use free model decay motion tests, as first described by Scanlan and Sabzevari (Scanlan and Sabzevari, 1969). For each velocity setting V the various aerodynamic derivatives are plotted as functions of the reduced velocity $\hat{V} = V_n / \left[B\omega_j(V_n) \right]$, $j = z$ or θ, where $\omega_j(V_n)$ is the in-wind eigenfrequency of the combined flow and structural system, each depending on which degree of freedom (z or θ) is responsible for the motion, as illustrated in Fig. 5.12, where the properties of a flat-plate type of cross section have been plotted.

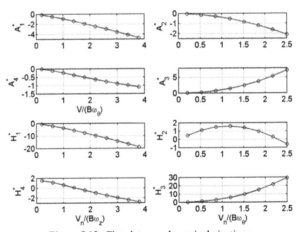

Figure 5.12. Flat plate aerodynamic derivatives.

5.4.2 Determination of AD's by Tests with Forced Vibrations

In forced vibration tests, the equilibrium conditions are given by

$$\mathbf{F}(V,\mathbf{r},t) = \begin{bmatrix} F_z \\ F_\theta \end{bmatrix} = \begin{bmatrix} \tilde{M}_z & 0 \\ 0 & \tilde{M}_\theta \end{bmatrix} \begin{bmatrix} \ddot{r}_z(\omega,t) \\ \ddot{r}_\theta(\omega,t) \end{bmatrix} + \begin{bmatrix} \tilde{C}_z & 0 \\ 0 & \tilde{C}_\theta \end{bmatrix} \begin{bmatrix} \dot{r}_z(\omega,t) \\ \dot{r}_\theta(\omega,t) \end{bmatrix} - \int_L \begin{bmatrix} q_z((V,t)) \\ q_\theta(V,t) \end{bmatrix} dx$$

$$- \int_L \frac{\rho B^2}{2} \omega \begin{bmatrix} H_1^* & BH_2^* \\ BA_1^* & B^2 A_2^* \end{bmatrix} dx \begin{bmatrix} \dot{r}_z(\omega,t) \\ \dot{r}_\theta(\omega,t) \end{bmatrix} - \int_L \frac{\rho B^2}{2} \omega^2 \begin{bmatrix} H_4^* & BH_3^* \\ BA_4^* & B^2 A_3^* \end{bmatrix} dx \begin{bmatrix} r_z(\omega,t) \\ r_\theta(\omega,t) \end{bmatrix}$$

$$(5.61)$$

where again, all quantities are at model scale, and where F_z and F_θ are the external exciting forces that are applied to the corresponding degrees of freedom in the system. It is taken for granted that modal masses are known, and that the damping properties have been obtained from simple decay recording at $V = 0$. The tests are usually performed in smooth flow, but there is no reason why the tests cannot be performed in turbulent flow. The test procedure is as follows: First, the structure in kept at rest, i.e., $\mathbf{r} = 0$, while V varies across the desired range of mean wind velocities, from which $\mathbf{F}(V, \mathbf{r} = 0, t)$ is obtained. Then, without wind (i.e., at $V = 0$), a suitable range of tests with $\mathbf{r} = \begin{bmatrix} a_z & a_\theta \end{bmatrix}^T e^{i\omega jt} \neq 0, j = z$ or θ are performed, first with purely vertical motion, followed by purely torsion motion, from which $\mathbf{F}(V = 0, \mathbf{r}, t)$ is obtained. Finally, tests at suitable range of mean wind velocities settings are performed, again, first with purely vertical motion, followed by purely torsion motion, and finally in a combined motion, from which $\Delta\mathbf{F} = \mathbf{F}(V, \mathbf{r}, t) - \mathbf{F}(V = 0, \mathbf{r}, t) - \mathbf{F}(V, \mathbf{r} = 0, t)$ is determined. It is taken for granted that $\mathbf{r}(\omega_j, t) = \mathbf{a}_r(\omega_j) e^{i\omega jt}$, where $\mathbf{a}_r(\omega_j) = \begin{bmatrix} a_{r_z} & a_{r_\theta} \end{bmatrix}^T$ is real, and that $\Delta\mathbf{F}(V, \mathbf{r}, t) = \mathbf{A}_F(V, \mathbf{r}, \omega_j) e^{i\omega jt}$ where $\mathbf{A}_F(V, \mathbf{r}, \omega_j) = \begin{bmatrix} A_{F_z} & A_{F_\theta} \end{bmatrix}^T$ is complex. It is also assumed that the aerodynamic derivatives are constants along the span L of the model.

Defining
$$\hat{\mathbf{A}}_F(V, \mathbf{r}, \omega_j) = \frac{2}{\rho B^2 L \omega_j^2} \begin{bmatrix} A_{F_z}/a_{r_z} \\ A_{F_\theta}/a_{r_\theta} \end{bmatrix} \tag{5.62}$$

then the aerodynamic derivatives may be obtained from

$$i \cdot \hat{\mathbf{c}}_{ae}(V, \omega_j) + \hat{\mathbf{k}}_{ae}(V, \omega_j) = \hat{\mathbf{A}}_F(V, \mathbf{r}, \omega_j) \tag{5.63}$$

where $\hat{\mathbf{c}}_{ae}(V, \omega_j) = \begin{bmatrix} H_1^* & BH_2^* \\ BA_1^* & B^2 A_2^* \end{bmatrix}$ and $\tilde{\mathbf{k}}(V, \omega_j) = \begin{bmatrix} H_4^* & BH_3^* \\ BA_4^* & B^2 A_3^* \end{bmatrix}$ $\quad$ (5.64)

As can be seen, aerodynamic derivatives derived from this type of tests will be functions of the testing amplitude as well as the force frequency of motion. It cannot be ruled out that in some cases this procedure will render results that are reliable enough, but principally the procedure cannot be recommended for the following reason: Since the helical springs have been removed, the model has lost one of its dynamic properties, i.e., its flexibility. Thus, it no longer fully represents the full scale structure it is meant to be a model of. Therefore, the set-up for such a test procedure violates one of the main similarity requirements in the model laws of the experiment, i.e., the similarity requirement that the system also has a flexibility to move as it chooses itself. This violation will be the case even if the force frequency coincides with the eigenfrequencies obtained at zero mean wind velocity, because it will still not allow for changes to the in-wind resonance frequency of the combined model and flow system. The model is simply not allowed to respond according to its own preference, as defined by the shape of its cross section, its mass and its stiffness properties. The procedure of forced vibration tests comes from the idea that the aerodynamic derivatives are cross sectional properties alone. That assumption is principally incorrect. Of course, the shape of the cross section will most often play a major part, but a cross section alone cannot possess all the dynamic properties of the combined structure and flow system, i.e., the motion induced forces comes from the motion preferences of the system as well as the shape of its cross section.

5.4.3 *Full Scale Application of Aerodynamic Derivatives*

Thus, it is an inescapable fact that the vertical and torsion degrees of freedom that are modelled in the aeroelastic section model do not represent any of the physical degrees of freedom of the full scale structure itself; the model set-up is modal, it represents the modal coordinates of the two modes deemed critical for the stability limit of the full scale structure, from which it follows that all the test results coming out of the dynamic tests (e.g., the aerodynamic derivatives) are also modal quantities. The implication of this is that aerodynamic derivatives from these tests cannot be used for full scale structural response calculations in the original finite element degrees of freedom, neither in time domain nor in frequency domain. However, in a modal approach they may be used in time domain as well as in frequency domain. If a solution in original degrees of freedom is pursued, it will be necessary to use the motion induced forces from the quasi-steady theory (see Eqs. (5.3) and (5.35) or to determine the indicial functions [(see Eq. (5.4))], a task which presently is at the research stage.

It should also be noted that the aerodynamic derivatives are functions of the reduced velocity, $V/[B\omega_i(V)]$, $i = z$ or θ. They are not functions along a continuous frequency axis, although they are often seen in the literature as functions of $V/(B\omega)$ or $V/(Bf)$, or even $K = \omega B/V$. In a Buckingham π type of modelling the requirement is that all entries in a non-dimensional number must be physical quantities, and at each velocity setting V, B and ω_i are all fulfilling this requirement, while ω or f are not. Thus, for the purpose of full scale response calculations the application of the aerodynamic derivatives will only fulfil the similarity requirements if they are extracted as functions of the same reduced velocity, either $\hat{V}_z = V/(B\omega_z)$ or $\hat{V}_\theta = V/(B\omega_\theta)$, depending on how they have been normalised, where V is the mean wind velocity setting, and where ω_z and ω_θ are the in-wind resonance frequencies (as affected by aerodynamic stiffness contributions). Hence, in any computational procedure iterations will be required.

Because the two-degrees-of-freedom experimental section model setup represents the full scale bridge structure in two particularly singled out modal coordinates, it should also be noted that in any full scale response calculation the aerodynamic derivatives will, strictly spoken, only be applicable to the corresponding modal coordinates of the system. As to whether or not they may also be representative for other combinations of modes is an open question, hopefully yes but most likely not. In a mean wind velocity range well below a stability limit the motion induced forces are playing a minor part as compared to the wind buffeting load, and thus, in this velocity range any restriction in the use of the aerodynamic derivatives is likely to be excessive. However, in the vicinity of a stability limit the motion induced forces play a major part, and therefore, without any further investigations, full scale response calculation in the vicinity of a stability limit can at the moment only include the two modelled mode shapes.

5.5 Concluding Remarks and Further Studies

The theory of wind induced dynamic response calculations has gone through considerable development over the last two decades. In modal coordinates it has become possible to follow the development of the response up to mean wind values in the vicinity of the stability limit, which has enabled designers to assess the safety factor up structural collapse. Unfortunately, similar calculations are still not possible in the original finite element degrees of freedom. The reason for this is that the methods for experimental determination of motion induced forces have not made similar advances. It is widely accepted that at an incremental time domain approach, linearity between motion induced forces and the motion itself may still be adopted within each small time step. The problem is to determine this connection in an experimental set-up in the wind tunnel. This is where the main research effort should be.

NOTATION

Matrices are in bold upper case Latin or Greek letters. Vectors are in bold lower case letters.

diag[·] is a diagonal matrix whose content is written within the brackets.

det(·) is the determinant of the matrix within the brackets.

Super-script T indicates the transposed of a vector or a matrix.

Super-script * indicates the complex conjugate of a quantity.

Dots above symbols (e.g., $\dot{r}$, $\ddot{r}$) indicates time derivatives, i.e., d/dt, d^2/dt^2.

A prime on a variable (e.g., C'_L or ϕ') indicates its derivative with respect to a relevant variable, e.g., $C'_L = dC_L/d\alpha$ and $\phi' = d\phi/dx$. Two primes is then second derivativeand so on.

Line (−) above a variable (e.g., $\bar{C}_D$) indicates its average value.

A tilde (~) above a symbol (e.g., $\tilde{M}_i$) indicates a modal quantity.

A hat (∧) above a symbol (e.g., $\hat{B}_q$) indicates a normalised quantity.

Latin letters

A	Area, cross sectional area
A_n	Wind spectrum coefficient ($n = u$, v or w)
$A_1^* - A_6^*$	Aerodynamic derivatives associated with the motion in torsion
$\mathbf{A}, \mathbf{A}_m, \mathbf{A}_n$	Connectivity matrix (associated with element m or n)
a, $\mathbf{a}$	Fourier coefficient, Fourier coefficient vector

B	Cross sectional width
$\mathbf{B}_q$	Buffeting dynamic load coefficient matrix at cross sectional level
$\mathbf{b}_q$	Quasi-static load coefficient matrix
$C, \mathbf{C}$	Damping coefficient or matrix containing damping coefficient
$C_{ae}, \mathbf{C}_{ae}$	Aerodynamic damping, aerodynamic damping matrix
C_D, C_L, C_M	Force coefficients at mean angle of incidence
$\mathbf{c}, \mathbf{c}_0, \mathbf{c}_{ae}$	Damping matrix at cross sectional or element level, aerodynamic damping matrix
$Co, \mathbf{Co}$	Co-spectral density, co-spectral density matrix
$\mathbf{Cov}_{rr}$	Matrix containing covariance of response quantities
D	Cross sectional depth
$\mathbf{d}, d_k$	Element displacement vector, element end displacement $(k = 1, 2, \ldots, 12)$
$\mathbf{F}, F$	Element force vector, force
f	Frequency (in Hz)
$H_1^* - H_6^*$	Aerodynamic derivatives associated with the across-wind vertical motion
$\mathrm{H}, \mathbf{H}, \mathbf{H}_r$	Frequency response function, frequency response matrix
$\tilde{H}_\eta, \tilde{\mathbf{H}}_\eta$	Modal frequency response functions, modal frequency response matrix
I_u, I_v, I_w	Turbulence intensity of flow components u, v or w
$\mathbf{I}$	Identity matrix
i	The imaginary unit (i.e., $i = \sqrt{-1}$) or index variable
$K, \mathbf{K}$	Stiffness, stiffness matrix
$K_{ae}, \mathbf{K}_{ae}$	Aerodynamic stiffness, aerodynamic stiffness matrix
$\mathbf{k}, \mathbf{k}_{ae}$	Stiffness matrix at element level, aerodynamic stiffness matrix
L	Element length (assumed equal to wind exposed length)
^mL_n	Integral length scales $(m = y, z$ or $\theta, n = u, v$ or $w)$
$m, \mathbf{M}$	Mass or mass matrix at system level
$\mathbf{m}_0, \mathbf{m}$	Mass matrix at a cross sectional level, mass matrix at element level
$\bar{N}$	Time invariant (mean) normal force
N, N_r	Number, number of degrees of freedom
$\mathbf{N}, N_i(x)$	Shape function matrix, polynomial shape function $(i = 1, 2, \ldots, 12)$
p, k	Node numbers
$P_1^* - P_6^*$	Aerodynamic derivatives associated with the along-wind motion
q_D, q_L, q_M	Drag, lift and moment cross sectional loads in flow axes
q_x, q_y, q_θ	Drag, lift and moment cross sectional loads in element axes

$\mathbf{q}, \mathbf{q}_{tot}$	Wind load, buffeting wind load or total wind load vector at cross sectional level
$q_{ae}, \mathbf{q}_{ae}$	Aerodynamic (motion induced) load at cross sectional level
$R, \mathbf{R}, \mathbf{R}_n$	External load, external load vector at system level, at element level
$\tilde{R}, \tilde{\mathbf{R}}$	Modal load, Modal load vector
$r_x, r_y, r_\theta, \mathbf{r}$	Cross sectional displacements, cross sectional displacement vector
$s = t - \tau$	Relative time
$S, \mathbf{S}$	Auto or cross spectral density, cross-spectral density matrix
$\mathbf{S}_{rr}, \mathbf{S}_{RR}$	Cross spectral density matrix associated with response or load
$\mathbf{S}_{\eta\eta}, \mathbf{S}_{\tilde{R}\tilde{R}}$	Cross spectral density matrix associated with modal response or load
$t, \Delta t$	Time, incremental time step
u, v, w	A long-wind, across wind horizontal and vertical wind velocity components
V	Mean wind velocity
$\mathbf{v}, \hat{\mathbf{v}}_0, \hat{\mathbf{v}}_n$	Wind velocity vector or reduced wind velocity vector
X, Y, Z	Cartesian structural global axis
x, y, z	Element axis (origo in the shear centre, x in span-wise direction and z vertical)
x_f, y_f, z_f	Cartesian flow axis (x_f in main flow direction and z_f vertical)
W_{ext}, W_{ext}	External and internal work

Greek letters

α	Coefficient, angle of incidence
$\boldsymbol{\beta}_0$	Matrix containing element orientation properties
ζ or $\boldsymbol{\zeta}$	Damping ratio or damping ratio matrix
η or $\boldsymbol{\eta}$	Generalised coordinate or vector containing $N_{mod}\eta$ components
θ	Index indicating cross sectional rotation (about shear centre)
$\boldsymbol{\kappa}_{ae}$	Matrix containing aerodynamic modal stiffness contributions
ρ	Coefficient or density (e.g., of air)
σ, σ^2	Standard deviation, variance
τ	Dummy time variable
Φ	$3 \cdot N_{mod}$ by N_{mod} matrix containing all mode shapes φ_i
$\phi_y, \phi_z, \phi_\theta$	Mode shapes associated with horizontal, vertical and torsion motion
Φ_{mn}	Indicial function ($m = Drag$, $Lift$ or $Moment$, $n = y, z$ or θ)
$\boldsymbol{\varphi}, \boldsymbol{\varphi}_n$	Mode shape vector, mode shape vector number n
$\boldsymbol{\psi}, \boldsymbol{\psi}_n, \boldsymbol{\psi}_0$	Element displacement shape function matrix

ω	Circular frequency (rad/s)
ω_i	Still air eigenfrequency associated with mode shape i
$\omega_i(V)$	Resonance frequency assoc. with mode i at mean wind velocity V

References

Cook, R.D., Malkus, D.S., Plesha, M.E. and Witt, R.J. 2002. Concepts and applications of finite element analysis, 4th edn. John Wiley & Sons Inc.

Davenport, A.G. 1962. The response of slender line-like structures to a gusty wind. Proceedings of the Institution of Civil Engineers 23: 389–408.

Davenport, A.G. 1978. The prediction of the response of structures to gusty wind. Proceedings of the International Research Seminar on Safety of Structures under Dynamic Loading; Norwegian University of Science and Technology, Tapir, pp. 257–284.

Deodatis, G. 1996. Simulation of ergodic multivariate stochastic processes. Journal of Engineering Mechanics, ASCE 122(8): 778–787.

Dyrbye, C. and Hansen, S.O. 1988. Calculation of joint acceptance function for line-like structures. Journal of Wind Engineering and Industrial Aerodynamics 31: 351–353.

Dyrbye, C. and Hansen, S.O. 1999. Wind loads on structures. John Wiley & Sons Inc.

ESDU International, 27 Corsham St., London N1 6UA, UK.

Jakobsen, J.B. 1995a. Estimation of motion induced wind forces by a systems identification method. Proc. of 9th IAWE Int. Conf. on Wind Engineering, Wiley Eastern Ltd., New Delhi.

Jakobsen, J.B. 1995b. Fluctuating wind load and response of a line-like engineering structure with emphasis on motion-induced forces, Ph D thesis. Department of Structural Engineering, Norwegian University of Science and Technology, Trondheim, Norway.

Kaimal, J.C., Wyngaard, J.C, Izumi, Y. and Coté, O.R. 1972. Spectral characteristics of surface-layer turbulence. Journal of the Meteorological Society 98: 563–589.

Krenk, S. 1995. Wind field coherence and dynamic wind forces. Proceedings of Symposium on the Advances in Nonlinear Stochastic Mechanics. Næss, A. and S. Krenk (eds.). Proceedings of the IUTAM Symposium held in Trondheim, Norway, 3–7 July 1995.

Millkan, C.B. 1938. A critical discussion of turbulent flows in channels and circular tubes. Proceedings of the 5th International Congress of Applied Mechanics, Cambridge, MA, USA, pp. 386–393.

Salvatori, L. and Borri, C. 2007. Frequency- and time domain methods for the numerical modelling of full-bridge aeroelasticity. Computers and Structures 85: 675–687.

Scanlan, R.H. and Tomko, A. 1971. Airfoil and bridge deck flutter derivatives. Journal of the Engineering Mechanics Division, ASCE, (EM6), Dec. Proc. Paper 8609 97: 1717–1737.

Scanlan, R.H. and Sabzevari, A. 1969. Experimental aerodynamic coefficients in the analytical study of suspension bridge flutter. Journal of Mechanical Engineering Science 11(3): June.

Selberg, A. 1961. Oscillation and aerodynamic stability of suspension bridges. Acta Polytechnica Scandinavica, Civil Engineering and Building Construction Series No. 13, Oslo.

Simiu, E. and Scanlan, R.H. 1996. Wind effects on structures, 3rd edn. John Wiley & Sons.

Solari, G. and Piccardo, G. 2001. Probabilistic 3 – D turbulence modelling for gust buffeting of structures. Journal of Probabilistic Engineering Mechanics 16: 73–86.

Strømmen, E. 2010. Theory of bridge aerodynamics. 2nd edn. Springer-Verlag Berlin Heidelberg, 302 pp.

Strømmen, E. 2014. Structural dynamics. Springer Series in Solid and Structural Mechanics. Springer Verlag.

Shinozuka, M. 1972. Monte carlo solution of structural dynamics. Computers and Structures 2: 855–874.

6

Neo-Deterministic Scenario-Earthquake Accelerograms and Spectra
A NDSHA Approach to Seismic Analysis

Paolo Rugarli,[1] *Claudio Amadio,*[2] *Antonella Peresan,*[3]
Marco Fasan,[2] *Franco Vaccari,*[4] *Andrea Magrin,*[3]
Fabio Romanelli[4,*] and *Giuliano F. Panza*[5]

"I had", said he, "come to an entirely erroneous conclusion, which shows, my dear Watson, how dangerous it always is to reason from insufficient data."

Conan Doyle
The adventure of the Speckled Band, 1892

6.1 Introduction

The concept of Performance Based Seismic Design (PBSD) introduced in was first translated into design guidelines in 1978 with the ATC3-06 publication (ATC, 1978).

[1] CASTALIA S.r.l., Via Pinturicchio, 24, 20133, Milano, Italy.
  Email: paolo.rugarli@castaliaweb.com
[2] Department of Engineering and Architecture, University of Trieste, Via Alfonso Valerio 6/1, 34127, Trieste, Italy.
  Emails: amadio@units.it; marcofasan@hotmail.it
[3] OGS, Istituto Nazionale di Oceanografia e di Geofisica Sperimentale, Via Treviso 55, 33100, Udine, Italy.
  Emails: aperesan@inogs.it; amagrin@inogs.it
[4] Department of Mathematics and Geosciences, University of Trieste, Via Weiss 4, 34128, Trieste, Italy.
  Email: vaccari@units.it
[5] Accademia Nazionale dei Lincei, Roma, Italy; Institute of Geophysics, China Earthquake Administration, Beijing, China; Beijing University of Civil Engineering and Architecture, Beijing, China; International Seismic Safety Organization (ISSO), Arsita, Italy. 2018 AGU International Award winner.
  Email: giulianofpanza@fastwebnet.it
* Corresponding author: romanel@units.it
NB: Order of authors, except the last one, determined by lot

The assessment of seismic hazard was based on a single map and the achievement of better performance for buildings with greater risk to the public was reached by classifying the buildings in four different Seismic Performance Categories, each requiring different levels of security and anti-seismic details. The hazard map was determined with a probabilistic approach since it was a *"second policy decision"* of the ATC3-06 committee that *"the probability of exceeding the design ground shaking should—as a goal—be roughly the same in all parts of the country"*. There was not full agreement with this policy decision: *"There is not unanimous agreement in the profession with this policy decision. In part this lack of agreement reflects doubt about as to how well the probability of ground motion occurrence can be estimated with today's knowledge and disagreement with the specific procedures used to make the estimates"*. So this *"second policy decision"* was not unanimous and yet at the time professionals were against. However, the document says that *"there is no workable alternative approach to the construction of a seismic design regionalization map which comes close to meeting the goal of the second policy decision"*(ATC 1978) even if it was recognized that the *"assumption* [of Poissonian occurrences of earthquakes] *is of limited validity"*. Hence, for consistency with a priori not unanimous decisions, the ATC3-06 Committee adopted a method known to be based on wrong assumptions. Of course, at that time deterministic seismic hazard approaches were available *"based upon estimates of the maximum ground shaking experienced during the recorded historical period without consideration of how frequently such motions might occur"* but *"considering the significant cost of designing a structure for extreme ground motions, it is undesirable to require such a design unless there is a significant probability that the extreme motion will occur"*. In other words, the committee decided to "cut down" the ground motion level in order to save in the cost of construction and the probabilistic method was used to give a semblance of rationality to the choice, also if the methods were labelled *"quite crude"* by the Committee itself. *"Not unanimous"*, *"quite crude"* unscientific as intrinsically not falsifiable, and so: everlasting. This decision to cut down ground motion was supported by the fact that a seismic hazard map was first drafted for ATC3-06 *"having literally been drawn by a committee"* based on expert judgement and subsequently since this map *"appeared to agree reasonably well with the level of acceleration determined by Algermissen and Perkins [...] their map was used as a guide for the rest of the country"*. It happened that the map of Algermissen and Perkins (Algermissen and Perkins 1976) was based on a not further defined *"mean recurrence interval (also referred to as return period)"* of 475 years, so a 10% in 50 years map was adopted in ATC3-06.These subsequently became standard numbers all over the world (such spreading is clearly due to the so called "genetic fallacy", that is, the alleged perfection of the stuff coming from authoritative sources, Rugarli 2014; Rugarli, 2015a). So the use of an "average life" of 50 years is "explained" as *"a rather arbitrary convenience"* and the 10% "probability" of exceedance as a number often taken, still recently, *"to be meaningful"*(Bommer and Pinho, 2006). However, the Committee, well aware of the issue, declared pre-emptively that *"a mean recurrence interval of 475 years does not mean that the earthquake will occur once, twice or even at all in 475 years. With present knowledge, there is no practical alternative to assuming that a large earthquake is equally likely to occur at any time, and quantities such as return period only indicate the likelihood that such event will occur."A likelihood,* however, is not enough to build safely, as the earthquakes in the last 40 years made clear. So why use it? The only use of "return period" was to make applicable the Poissonian methods, and to compute "something" with no physical or statistical meaning. *Compute something* was the apparent goal. Basically, a quite expensive mathematical joke, which opened the path to a huge set of physically groundless inventions, and dozen and dozen of papers. It must be underlined that the use of a set of misleading terms, like "probability", "return period", "exceedance", "collapse prevention", and the systematic use of complex math to hide simple baseless rules, has systematically deceived the population who was not told the real truth: that the building was designed for potentially unsafe actions, and that "rather crude" reasoning was applied to their lives.

This hiding of the truth is particularly visible in Italy, a country that has several unique features making it different from, for instance, California:

- Italy has one of the longest existing databases of historically mentioned earthquakes, probably unique.
- Italy has a huge set of ancient masonry buildings, whose value is simply too high to even think to destroy and rebuild. Existing irregular and layered masonry buildings pose problems totally different from those of steel or reinforced concrete new (regular) buildings.
- Italy has, as it is generally agreed, the largest slice of Art Heritage, and this should be protected against all threats.

No matter these very important and specific features, each middle intensity earthquake in Italy causes death and huge damage. This is not due to daredevil superficiality of the population, but to ignorance fuelled by laws that have systematically underestimated the seismic hazard, also using as a useful drug the computation of phantasmatic "probabilities" and "return periods".

Historically, the same target "probability" of exceedance, P_{EY}, of 10% in 50 years has been used worldwide as a reference to design ordinary buildings without any clear risk-based rationale and regardless of differences with respect to the USA in terms of seismicity, construction practices and economic prosperity (Bommer and Pinho, 2006). These values of ground motion, as it could be expected given their "probabilistic" nature, have been repeatedly and largely exceeded in real records as documented by Kossobokov and Nekrasova (2012). So the standards based on these methods are surely unsafe. Moreover, the comparison between different "probabilistic" hazard maps reveals how the peak values (e.g., Peak Ground Acceleration (PGA) with P_{EY} = 10% in 50 years) are not consistent from map to map, and large differences have been found (Nekrasova et al., 2014). These observations and other engineering considerations have led, in some countries (e.g., USA), to a change of the value of P_{EY} from 10% to 2% - 50 years, *"In part,*

2% - 50 years years was selected because USGS had already produced maps for this hazard level" (BSSC, 2015).

The main contribution to the development of the PBSD philosophy of design has been given by the Vision 2000 report (SEAOC, 1995) which firstly introduced a multi-performance levels check. This document defines a series of performances (in terms of acceptable damage) that a building should achieve during earthquakes of different strength. These performance levels are usually defined as: Operational Limit (OL), Immediate Occupancy (IO), Life Safety (LS) and Collapse Prevention (CP). The "return periods", P_R, arbitrarily associated with them are 43, 72, 475 and 970 years and correspond to a "probability" of exceedance of 69%, 50%, 10% and 5% in an interval of 50 years, respectively (Fig. 6.1). The choice of the "return periods" corresponding to the four performance levels has been arbitrarily selected for California (Bertero and Bertero, 2002) and it has never been motivated (Bommer and Pinho, 2006). Other "return periods" (depending on preset "nominal life" duration and "exceeding probabilities") would lead to other ground shaking values, so to what has been called "The Earthquake Supermarket" (Rugarli, 2015b): the ability to choose the ground shaking level according to one's needs.

This dangerous assumption was however useful for administrators. It led to the perverse adoption of the concept of *"remaining nominal life"* by Italian law makers, who, having the need to cut down arbitrarily downward the severity of the shaking in view of the retrofitting of the historical buildings which shape the country, adopted this trick to assess that a building may indeed have, *for a limited number of years*, i.e., for the *"remaining* nominal life", the same "probability" to collapse of a new building in 50 years. Some years are then "gained", *with no action*. Sons and nephews will solve.

Nowadays, the so-called "most advanced seismic codes" change the "reference average life" or also "nominal life" Y with the change of the importance of the structure (risk category) that is controlled by the (hypothetical) consequences of its failure (the more dangerous the consequences, the longer the "nominal life") (e.g.,

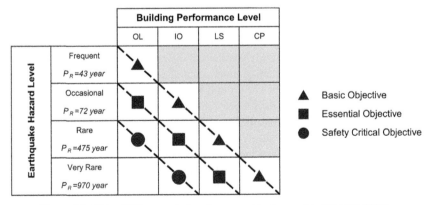

Figure 6.1. Vision 2000 conceptual performance objectives matrix (SEAOC, 1995).

NTC08 (C.S.L.P., 2008)). This leads the codes to increase the expected structural performance with increasing importance of the structure, by using a rather tricky procedure of extending the "nominal life" and so indirectly the "return period". Indeed, ideally, we would prefer better structural performances for earthquakes that occur frequently (i.e., for low intensities) and, on the other hand, we would accept higher damages for a very rare earthquake.

However, as it is totally groundless the ability to compute "probabilities" of ground motion exceedance, it is necessarily also groundless the ability to tune those "performance level" as a function of earthquake "probabilities". The words "Frequent", "Rare" and "Very Rare" are presently wishful thinking and are also dangerous as they may deliver to the population the wrong meaning of "probable" or "almost impossible". If it is agreed that the most severe earthquake could happen tomorrow, and that we cannot evaluate *recurrences* and probabilities, then, normative reasoning implies that there would be no need to distinguish these performances but as a rough tool, very likely already covered by good design rules and minimum horizontal forces.

The "probability" of exceedance P_{EY} is related with the structural performance level to check: the lesser the acceptable damage, the higher the "probability" of exceedance. Let us consider the following *exemplum fictum*: Using the Italian Building Code NTC08 (CSLP, 2008) a residential building should be designed to reach the Collapse Prevention Level for an earthquake with $P_R = 975$ years (i.e., a response spectrum whose accelerations are supposed to have $P_{EY} = 5\%/Y = 50$ years). On the other side, an Essential Building (e.g., a school) should be designed to reach this level when impacted by an earthquake with $P_R = 1462$ years (i.e., $P_{EY} = 5\%/Y = 75$ years). Focusing on the so called "Collapse Prevention Level", this means that if an earthquake consistent with a $P_R = 1462$ years happens, the residential building designed in accordance with the Italian Building Code could, in a notional sense, collapse. If an earthquake with $P_R = 2000$ years happens, even a school would fall down. At a first glance, some of these "probabilities" of occurrence could appear very low, but this concern is sitting on the erroneous and thus very misleading concept of "return period". Moreover, a 10% probability in 50 years is not so a low probability, after all. As it has been recently shown (Bizzarri and Crupi, 2013), physical roots for P_R are totally lacking and thus P_R represents a rather arbitrary choice and nothing more (see also Section 6.1.1). Actually, events that have never happened before happen every day (Taleb, 2007).

This procedure, basically physically rootless, is usually justified on the basis of an economic assessment of the cost to build structures in seismic areas. This idea was introduced in ATC3-06 (ATC, 1978) although the estimates of losses in that document are based entirely on expert judgment rather than modelling, and the tone is very much one of assessing and judging the chosen design basis as being reasonable and at least as stringent as the design basis in use at the time (Bommer and Pinho, 2006).

However, the standard breakdown of the overall cost of a modern building is: 8–18% for structural components, 48–62% for non-structural components and

20–44% for contents (Miranda and Aslani, 2003). The costs optimization using a probabilistic value of ground motions when evaluating the Collapse Prevention Level appears to be unreasonable, at least for these reasons:

- the fallacy of the "return period" concept and of all the recurrence models for earthquakes;
- the impossibility to evaluate objectively earthquake probabilities;
- the benefits (reduction of costs) due to a PSHA (Probabilistic Seismic Hazard Assessment) decrease of ground motion involve a very small percentage of the overall cost (the structural and not structural components);
- it does not take into account the post-earthquake recovery costs.

Ideally, a PBSD procedure should aim to build an earthquake "resilient system". An earthquake resilient system would be a system with the following features (Bruneau et al., 2003):

- reduced failure probabilities;
- reduced consequences from failures, in terms of lives lost, damage, and negative economic and social consequences;
- reduced time to recovery (restoration of a specific system or set of systems to their "normal" level of performance).

Indeed, recent earthquakes demonstrated that a PBSD approach based on PSHA is neither reliable nor cost effective. The P_R = 2500 years acceleration response spectra prescribed by the New Zealand seismic code was exceeded by the Christchurch earthquake (New Zealand, 22 February 2011, M_w = 6.2) that caused 181 deaths. It was estimated that at least 900 buildings in the business district and over 10,000 homes had to be demolished. The repairs cost was estimated in about US$ 15–20 billion, the highest cost ever caused by an earthquake in New Zealand (Morgenroth and Armstrong, 2012; Kaiser et al., 2012). In the Tohoku earthquake (Japan, 11 March 2011, M_w = 9), followed by a devastating tsunami, the cost to the government has reached US$ 260 billion (Iuchi et al., 2013). The Wenchuan (Sichuan) earthquake (China, 12 May 2008, M_w = 7.9) resulted in US $124 billion of direct losses and at least other US$ 100 billion of indirect losses to production and housing sectors (Wu et al., 2012). Italy, a seismic country but with relatively low maximum recorded magnitudes, has spent from 1944 to 2012 almost €181 billion (CRESME, 2012), only in public funding, because of earthquakes. A devastating series of earthquakes struck the country again between August and October 2016, and much higher spectral accelerations than those with a "return period" of 2475 years given by the Building Code have been recorded.

It seems clear that the statement of the ATC 3-06 committee *"considering the significant cost of designing a structure for extreme ground motions, it is undesirable to require such a design unless there is a significant probability that the extreme motion will occur"* is no longer acceptable in order to create a resilient system. In addition, the progresses of engineering knowledge and new technologies, such as the use of seismic isolation and/or dissipative systems, make this statement no longer true.

Since the end of the 70s, the ideas embedded in the cited ATC 03-06 document spread all over the world, and very many standards adopted them with almost no modification, no matter their lack of physical and statistical meaning. As happens in human-organized systems (as to structural engineering see, e.g., Rugarli, 2014, in general see, e.g., Catino, 2014) information between interfaces was gradually lost. No part of the original ATC 3-06 cautions in referring to probabilistic methods was saved. Year 2008 saw the birth of Italian Technical Standards (NTC08), the absolute champion of this memory-deletion process, where a meaningless interpolation of 3–4 digits PGAs over a 5.5 km grid was asked to assess, for each limit state, the design response spectrum *at bedrock*. Still today designers have to determine exact latitude and longitude (with less than 100 meters error), and then interpolate the four corner values of the pertinent grid, to "compute" the bedrock PGA related to the "nominal life" and exceeding "probability". On the boundary line between cells, first type discontinuity in PGA was systematically detected, so that sarcastic comments, indicators of a deep disagreement with the probabilistic kernel of seismic hazard assessment, and coming from engineers, suggested to compute it by a repeated throw of dices (Rugarli, 2008).

Since the end of the 70s, an impressive number of works criticizing PSHA from the statistical (e.g., Stark and Freedman, 2001), mathematical (e.g., Wang, 2010), physical (e.g., Bizzarri and Crupi, 2013; Panza and Peresan, 2016) and engineering point of view (e.g., Klügel, 2007a; Rugarli, 2008; 2014; 2015a; 2015b) did appear. Currently PSHA is still necessarily acknowledged (Junbo, 2018) as the most used method in the world.

However, a new *paradigm* is needed if Disaster Risk Mitigation is to succeed in fulfilling its very worthy goals. Mirroring the cautions and warnings of dozens of earlier papers (e.g., Molchan et al., 1997; Kossobokov and Nekrasova, 2012; Panza et al., 2012), most recently Geller et al. (2015) and Mulargia et al. (2017) have concluded: (i) that everyone involved in seismic safety concerns should acknowledge the demonstrated *shortcomings* of PSHA; (ii) that its use as a sacrosanct and unquestioningly-relied-upon *black box* for civil protection and public well-being must *cease*; and (iii) that most certainly a *new paradigm* is needed! Recent, quite convincing evidences reported by several authors (Bergen et al., 2017; Cowie et al., 2017; Dolan et al., 2016; Stockmeyer et al., 2017; Zinke et al., 2017) are very important local and regional examples (California-Garlock and Puente Hills faults; New Zealand—Awatere fault; China - Southern Junggar Thrust) of large time-changes in fault slip rates that provide a basic input for earthquake recurrence models used in PSHA. The results of these studies well support the PSHA pitfalls evidenced at global scale by Panza et al. (2012).

Therefore, a different method to define the seismic hazard should be defined and used in a PBSD framework. This chapter tackles this problem and offers an operative solution.

6.1.1 Some Remarks About the Use of the Term "Probability"

There is no general agreement about the meaning and implications of the term *probability*. Frequentist and subjectivist approaches are the most frequently used. The first requires a sufficiently wide amount of data to issue probability evaluations, otherwise by definition loses its meaningfulness. It raises frequencies to the rank of *probabilities* extrapolating their validity to the future. The second affirms the possibility to issue probability evaluations also with no data at all, and affirms that, eventually, every probability estimate is subjective. According to this approach the probability evaluations of PSHA are possible as such, but should be considered as (very much questionable) *opinions*, not as objective evaluations. It happens that these opinions are enforced by law, and that these opinions are dangerous.

This is not the place where to discuss these items, however it is necessary to underline that in this work the term "probability" has been used out of citations with inverted commas, to underline that the validity of its use as objective result to predict the future frequencies is denied. The every day probability based on a sufficient amount of experience has nothing to do with the "probability" computed by PSHA believers.

The authors of this paper do not deny the validity of probabilistic methods as such, however. The percentiles that are computed in this work, are computed using a large amount of numerically generated data, and assume no distribution *a priori*. No "probability" is computed in this work. The authors believe that the arbitrary promotion of uncertain numerical indices to "probability", and the publication of *laws* using these terms, are dangerous and misleading steps and should be avoided.

The authors believe that it is much better to declare the impossibility of a probability estimate than to use a surely wrong one. For more details see, e.g., Panza et al. (2014).

6.2 Fundamentals—NDSHA Theory

The Neo Deterministic Seismic Hazard Assessment (NDSHA) (Panza et al., 2001; 2012) does not use empirical equations such as Ground Motion Prediction Equations (GMPEs) to derive the Intensity Measure (IM) of interest (e.g., PGA or Spectral Acceleration (SA)). Instead, it is a multi-scenario-based procedure which supplies realistic time history ground motions calculated as the tensor product between the tensor representing in a formal way the earthquake source and the Green's function of the medium. NDSHA is based on the maximum magnitudes expected at a site regardless of their likelihood of occurrence. Physics-based synthetic seismograms can be computed through the knowledge of the earthquake generation process and of the seismic wave propagation in an inelastic medium. As any physical model NDSHA suffers uncertainties and limitations due to the uncertainty intrinsic in the basic data, chiefly earthquake catalogues, and lack of satisfactory theories about earthquake source. For this reason, hazard values at national/regional scale supplied by NDSHA are given as discrete ranges in geometrical progression close to 2 (Fig. 6.2), over areas whose extension is consistent with the information content of the basic data. In this way over-parameterization is avoided, as the inclusion of too high a level of detail to describe past earthquakes may lower the maps' ability to predict future shaking, as recently echoed by Brooks et al. (2017). More specific hazard estimates can be obtained at local scale by means of ad hoc studies, as shown, for example in Section 6.2.2.

In the original formulation (Panza et al., 2001; 2012) NDSHA is solving, in a first approximation, fundamental problems posed by an adequate description of the physical process of earthquake occurrence. It can take the largest event physically possible, usually termed maximum credible earthquake (MCE), whose magnitude M_{design} at a given site can be tentatively, until proven otherwise, set equal to the *maximum* observed or estimated magnitude M_{max} plus some multiple of its accepted global standard deviation (σ_M). In areas without information on faults or sparse data, historical and morphostructural data are used to estimate the maximum.

Specifically, no more than $1/k^2$ of a distribution's values can be more than k standard deviations away from the mean (or equivalently, at least $1-1/k^2$ of the distribution's values are within k standard deviations of the mean). If $k = 2$ at least 75% of the values fall within $2\sigma_M$ and if $k = 3$ at least 89% of the values fall within an interval of $3\sigma_M$ centered on the mean. The factor k can be considered a tunable safety factor that may be named consistently with the safety factors used in structural engineering, γ_{EM} (Earthquake Magnitude). So

$$M_{design} = M_{max} + \gamma_{EM}\ \sigma_M$$

where it is currently assumed

$\sigma_M = 0.2 - 0.3$ (e.g., Båth, 1973, p. 111)

and it is proposed to use

$\gamma_{EM} = 1.5 - 2.5$

A(g)

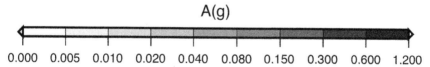

| 0.000 | 0.005 | 0.010 | 0.020 | 0.040 | 0.080 | 0.150 | 0.300 | 0.600 | 1.200 |

Figure 6.2. Typical discrete ranges of hazard values (units of g), in geometrical progression close to 2, consistent with the real resolving power of the worldwide available data (e.g., Cancani, 1904; Lliboutry, 2000).

It is underlined that the design value M_{design} is obtained adding to the *maximum* estimated magnitude M_{max} value a further tunable increment, ΔM, so it must be considered an envelope evaluated at the best of present day knowledge. This choice is consistent with Chebyshev's theorem: for a very wide class of probability distributions, no more than a certain fraction of values can be at more than a certain distance from the *mean* (but here the *maximum* estimated is used).

NDSHA aims to supply an envelope value, in other words a value that should not be exceeded, therefore it is immediately falsifiable/verifiable: if an earthquake occurs with a magnitude, M, larger than that indicated by NDSHA, M_{design}, then

$$\Delta M = M - M_{max} > \gamma_{EM}\, \sigma_M$$

and the product $\gamma_{EM}\sigma_M$ should be increased. Given the way M_{design} is defined this is expected as a rare condition. Similarly, the product $\gamma_{EM}\sigma_M$ should be raised if the peak values (e.g., PGA) recorded *at the bedrock* at the occurrence of an earthquake after the compilation of NDSHA maps exceed, within error limits, those given in the same maps. This would require installing stations over stiff soils, so as to avoid the local amplifications due to site effects (the most part of the stations of the Italian net, are, for instance, placed over soft soils).

The decision of the multiplier γ_{EM} to be applied to the standard deviation cannot be proved by equations (and it would be misleading to try) and today is partly heuristic. However, this heuristic can be falsified by natural experiments, and this multiplier can be gradually set to the minimum safe value. This is what has already been done with all the safety factors used in engineering (the 1.5 safety factor for material limit stresses was used well before the availability of reliable statistical measures, and the semi-probabilistic methods used in structural engineering are *de facto* tuned to confirm these already validated-by-experience values).

The computed seismograms are used to estimate engineering relevant parameters such as Peak Ground Acceleration (PGA), Displacement (PGD), Velocity (PGV) and spectral values, and can be used directly as input for Non-Linear Time History Analysis of structures.

In the NDSHA framework the computations of physics-based synthetic seismograms is performed with different levels of details, depending on the purpose of the analysis. For national-scale seismic hazard mapping, a "Regional Scale Analysis" (RSA) is carried out using many possible sources and simplified structural models representative of bedrock conditions. When a detailed analysis is needed, a "Site-Specific Analysis" (SSA) can be performed. A SSA can consider structural and topographical heterogeneities, but also the influence of the source rupture process on the seismic wave field at a site. So far the NDSHA method has been applied in several countries at different levels of detail (PAGEOPH Topical Volume 168 (2011), Panza et al., 2012). Some features of NDSHA can be tested thanks to the development of a web application (http://www.xeris.it/index.html) (Vaccari, 2016).

The steps required to perform a RSA and a SSA, starting from NDSHA computations, are described in the following, with a focus on the Italian territory. In particular, with respect to the procedure described by Panza et al. (2001; 2012), in order to better fit engineering needs, upgrades in the seismograms computation are described. These upgrades are described by Fasan et al. (2015; 2016; 2017), Fasan (2017) and Magrin et al. (2016).

6.2.1 Regional Scale Analysis (RSA)

The properties of the sources and structural models of the Earth are needed in order to perform NDSHA. As a rule, NDSHA allows for the use of all the available information about the spatial distributions of the sources, their magnitudes and focal mechanisms, as well as about the properties of the inelastic media crossed by earthquake waves. The procedure can be divided into three steps:

- identification of possible seismic sources;
- characterization of the mechanical properties of the medium in which the seismic waves propagate;
- computation of the seismograms at sites of interest.

6.2.1.1 Seismic Sources

The objective of NDSHA is to incorporate all possible seismic sources. The potential sources are defined combining all the available information about historical and instrumental seismicity (Fig. 6.3), seismotectonic models (Fig. 6.4) and morphostructural analysis (Fig. 6.5). As far as the Italian territory is concerned, the magnitudes are derived from:

- the parametric catalogue of Italian earthquakes CPTI04 (CPTI Working Group, 2004);
- the earthquakes catalogues for Slovenia and Croatia (Markušić et al., 2000; Živčić et al., 2000);
- the ZS9 seismogenic zones (Meletti et al., 2008), i.e., seismotectonic homogeneous areas capable of generating earthquakes (Fig. 6.4);
- the seismogenic nodes, i.e., zones prone to strong earthquakes identified through a morphostructural analysis (Gorshkov et al., 2002; 2004; 2009) (Fig. 6.5).

The seismogenic nodes are placed at the intersection of lineaments, identified by morphostructural analysis (Gelfand et al., 1972). In Fig. 6.5, for example, the nodes are represented as circles of radius R = 25 km within which earthquakes have magnitude $M_N \geq 6$ or $M_N \geq 6.5$. The choice of the radius is consistent with the average source dimension of earthquakes within the same range of magnitudes (Wells and Coppersmith, 1994) and with the uncertainty in their location. The use of seismogenic nodes allows for the inclusion of the effects of earthquake prone

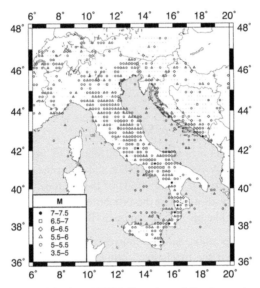

Figure 6.3. Discretized seismicity from CPTI04, Slovenian and Croatian catalogues (Markušić et al., 2000; Živčić et al., 2000; CPTI Working Group, 2004).

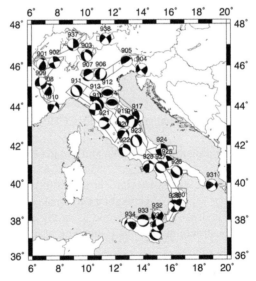

Figure 6.4. ZS9 Seismogenic zones and associated focal mechanisms (Meletti et al., 2008).

areas not yet affected in historical time by strong events (and hence not reported in catalogues) (Peresan et al., 2009).

Consistently with the level of detail adopted and required at regional scale, possible epicentres over the territory are discretized into $0.2° \times 0.2°$ cells (about a 20×20 km grid).

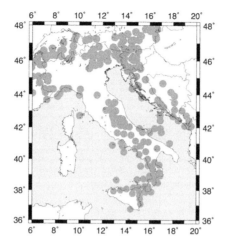

Figure 6.5. Seismogenic nodes identified by morphostructural analysis ($M_N \geq 6$ or $M_N \geq 6.5$) (Gorshkov et al., 2002; 2004; 2009).

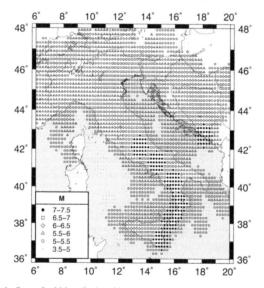

Figure 6.6. Smoothed historical and instrumental seismicity (Panza et al., 2001).

The first step is to elaborate the information contained in historical catalogues. Magnitudes derived from historical catalogues are grouped into each cell and only the maximum magnitude recorded within each cell is retained. This step results in a discretization of the historical and instrumental seismicity, as reported in Fig. 6.3. The second step consists in applying a smoothing procedure (Panza et al., 2001) to roughly account for the spatial uncertainties and the source dimensions (see Fig. 6.6). The discretized magnitudes are spread within a circle, centred on their

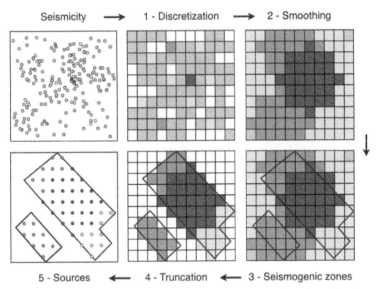

Figure 6.7. Procedure for the choice of the magnitude to be assigned to each cell.

original position, of radius equal to three times the cell's side. After this smoothing, only the sources falling into the seismogenic zones and the seismogenic nodes are retained. The procedure is summed up in Fig. 6.7. The aim of the procedure is to get an envelope of what can be expected.

The magnitude to be assigned to each cell, which will represent the magnitude used in the computation of seismograms, is chosen as *the maximum* between:

- the magnitude M_N of the seismogenic nodes;
- the magnitude resulting from the smoothing procedure;
- a minimum magnitude of 5.

The resulting map of seismic sources for the Italian territory is shown in Fig. 6.8. The reason for assigning a minimum magnitude of 5 to any cell falling within a seismogenic area (thus potentially capable of generating earthquakes) is that 5 is the value after which one begins to observe structural damage (D'Amico et al., 1999).

6.2.1.2 Structural Models

At a regional scale, consistently with the approximations in the computational method and with the required level of detail, structural models are represented by flat, parallel inelastic layered media. The physical properties of the source-site paths are defined using a set of cellular structures (Fig. 6.9) obtained through an optimized nonlinear inversion of surface wave dispersion curves (Brandmayr et al., 2010). Every cell has a dimension of $1° \times 1°$ (about 100×100 km) and represents the average structural properties of the lithosphere at regional scale, as

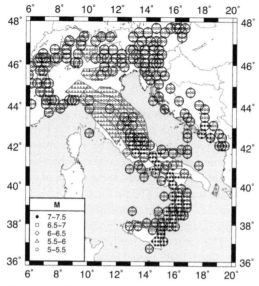

Figure 6.8. Final sources configuration used in NDSHA computations (Panza et al., 2012).

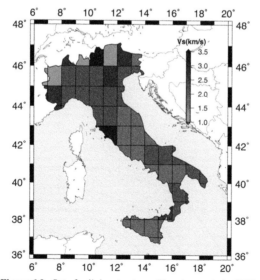

Figure 6.9. Set of cellular structures (Brandmayr et al., 2010).

a rule bedrock conditions. The properties of the medium assigned to each cell are the result of knowledge gained over the last two decades in the Italian area, mostly in the framework of the project "Determinazionedel potenziale sismogenetico in Italia per il calcolo della pericolosità sismica" (INGV-DPC 2007–2009 agreement).

6.2.1.3 Computations of Physics-Based Synthetic Seismograms

The computation of seismograms by means of NDSHA is done into two steps:

- simulation of the rupture process on the faults;
- simulation of the propagation of seismic waves through the definition of a transfer function (Green's function).

The starting point for the upgrade of the methodology is represented by "Model 6" of Panza et al. (2012). The upgrades are described in Fasan et al. (2015; 2016; 2017), Fasan (2017) and Magrin et al. (2016). A double-couple, a tensor that represents a dislocation consistent with the tectonic character of the seismogenic zone or of the seismogenic node, is placed at the centre of each cell. The depth is chosen as a function of the magnitude (10 km for $M \leq 7$, 15 km for $M > 7$) to account for the existing magnitude—depth relationship (Caputo et al., 1973; Molchan et al., 1997; Doglioni, 2016). The moment-magnitude relation chosen is that given by Kanamori (1977). The sources are modelled as point sources scaled in size and duration (STSPS). The STSPS model is based on an extended source model provided by the PULSYN06 algorithm (Gusev, 2011) and considers a reference scaling law for source spectra (SLSS). The SLSS used in "Model 6" of Panza et al. (2012) is the G83 (Gusev, 1983) that reasonably represents seismic source data at a global scale, as successfully tested in particular by Boore (1986). Magrin et al. (2016) updated the SLSS focusing on the Italian region (Fig. 6.10).

A further upgrade of the procedure, adopted to build the "Model 6" of Panza et al. (2012), is the generation of different stochastic realizations of the source model (slip distribution and rupturing velocity), for each source-to-site path, using the PULSYN algorithm (Gusev, 2011). This is done to account statistically

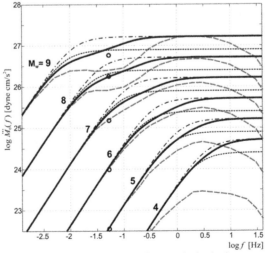

Figure 6.10. Updated SLSS for the Italian region, for magnitudes in the range 4–9 (Magrin et al., 2016).

for the variability of the ground motion at a site due to unpredictable variations in the rupture process, which can have a strong impact on ground motion critical features. In fact, many rupture parameters cannot be predicted in a deterministic way (it is impossible to predict the precise style of the next rupture); therefore, in line with the method's choice to envelop possible future scenarios, a Monte-Carlo simulation of these parameters is needed.

Moreover, in the standard NDSHA, the ground motion at a site is computed using the Modal Summation (MS) technique (Panza, 1985; Florsch et al., 1991; Panza et al., 2001). The MS technique is computationally very fast and provides an adequate simulation of ground motion in the far field, while the Discrete Wave Number technique (DWN) is used in near field conditions (short paths). The DWN in the implementation of Pavlov (2009) gives the full wave field, including all body waves and near field. The computational cost of DWN increases with epicentral distance-source depth ratio since the number of wavelengths to be calculated for the series convergence depends on the angle formed with the vertical: the more a radius is vertical, the lesser are the terms to be calculated, hence the greater is the ratio between the epicentral distance and the depth of the source, the more the calculation time grows (Magrin, 2013). A good compromise between accuracy and CPU time is to use DWN in computations for epicentral distances less than 20 km and MS for larger distances, routinely up to 150 km. Synthetic seismograms are then computed over the Italian territory with a frequency cut off at 10 Hz for each node of a grid of $0.2° \times 0.2°$ shifted by $0.1°$ from the grid of the sources.

6.2.2 Site-Specific Analysis (SSA)

The results of the RSA are valid as long as the site of interest is placed on bedrock soil. This condition is quite rare and usually the ground motion at a site is strongly controlled by the interaction between source radiation and lateral heterogeneities, whether topographical or due to the presence of soft-sedimentary soil.

Routinely, local "amplifications" are evaluated in a very simplified manner by modifying the shape of the response spectra at the bedrock using different coefficients. These coefficients, introduced in the seismic codes, are functions of the mechanical properties of the surface layer and of its topography. A popular computation of the local effects might be carried out using the ratio between the horizontal and the vertical response spectra (H/V ratio) (Nakamura, 1989). This widely used factor is obtained from seismic noise, and it can be acceptable under the stringent condition, very rarely fulfilled, that the vertical ground motion is not affected by the superficial layer(s). This method has been demonstrated to be unable, as a rule, to give correct local effects (Panza et al., 2012). In fact, the vertical component of motion can be severely affected by local soil mechanical conditions, too, as demonstrated in a pioneering paper by Panza et al. (1972). "Amplifications" of both vertical and horizontal components of motion are strongly dependent not only on the soil and topography characteristics, but also on the incidence angles of the radiated wave field.

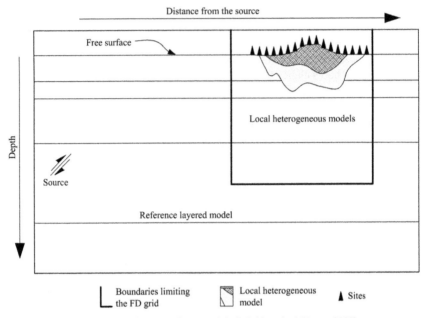

Figure 6.11. Schematic diagram of the hybrid method (Fasan, 2017).

To overcome these limitations, a method based on computer simulations exploiting the knowledge about the source process, the path source-to-site and the local site conditions has been developed. This hybrid method combines MS with the finite-difference technique (Fäh and Panza, 1994). The wave field generated by MS is introduced in the mesh that defines the local heterogeneous area and it is propagated according to the finite-differences scheme as shown in Fig. 6.11.

To reduce the CPU time costs, the procedure is applied only with a small number of sources (for the path source to site), i.e., those that give the largest bedrock hazard for the IM of interest (e.g., Spectral Acceleration (SA)). In other words, the SSA is an RSA carried out only for the most hazardous sources for the site of interest but considering the local soil conditions.

From an engineering point of view, in addition to accounting for realistic site amplifications, an SSA provides realistic and site-specific synthetic seismograms. This possibility is truly important given that the number of available recorded ground motion is very low, particularly for large earthquakes. A preliminary SSA is then essential to run time history structural analysis using seismograms representative of the dynamic characteristics of the site of interest.

6.2.3 Maximum Credible Seismic Input

In Sections 6.2.1 and 6.2.2, the computation of synthetic seismograms via the NDSHA method with its upgrades is shown. NDSHA-MCSI (henceforth called MCSI) can be defined as a Response Spectrum or as a set of accelerograms. In engineering analysis, all the accelerograms generated by NDSHA can be used to perform nonlinear analysis of a structure. However, since thousands of ground motions are simulated, the available information needs to be summarized, in order to reduce the analysis time.

Following the NDSHA method, MCSI can be defined at a given site at two levels of detail (Fasan et al., 2017). The first level uses the results computed with an RSA, as explained in Section 6.2.1. It provides the "Maximum Credible Seismic Input at bedrock" ($MCSI_{BD}$), without considering the site effects. At the second level of detail, the RSA is used as a reference to choose the most dangerous sources for the site and ground motion parameter of interest. As for these sources, a detailed SSA that considers the local structural heterogeneities is then carried out for each source-to-site path as described in Section 6.2.2. The SSA allows us to determine the "Maximum Credible Seismic Input Site Specific" ($MCSI_{SS}$).

The use of source spectra computed by PULSYN06 (Gusev, 2011) introduces a stochastic element in NDSHA. In order to define the MCSI, its relevance must be evaluated to enable realistic estimates of seismic hazard and their uncertainty. For this purpose, the procedure described in Fasan et al. (2015; 2017) is applied.

In the first step, for each path and ground motion parameters, the distribution due to the different realizations of the rupture process is determined. This step does not consist in the assumption of the distribution "a priori" (e.g., lognormal). Instead, the distribution of percentiles is retrieved directly from the Monte-Carlo simulations, treating them as observations.

In the second step, for each site, the median of these distributions is compared and the distribution of the path (i.e., the source) that gives the maximum median value is chosen for the ground motion parameter of interest. In this way, only the source that gives the "worst case" scenario is considered. In fact, if the distribution of parameters due to all sources was chosen, the values corresponding to certain percentiles would be reduced from the sources that give lower estimates of the selected parameter.

The steps can be repeated for each selected ground motion parameter (e.g., PGA, PGV, SA, etc.). The procedure to construct the MCSI response spectrum is summarized in Fig. 6.12. At each site and at each period, SA values computed from different scenarios are compared and the maximum is chosen. In other words, as a rule, the MCSI at different periods can be controlled by different scenarios (e.g., in terms of magnitude, epicentral distance, earthquake focal mechanism). In fact, MCSI represents a sort of Uniform Hazard Spectra (UHS) (Trifunac, 2012), where, at each site, the hazard is controlled by the maximum magnitude of every source that could affect the site of interest.

MCSI response spectrum should be set equal to the envelope response spectrum of all the simulated response spectra (100th percentile). However, due to the stochastic nature of the algorithm used to account for different rupture process, the 100th percentile is initially very sensitive to the number of simulations. Based on the limited experience gained so far, an acceptable compromise between computational costs and accuracy of results is the use of the 95th percentile of at least one hundred

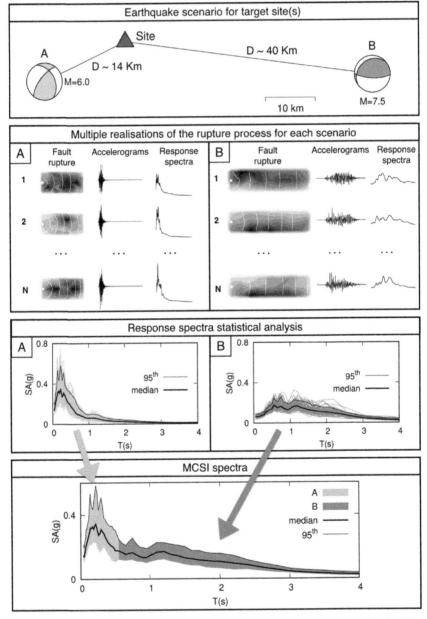

Figure 6.12. Description of the NDSHA-based MCSI response spectrum construction (modified from Fasan et al., 2016).

realizations of the rupture process. The number of simulations can be increased at will in order to stabilize the value of the 100th percentile with a linear increase of computational time with increasing number of simulations: should it be required by particularly sensible objects (e.g., hazardous power plant), one can increase the product between γ_{EM} and σ_M to a value that represents the uncertainty and run "one million" simulations as USGS claims makes its UCERF3 forecast of earthquake "probabilities" in California "plausibly scientific". It is worth noting, however, that even the choice of the 100th percentile is still arbitrary since it represents just the upper value of the simulations and not the, unpredictable with precision, *maximum maximorum* that could happen at a site. But this is very likely far from human capabilities, and will be so for quite some time to come. The uncertainty that is still present in MCSI is reflected in the word "credible" adopted in the definition of the acronym and places major responsibility in the engineering choices, which are still the fundamental part of a reliable design process.

As an example we describe the procedure applied to the city of Trieste (Italy). The resultant response spectrum has been found to be a meaningful parameter for structural design (BSSC, 2009). Therefore, the MCSI response spectrum is calculated as the resultant (herein called "Res", see Fig. 6.13) rather than the maximum between the components of ground motion in the horizontal plane (herein called "Max_xy"), which are dependent on the reference system (see Fig. 6.15a).

Both "Res" and "Max_xy" have been calculated. This has been done to show the dependency of the Max_xy response spectrum on the choice of the reference system. Furthermore, it is necessary for a comparison with the response spectra of the Italian Building Code that represents the Max_xy and not the resultant. This variability (shown in Fig. 6.14a and b) of the spectral accelerations, respectively for the Max_xy and Res response spectra, as resulting from the RSA, is due to the contribution of different sources, magnitudes, focal mechanisms and rupture processes. The variability at each period represents the variability of the spectral values for the source that gives the highest hazard at that period.

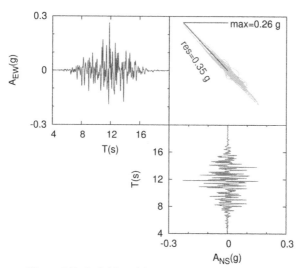

Figure 6.13. Definition of the resultant response spectrum.

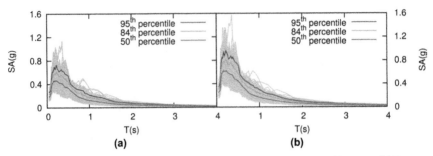

Figure 6.14. Variability of response spectra shape at the site of interest: (a) Max_xy; (b) Res.

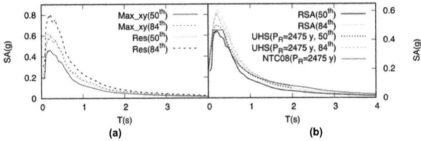

Figure 6.15. (a) Comparison between Res and Max_xy (RSA); (b) Comparison between Max_xy resulting from a RSA and the Italian building code response spectra.

An RSA allows for the identification of the seismic input at the bedrock (i.e., soil class "A" as per Italian Building Code). The Uniform Hazard Spectra (UHS) given by the Italian seismic hazard map represent the Max_xy response spectra at bedrock for a given "probability" of exceedance. In Fig. 6.15b it is shown a comparison between the Max_xy response spectra resulting from the RSA (50th and 84th percentile) and the Uniform Hazard Spectra of the Italian seismic hazard map for a "mean return period" $P_R = 2475$ years (50th and 84th percentile).

The "return period" of 2475 years is the highest value used to calculate the Italian seismic hazard map. Consequently, it is gratuitously assumed that for

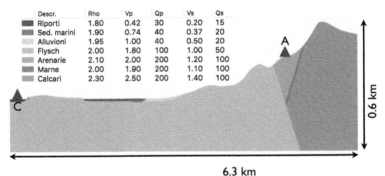

Figure 6.16. Profile and sites of interest used for the SSA.

this value of "probability" of exceedance (2% in 50 years) the resulting spectral accelerations are still realistic, even if the length of the reliable earthquake catalogue, including the pre-instrumental macroseismic intensities, is little above a millennium.

As it can be seen, MCSI caps the spectral acceleration when using small values of "probability" of exceedance. In addition, in Fig. 6.15b the response spectrum of the Italian Building Code for a "mean return period" P_R = 2475 years is reported, too. This response spectrum represents a "code fit" of the median UHS.

As a second step, an SSA has been performed. To consider the effects of soil and topographic characteristics on both vertical and horizontal components of the earthquake ground motion, a laterally heterogeneous profile representative of the local conditions has been composed using data from literature and fieldwork.

To save computer time costs, the scenario used for the SSA has been chosen from the disaggregation of the MCSI$_{BD}$ response spectrum calculated with the RSA (i.e., the source-distance combination that gives the largest spectral acceleration at periods of interest). The controlling event, for the range of periods from 0 s to 4 s, has been found to be a magnitude 6.5 earthquake at epicentral distance in the range from 15 to 20 kilometres (Fig. 6.17a).

This scenario is consistent with the seismic potential of the Branik-Ilirska Bistrica fault (SICS004) (DISS Working Group, 2015) and falls within a seismogenic node (Fig. 6.17a). In agreement with the local dominant tectonic style (DISS Working Group, 2015), the assumed focal mechanism parameters are: depth =10 km, strike = 281°, dip = 79°, rake = 16°.

Two sites have been selected as representative for the analysis of the results of the entire study: A—representative of a soil type "A" ($V_{S,30} \geq$ 800 m/s), and C— representative of a soil type "C" (180 < $V_{S,30} \leq$ 360 m/s), as defined by the Italian Building Code. Figure 6.18 shows a comparison between the Max_xy response spectrum (50th and 84th percentile) and the resultant response spectrum (50th and 84th percentile) for the sites resulting from the SSA. The same comparison has been done in Fig. 6.15a for the response spectra resulting from the RSA. As it can be seen, the ratio between the two response spectra varies from about 1.4 in

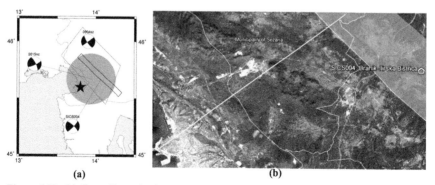

Figure 6.17. (a) Controlling seismic sources resulting from a RSA; (b) Source to site path used in the SSA.

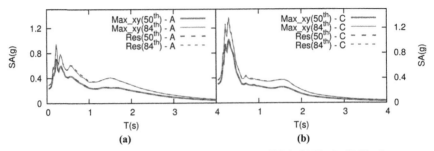

Figure 6.18. Comparison between Res and Max_xy (SSA): (a) Site A; (b) Site C.

the RSA to about 1.0 in the SSA where Res and Max_xy are almost overlapped. This comparison confirms that the Max_xy response spectrum is dependent on the orientation of the reference system and therefore it isn't a valuable tool for seismic hazard definition and consequently seismic design (i.e., the same earthquake, depending on the orientation of the instrument used to record it, can have different Max_xy response spectra and therefore using a Max_xy response spectrum in a structural design could lead to an underestimation of the seismic load).

Figure 6.19 shows a comparison between the maximum response spectrum provided by the Italian code ($P_R = 2475$ years), the input associated with the code to the Collapse Prevention Level for a standard residential building ($P_R = 975$ years) and the response spectrum adopted by the code for a standard design associated with the Life Safety Level ($P_R = 475$ years). As it can be seen, both for site A and for site C the structural lateral heterogeneities have a strong effect on the shape and amplitude of the response spectrum. In particular, the use of standard soil coefficient provided by codes can lead to a strong underestimation of the local amplification. In fact, even though the median UHS is very close to the median Max_xy response spectrum resulting from the RSA (i.e., the response spectrum at the bedrock without considering local soil and topographic conditions) (Fig. 6.19b), the differences increase when adopting an SSA.

Figure 6.19 is a paradigmatic example of the inadequacy of simplified approaches based on scalar correcting coefficients (e.g., for site conditions): the

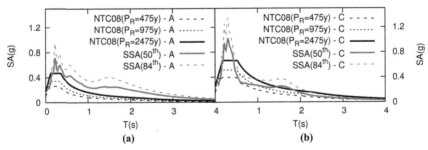

Figure 6.19. Comparison between Max_xy resulting from a SSA and the Italian code response spectra.

MCSIs evidence that local effects can lead to spectral values that are, at different periods, well above or below those of the elastic design spectra given by the Italian code (into force in 2017) on account of local conditions. Therefore, the seismic source process, the propagation of earthquake waves and their combined interactions with local effects should be duly taken into account to safeguard the tensorial nature of earthquake ground motion.

6.2.4 NDSHA and Long-Term Properties of Earthquake Occurrence

NDSHA, in its standard form, supplies time-independent maps that do not depend on temporal information about earthquakes occurrence. However, the flexibility of NDSHA naturally permits to account for the properties of earthquake occurrence, eventually providing information about the rates of occurrence of the expected ground shaking and allowing for the generation of maps at specified long-term average occurrence times, as described by Magrin et al. (2017). Remarkably, the resulting maps do not vary with time and do not imply any assumption on the probability model of earthquakes occurrence (e.g., not validated Poissonian model assumption), since information is provided only in terms of *average occurrence rates*. The transformation of (necessarily rough) average occurrence rates into "probabilities" is willingly avoided. Clearly, including additional temporal information, which is very uncertain, increases uncertainties in the resulting estimates and thus it should be considered only if and when strictly necessary. The authors believe that the engineering use of these average occurrence rates is questionable, and do not suggest their use.

The characterization of the frequency-magnitude relation for earthquakes in the Italian region is performed according to the multi-scale seismicity model

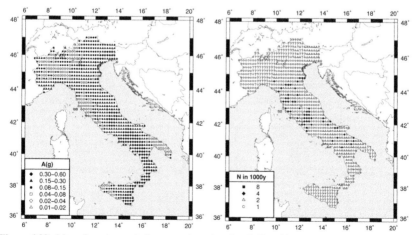

Figure 6.20. Map of maximum design ground acceleration (left) and its occurrence (right) expressed, to be conservative, as the upper bound of the (integer) number of times the ground motion is likely to be observed in 1000 years. Question marks represent sites where occurrence estimate of maximum ground motion is unreliable due to the lack of sufficient data (modified after Magrin et al., 2017).

(Molchan et al., 1997) and an average long-term occurrence is associated to each of the modeled sources. Since the occurrence rate of the source is associated to the related seismograms, a standard map of ground shaking is obtained along with the map of the corresponding average occurrence time (Fig. 6.20). Even though the concept of "return period" is totally rootless for earthquakes, if one wishes to have some kind of comparison with existing methods the transparent introduction of occurrence estimates in NDSHA allows for the generation of ground shaking maps that can be compared in a straightforward way with PSHA maps for specified return periods (Fig. 6.21).

The estimates of seismic hazard obtained according to the NDSHA and according to the "probabilistic" (PSHA) approaches have been compared for the Italian territory (Zuccolo et al., 2011). Remarkably, the PSHA expected ground shaking estimated with 10% "probability" of being exceeded in 50 years (corresponding to a "return period" of 475 years) appears underestimated (by about a factor 2) with respect to NDSHA estimates, particularly for the largest values of PGA. When a 2% "probability" of being exceeded in 50 years is considered (i.e., "return period" of 2475 years) PSHA estimates in high-seismicity areas become comparable with NDSHA; in this case however, the "probabilistic" maps overestimates the hazard in low-seismicity areas.

Thus, the predictive capability of the PSHA maps turns out quite unsatisfactory, as shown by Nekrasova et al. (2015). On the other side, the maps developed according to the NDSHA approach describe quite satisfactorily the ground shaking and even the occurrence rates at specific sites, as shown by Magrin et al. (2017).

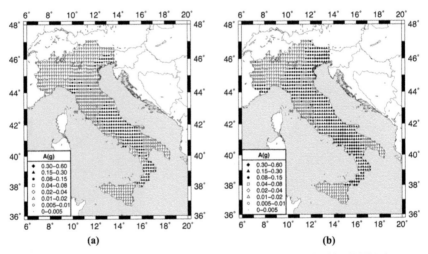

Figure 6.21. Maps of maximum design ground acceleration determined with NDSHA for return periods typical of PSHA: (a) T = 475 years and (b) T = 2475 years. Question marks represent sites where occurrence estimate of maximum ground motion is unreliable due to the lack of sufficient data (modified after Magrin et al., 2017).

6.2.5 Time Variable NDSHA Scenarios at National Scale

Besides the time-independent NDSHA maps described so far, time-variable maps can also be generated, when formally defined and tested information about earthquakes occurrence is available. Earthquakes cannot be predicted with precision, but algorithms exist for intermediate-term middle-range prediction of main shocks above a pre-assigned threshold, based on seismicity patterns (Keilis-Borok and Soloviev, 2003). Based on NDSHA approach, an operational integrated procedure for seismic hazard assessment has been developed (Peresan et al., 2011) that allows for the definition of time-dependent scenarios of ground shaking, through the routine updating of earthquake predictions, performed by means of the algorithms CN and M8S (Peresan et al., 2005). The integrated NDSHA procedure for seismic input definition, which is currently applied to the Italian territory, combines different pattern recognition techniques, designed for the space-time identification of strong earthquakes, with algorithms for the realistic modelling of ground motion (Peresan, 2018). Accordingly, a set of deterministic scenarios of ground motion at bedrock, which refers to the time interval when a strong event is likely to occur within the alerted areas, can be defined by means of full waveform modelling, both at regional and local scale. The NDSHA ground motion scenarios at bedrock for alarmed Central Italy CN region are shown in Fig. 6.22 in terms of PGV. By the way, the PGV values observed by Rete Accelerometrica Nazionale–Dipartimento Protezione Civile (RAN–DPC) at Amatrice (up to 31 cm/s) and Norcia (up to 56 cm/s) fit very nicely the values predicted by NDSHA ground motion scenarios (30–60 cm/s).

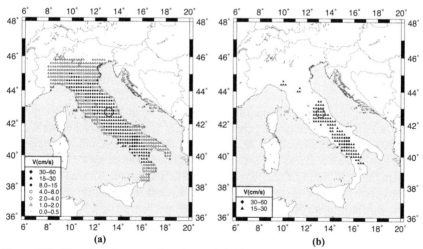

Figure 6.22. Time-dependent scenarios of ground shaking associated to the alarm in CN Central Region. On the left (a) maps of peak ground velocity (PGV) are shown, computed using simultaneously all of the possible sources within the alarmed area and for frequencies up to 10 Hz. On the right (b), the same maps are provided, but for PGV > 15 cm/s (modified after Peresan et al., 2015). The circle on maps (a) and (b) evidences the area within 30 km distance from the epicenter of the Central Italy earthquake (M6. 2, 2016).

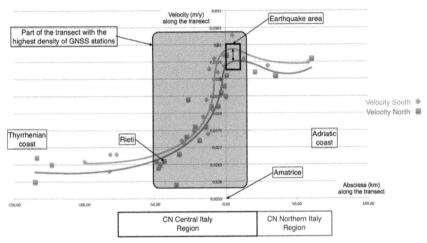

Figure 6.23. Velocity pattern along the Amatrice transect (Central Italy), located across the Apennines (time range: January 2005—middle of August 2016). The velocity gradient along the transect displays a gradient peak (geodetic signature) stable over time in the Amatrice earthquake area (24 August 2016). Remarkably, the velocity gradient is higher in the southern than in the northern side of the transect (after Panza et al., 2017).

The proper integration of seismological and geodetic information, also supplies a significant contribution to the reduction of the space extent of predictions of earthquakes (Panza et al., 2017). GPS data are not used to estimate the standard 2D velocity and strain field in the area, but to reconstruct the velocity and strain pattern along transects, which are properly oriented according to the a priori information about the known tectonic setting. As shown for the 2016–2017 seismic crisis in Central Italy and the 2012 Emilia sequence, the analysis of the available geodetic data indicates that it is possible to highlight the velocity variation and the related strain accumulation in the area alarmed by CN. Some counter examples, across CN alarmed and not-alarmed areas, do not falsify the predictive pattern (Panza et al., 2017).

The integrated routine monitoring, CN and GPS, could be more widely applied in the near future, since dense permanent GNSS networks could be established using low-cost GNSS receivers. An example of the possible significant reduction of the size of the CN alarmed areas, by the integrated monitoring, CN and GPS, is given in Fig. 6.23. In Fig. 6.23, the rectangle delimits the area covered by the GPS stations showing large anomalies in the velocity and strain pattern and encloses the seismic hazard scenario simultaneously defined/alerted by CN and GPS data that extends for about only 5000 km^2.

6.3 Basic Suggested Methodologies

6.3.1 Recommended Practice Using NDSHA

The problem of what ground motion level and hence what method of SHA is to be preferred becomes a matter of public safety when it is applied to the design of structures. When talking about strong earthquakes, the "rarity" argument (i.e., thinking that it is too expensive to design for strong earthquakes) is not acceptable for public safety because rare events are sporadic, therefore can occur at any time and it is not possible to predict them with precision (Taleb, 2007). The best way to reduce potential losses from earthquakes is to build seismic-resilient communities, which inevitably results in designing or retrofit buildings to withstand very strong earthquakes.

When assessing the so-called "Collapse Prevention Level", the situation that could involve the loss of the structure is dealt with. Given the fact that an engineer cannot control the earthquakes phenomena (so far nobody can tell with precision when and where an earthquake will take place) but can govern the building performance through the designing procedure, the least we can do is to use an upper-bound ground motion to design or retrofit buildings against the collapse. As a rule, an upper-bound ground motion should be used to assess every structural performance that involves the highest level of damage eligible for the building under design (e.g., "Collapse Prevention Level", CP, for Ordinary Buildings or "Immediate Occupancy", IO, for Hazardous Buildings).

The proposed procedure aims to address the following facts, evidenced by the analysis of seismic phenomena:

- any structure at a given location, regardless of its importance, is subject to the same bedrock shaking as a result of a given earthquake;
- it is impossible to determine with precision when a future earthquake of a given intensity/magnitude will occur;
- insufficient data are available to develop reliable statistics for natural earthquakes.

Structural performance levels depend on the damage (usually quite notionally defined in terms of acceptable storey drift or acceptable strain or rotation of plastic hinges) that is accepted to occur in the elements of a building when subjected to a certain level of ground motion. All the performances represent exclusively a conventional state of damage. We cannot be sure that the notional relationship between the computable damage and the true structural behaviour holds true for the single building under examination. This is particularly true for existing buildings and historical ones in particular. The definition of the ground motion used to check whether a performance level has been reached is a crucial step.

Imagine having to design a building of strategic importance, like a hospital. In case of earthquake occurrence, the hospital must be able to receive and treat the injured people. Therefore, is it reasonable to accept that it may be impracticable after the earthquake? It is believed not. In the light of the common-sense considerations made so far it is proposed to identify, according to the importance of the structure, a Target Performance Level (TPL) that is the highest level of damage acceptable for the building. The achievement of this level of performance will always be checked using the MCSI. In this way, the seismic input used to check the TPL becomes unrelated to the choice of the "nominal life" and the "probability of exceedance", which are arbitrary thresholds. Choosing to associate MCSI with different TPLs requires engineering judgment in the decision and evaluation of the TPL.

The definition of MCSI is affected by uncertainties, which cannot be removed using a percentile of the NDSHA simulations. Indeed uncertainness cannot be removed at all and sometimes it must be confessed the impossibility to express it with an agreeable number. This is one of the cases. MCSI represents an envelope of what could be expected at a site in the case that a MCE level scenario ground motion occurs, and may be invalidated by stronger-than-estimated earthquakes. The invalidation, however, due to how intrinsically the method is conceived, appears to be much less likely than the numerous and bloody ones experimentally observed for PSHA. That is what civil engineering should aim at.

Also the engineering procedure plays a major role in achieving a safe seismic design. Once the TPL is chosen, levels of performance that involve a lower percentage of damage assume a minor importance in terms of potential adverse consequences. These levels of performance are defined Lower Performance Levels (LPLs). By definition of LPLs, it is acceptable that they may be exceeded during the life of the structure, as they involve less damage than TPL. Consequently, the acceleration response spectra associated with them must be less than MCSI (which should be a reasonable upper limit) and, given the conventionality of this procedure, their choice is completely arbitrary, therefore not unique. Such levels could be transparently and notionally defined as a fraction of $MCSI_{ss}$ response spectra (*for example* 2/3 of $MCSI_{ss}$ for medium seismic input level and 2/5 of $MCSI_{ss}$ for low seismic input level). Behind this proposal is the refuse of computing with apparently complex math things which are not computable. One much needed and wished side-effect is a dramatic simplification of the design process. The proposed procedure, summed up in Fig. 6.24 whereby two values are suggested as examples, can be summarized as follows:

- step 1: identification of the Risk Category of the building (e.g., Ordinary Building, Essential Building or Hazardous Building);
- step 2: as a consequence of Step 1, choice of the Target Performance Level associated with the $MCSI_{ss}$ response spectrum;
- step 3: as a consequence of Step 2, choice of the Lower Performance Levels and the associated ground motions.

It is worth noting that the uncertainties, both structural and related to the seismic input, are such that it is impossible to predict exactly the seismic behaviour of a structure. In a nutshell, on account of the large uncertainties about the structural properties of the built environment and detailed characteristics of seismic input, and in order to avoid the illusory idea of optimizing costs "probabilistically" reducing the earthquake ground shaking, a reliable approach to be followed is believed to be that based on the use of the MCSI response spectrum, that represents a reasonably safe estimate of the worst possible case consistent with present day knowledge. Therefore, the procedure proposed in Fig. 6.24 should be used just as a minimum requested performance to assess the building during its design stage.

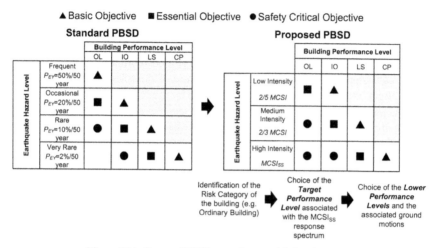

Figure 6.24. Proposed PBSD procedure considering the MCSI.

It is evident that the proposed procedure represents somehow an increment in the design criteria. Given the deaths and the costs due to rebuilding, and given that the cost variation is absolutely not excessive, this is a welcome and much needed result.

The discussion that has been carried out so far focuses mainly on the design of new buildings, however the same consideration about the earthquake phenomena must be applied to existing structures. Therefore, a retrofit project should be based on an analysis carried out using a MCSI level of seismic input at least to assess the gap between the required and the expected performance of the existing structure. The dangerous and people-kidding concept of "remaining nominal life" should be removed from all the standards and considered for what it is: a mean to edulcorate the truth.

6.4 Time History Selection

Non-linear Time History Analysis (NLTHA) is, so far, the best tool to assess dynamic seismic performance of structures (ATC, 2012; FIB, 2012). Since NLTHA belongs to the class of "garbage in, garbage out" procedures, the reliability of the outcomes of NLTHAs depends on how well the mathematical model represents the real behaviour of structures and on the reliability of the seismic input. In a Time-History Analysis (Linear or Non-Linear) of 3D structures the input is represented by a triplet of accelerograms (two horizontal and one vertical, the rotational components are not used). In the currently accepted procedures the selected ground motions should correspond as to magnitude, source distance and focal mechanism associated with the scenario controlling the hazard at the site of interest. No mention is usually paid to the specificity of source-to-site strata, which, instead, play a fundamental role. Their effect is kept into account by attenuation "laws" whose effectiveness is usually shown in a much flattening lognormal plane. Indeed these relationships have no physical ground. Appropriate site soil conditions and near fault effects (directivity and fling-step) should be accounted for (NIST, 2011; Haselton et al., 2017). The selection of ground motions is usually guided by the consistency between an Intensity Measure of the selected ground motion and a Target Intensity Measure, often identified by the 5% damped elastic SA at a selected period or over a range of periods (Katsanos et al., 2010). More complex criteria are based on Vector-based Intensity Measures, i.e., a couple of different IMs such as spectral acceleration at the first vibrational period and other parameters representative of the spectral shape (Baker and Cornell, 2008; Bojórquez and Iervolino, 2011; Theophilou et al., 2017).

The method described in Section 6.2 to define the MCSI is instead rooted in the NDSHA method. The NDSHA method estimates the seismic hazard by means of the computation of broadband physics-based synthetic accelerograms that account for known seismological, geophysical and geological characteristics from source to the site of interest. Therefore, it is a natural consequence to implement such accelerograms in the dynamic analysis of buildings and to create a direct bridge between seismology and structural analysis.

In this section: (a) suggestions based on the latest information available in literature are given on how to properly select the accelerograms using MCSI spectrum as target considering the natural periods of the structure at hand; (b) a procedure to select the accelerograms directly from the NDSHA simulations that contribute to the definition of MCSI is illustrated.

6.4.1 Accelerograms Selection: Current Issues and Suggestions

To properly estimate the median response of a structure, several ground motions need to be selected and employed. If an appropriate number of analyses using different ground motions consistent with a target response spectrum are performed, the value of a selected Engineering Demand Parameter (EDP, e.g., Story Drift Ratio) can be set equal to the average of the maximum values it has assumed in any analysis (ATC, 2012).

The response spectrum method is not precise and cannot be considered the best method to assess the structural behavior. This is due to its inability to consider the phases and the duration of the signal. However, it is still today the most frequently used method because of its relative ease, and because it may deliver important information about the maximum response of the single oscillators the structure is made of during the elastic phase. For engineering purposes, the seismic hazard is still today defined by a response spectrum that takes indirectly into account the variability of the possible signals. The response spectra used by the norms, however, have notional shapes and no true direct link to the site under examination. Usually, a set consisting of n accelerograms is deemed representative of the seismic hazard if the average spectrum of the n selected accelerograms is compatible with the target spectrum for the site of interest (CEN, 2004; CSLP, 2008; BSSC, 2015). This compatibility criterion is often translated into practice requiring that the values of the average spectrum of the selected accelerograms fall within a range of spectral accelerations between 90% and 130% of the reference spectrum for the site of interest (e.g., EC8 (CEN, 2004) or Italian Building Code NTC08 (CSLP, 2008)). All these are reasonable notional choices which have only an empirical ground.

Even if a set of accelerograms is spectrum compatible, records can have a significant duration which is not consistent with the scenario which governs the hazard at the site of interest (Bommer and Acevedo, 2004). Therefore, in addition to spectrum compatibility, accelerograms should be chosen among those representative of the seismological conditions at the site of interest (e.g., source and site effects). Then, the input selection for nonlinear dynamic analysis must be made considering, at least, the following parameters:

- the target response spectrum;
- the estimated periods of the structure for spectrum compatibility check;
- the minimum number of analyses to perform;
- source and site effects;
- source-to-site effect;
- the availability of accelerograms.

Once the accelerograms have been selected, another issue is to choose the orientation in which they are applied to the mathematical model of the structure with respect to the principal direction of the building plan (Beyer and Bommer, 2007).

6.4.2 Target Response Spectrum

Attention should be paid to which spectral acceleration refers to the target spectrum, since this affects the structural analysis results (Baker and Cornell, 2006a; Beyer and Bommer, 2007). For example, the target spectrum of the ASCE 7–10 code (ASCE, 2013) represents the maximum direction spectral acceleration for any possible orientation (RotD100), whereas the target spectrum of the Italian Building Code NTC08 represents the maximum spectral acceleration between two orthogonal directions (Max_{NS-EW}). When selecting accelerograms for the analysis of 3D structures, the use of the maximum direction spectrum is suggested since it automatically takes into account the bidirectional effects of ground motion (Huang et al., 2008; NIST, 2011). Therefore, the maximum direction spectrum of each pair of orthogonal accelerograms should be computed and then the average of the n selected maximum direction spectra should be compared with the target spectrum.

The maximum direction spectrum does not consider the vertical component of the ground motion. Historically, little attention has been given to the vertical component, as it was misbelieved that few structures are sensitive to it (e.g., structures with very long spans). This is simply not true, e.g., for the very much diffused in South Europe, but almost lacking in USA, masonry old buildings (as the recent Italian earthquakes have once more shown). Sites close to the epicentral areas can exhibit strong vertical accelerations. For example, in the event of October 30,

2016 of the central Italy (2016–2017) earthquakes sequence, a spectral acceleration larger than 1 g was recorded for a wide range of periods, including those typical of vertical modes of vibration. Therefore, once a set of orthogonal ground motions is selected, its compatibility with the vertical target response spectrum should also be evaluated, and all the three components of motion should be applied simultaneously to the model. This could lead to practical problems; in Section 6.4.7.1 a possible procedure using the MCSI target spectrum is suggested.

6.4.3 Range of Periods

The range of periods used to define the compatibility criteria should account for the free vibration periods which strongly affect the dynamic response of the structure (Haselton et al., 2017). It must be underlined that the evaluation of the free structural periods is itself a quite difficult procedure, especially for existing buildings. It is not infrequent the need to retrofit buildings for which no original design drawings are available. Moreover, the use of elastic methods to evaluate periods is questionable and probably wrong for ancient masonry and highly irregular buildings. For them, all is yet to come.

Traditionally, seismic codes rather crudely suggest to assess the spectral compatibility over a *range of periods* going from $0.2T$ to $1.5T$, where T is the fundamental translational free period of the structure (e.g., ASCE 7–10 (ASCE, 2013), EC8 (CEN, 2004) or the Italian Building Code NTC08 (CSLP, 2008)). This range of periods should assure to account for the effects of higher modes and for the increase of natural periods due to inelastic degradations of strength and stiffness; however, even if this range or periods was set for 2D planar analysis of regular frames (first mode dominated) it is taken as appropriate for 3D analysis (NIST, 2011) as well. Latest indications tackle this generalization and suggest that (Haselton and Baker, 2006; BSSC, 2015):

- the upper bound limit should be greater than or equal to $2T$, where T is the largest fundamental period of the building among translational directions and in torsion;
- the lower bound limit should be less than or equal to 20% of the period of the smallest first-mode between the two orthogonal horizontal directions of the response and such that the period range includes at least the number of *elastic* modes necessary to achieve 90% mass participation in each orthogonal horizontal direction.

6.4.4 Number of Analyses

The number of analyses to be performed, and therefore the number of accelerograms to be selected, varies from code to code. As a rule, a minimum of 3 analyses is required. This number is to be considered absolutely insufficient. If only 3 analyses are performed, the value of the EDP of interest must be set equal to the maximum value that it reached in the three analyses. If more than 7 (e.g., EC8 (CEN, 2004)) or 11 (e.g., FEMA P-58 (ATC, 2012) or FEMA P-1050 (BSSC, 2015)) analyses are performed, the value of the EDP of interest could be set equal to the average of the maximum values reached in all analyses.

We believe that these numbers are only due to the relative complexity related to managing large amount of data. A significant increase in the number of the time histories is the next future and definitely the only reliable tool to assess the structural behavior with some hope of reaching an envelope. Linear time histories may be run very quickly and may provide significant information also in order to select the signals (see below).

6.4.5 Geophysical and Geological Parameters

Since the selected accelerograms should be representative of what may be experienced at the site under analysis, the following scenario parameters must be considered (Bommer and Acevedo, 2004; Molchan et al., 2011; NIST, 2011; BSSC, 2015):

- *Source Mechanism*: the tectonic regime and the rupture mechanism should be the same as the ones of the scenarios controlling the hazard at the site;

- *Magnitude and Distance*: the magnitude and source-to-site distance of the accelerograms should be close to those governing the response spectrum at the structural period of interest since the strong-motion duration and the frequency content are strongly affected by these parameters;

- *Source-to-site Path and Site Soil Condition*: site condition can have a strong impact on the characteristics of ground motions. This is usually accounted for selecting records with the same soil classification of the site. The site classification is based on the mean shear velocity of the first 30 meters from the free surface ($V_{S,30}$) (e.g., Panza and Nunziata, 2018). This parameter is rather simplistic and does not consider the influence of the source-to-site path. Different geological features and mechanical properties of the Earth's crust can lead to different frequency content and attenuation. Moreover, in presence of sedimentary basins, local site effects (very often amplifications) depend on the relative position of the source with respect to the site (Molchan et al., 2011; NIST, 2011);

- *Near Source Effects*: Sites close to active faults can experience pulse-type ground motions in which most of the energy released by the fault rupture is concentrated in one or two pulses of motion that occur at the beginning of the record (Archuleta and Hartzell, 1981). These effects are attributable to the direction of propagation of the rupture, called *forward directivity* if the rupture propagates towards the site or *backward directivity* on the contrary, and to the static displacement of the ground surface due to the relative movement of the two side of the fault, called *fling-step*. The possibility to experience such phenomena depends on the source-to-site distance and on the site-source (Molchan et al., 2011; NIST, 2011) azimuth. A site can be classified as near-fault if the distance from the source is less than 10–20 km, but this depends on the source magnitude (Shahi and Baker, 2011; Haselton et al., 2017). When the rupture propagates toward the site (forward directivity), double-side pulses in the velocity ground motions can be observed and the significant duration of motion is usually shorter than the cases of backward directivity. Moreover, a polarization of the ground motion in the fault-normal component is usually observed up to 5 km from the source (Watson-Lamprey and Boore, 2007). Fling-step effect, instead, results in a monotonic step in the displacement ground motion, therefore in a single side pulse in the velocity ground motion. The effect of these pulse-type ground motions on the dynamic response of structures depends on the ratio between the pulse period and the fundamental period of the structure, with higher demand as this ratio approaches one (Kalkan and Kunnath, 2006). The period of the pulse is a function of the magnitude and lower magnitudes cause pulses with lower periods, therefore, near-fault earthquakes with moderate magnitudes could have spectral acceleration at intermediate periods higher than those due to larger earthquakes (NIST, 2011).

Several simplified models have been proposed to compute the period of the pulse and the peak ground velocity or the spectral acceleration amplification and the fling amplitude as a function of soil, magnitude and source-to-site distance (Bray and Rodriguez-Marek, 2004; Baker, 2007; Bray et al., 2009; Shahi and Baker, 2011; Burks and Baker, 2016). If a site can experience near-fault effects, the selected accelerograms should contain directivity and fling-step pulses to account for the energy they carry (e.g., Mollaioli et al., 2003).

6.4.6 Availability of Accelerograms

Usually, accelerograms are selected from online databases of natural recordings such as the NGA-West 2 database (Ancheta et al., 2014) in the U.S. or the Engineering Strong-Motion database (Luzi et al., 2016) in Europe. Since the number of available natural records is still very limited, especially in Europe, it is difficult to find a spectrum compatible set of accelerograms, which is strictly pertinent to the geological and geophysical properties of the site. By definition, moreover, selecting a signal recorded in a different part of the world, totally neglects the source-to-site bedrock strata which are typical of the sources-to-site paths of interest.

Therefore, in practice it is common to relax the allowable range of magnitudes, distances and site soil conditions and to allow for the use of different records of the same event in the same set (usually limited to a maximum of three records (Zimmerman et al., 2017)). Moreover, accelerograms are linearly scaled in amplitude in order to obtain SA similar to those of the target spectrum.

The procedure of amplitude scaling has been strongly debated, the main concern being that it can lead to unrealistic frequency contents and thus biased structural responses (Bazzurro and Luco, 2006; Luco and Bazzurro, 2007; Grigoriu, 2011).

Other authors argued that if spectral compatibility is accounted for the scaling procedure is acceptable (Iervolino and Cornell, 2005; Baker and Cornell, 2005; 2006b; Hancock et al., 2008). Even if the procedure of scaling is useful to overcome the lack of available data, altering the amplitudes without considering the change in frequency, duration and energy content is for sure an arbitrary procedure with no physical meaning.

Another source of accelerograms could be the use of programs such as SIMQKE (Gasparini and Vanmarke, 1976), which generate artificial accelerograms adding up sinusoidal functions with random phase angles and amplitudes in order to construct a response spectrum which matches the target one. However the use of artificial seismograms should be avoided since they overestimate the cyclic response and, on the contrary to what could be expected, underestimate the peak ductility demand (Bommer and Acevedo, 2004; Schwab and Lestuzzi, 2007; Iervolino et al., 2010).

A similar technique is the *"response spectrum matching"* where a real accelerogram (the seed) is altered in the time domain by adding adjustment functions to match the record response spectra with the target one, which should lead to preserve the non-stationary characteristic of the seed motion (Al Atik and Abrahamson, 2010; Grant and Diaferia, 2013). Even if this technique could lead to somehow "realistic" accelerograms, there are concerns that their use could lead to an underestimation of the response variability of the structures (Reyes et al., 2014) and to non conservative demand (Bazzurro and Luco, 2006; Iervolino et al., 2010). Again, this procedure has no physical meaning.

A sound option is the use of physics-based broadband synthetic seismograms. The intrinsic advantage of using accelerograms simulated in such a way is that they simultaneously account for the local site condition, the site-to-source path and the source properties including directivity and fling-step (Molchan et al., 2011; NIST, 2011). NDSHA method (Panza et al., 2001; 2012; Fasan et al., 2015; 2017; Magrin et al., 2016) is among the available simulation techniques capable of incorporating such information. Similar techniques are those recently proposed, for example, by Graves et al. (2011) and Graves and Pitarka (2010).

6.4.7 *Selection Using NDSHA*

6.4.7.1 *Selection Using MCSI Spectra*

The MCSI response spectrum is computed directly from NDSHA physics-based broadband simulations of the seismic process. This means that sets of spectrum compatible accelerograms can be found looking directly into the simulations used to define the target 5% damped MCSI response spectrum itself. The procedure described in Section 6.2.1 to define the seismic hazard at the bedrock, obviously does not consider the site-specific characteristics and it is intended just to give a lower bound of the seismic hazard over the country, that can be eventually compared with "probabilistic" maps (Zuccolo et al., 2011; Nekrasova et al., 2014).

NDSHA permits, if really necessary, to account for earthquake occurrence rate (Peresan et al., 2013 and references therein; Peresan et al., 2014; Magrin et al., 2017). Peresan et al. (2013) have performed the characterization of the frequency-magnitude relation for earthquake activity in Italy according to the multi-scale seismicity model (Molchan et al., 1997), so that a robust estimated occurrence is associated to each of the modeled sources. The occurrence assigned to the source is thus associated to the pertinent synthetic seismograms, coherently with the physical nature of the problem. Accordingly, two separate maps are obtained: one for the ground shaking, one for the corresponding occurrence. In fact when considering two sites prone to earthquakes with the same magnitude *M*, given that all the remaining conditions are the same, the parameters for seismic design must be equal at the two sites, since the magnitude we have to defend against is the same, independently from the sporadic occurrence of the earthquake. NDSHA maps and the related occurrence rates (not "probabilities") are obviously time-independent. Magrin et al. (2017), in their Figs. 6.4 and 6.5, duly evidence that, for a given territory, the additional information about occurrence rates can be considered, if necessary, together with the NDSHA hazard maps, but with unavoidable and severe uncertainties, represented by question marks (?).

Therefore, the regional-scale analysis can be useful to identify the most hazardous sources. The simulated accelerograms should be selected from a SSA as described in Section 6.2.2 and the target spectrum should be the $MCSI_{ss}$. In a SSA it is possible to take into account near-fault effects and, if necessary, the fault can be modelled as an Extended-Source (ES), e.g., Magrin (2013). When selecting accelerograms for bidirectional analysis the $MCSI_{ss}$ should be calculated accounting for the maximum direction spectral acceleration (RotD100) and the average of the *n* maximum direction spectra, chosen from the simulations database, should be spectrum compatible.

There is no need for filtering by magnitudes, distances or site classifications, as done with natural records since the accelerograms are all representative of the same site (site-specific). Moreover, there is no need for linearly scaling the accelerograms to match the target spectra. This procedure repeats what is usually done with natural records, replacing them with synthetic ones and replacing the Uniform Hazard Spectrum (UHS) prescribed by the codes with the $MCSI_{ss}$ response spectrum as suggested in Section 6.3.

If the, too often guiltily overlooked, vertical component of ground motion is needed, it is suggested to develop at least two sets of spectrum compatible accelerograms. The first one selecting the accelerograms on the maximum direction MCSI acceleration response spectrum (which considers the NS and EW components) and using in the structural analysis also the vertical components of the selected accelerograms. The second set could be defined selecting the accelerograms on the vertical MCSI acceleration response spectrum and then using in the structural analysis the associated NS and EW components of the ground motion. Since the accelerograms are all calculated for the specific site of interest, there is no chance to overestimate the seismic demand using this procedure (i.e., the accelerograms come out from a physic-based analysis so their characteristics are consistent with the generating scenarios) and the effects of all three components of motion can effectively be considered. Of course, these suggestions only are needed to optimize (reduce) the number of structural analyses to perform.

When looking for spectrum compatible sets of accelerograms using the MCSI as a target spectrum, although the simulations database contains the same accelerograms with which it was created, it is not uncommon to have difficulties in finding sets that match the spectrum, above all if an elevated number of accelerograms is required. This may sound strange but it is a direct consequence of the procedure followed to construct the MCSI spectrum. As explained in Section 6.2.3, the MCSI spectrum is a sort of UHS. It is built selecting firstly the most hazardous source at each period on the base of the median spectral acceleration of the simulations of every source and secondly, at each period, selecting as a reference value for the MCSI spectrum the 95th percentile value, not assuming any distribution, of the simulated spectral acceleration for the source that governs that specific period (as reported this is a minimum suggested value).

Hence, MCSI represents an envelope of different scenarios and cannot in general be matched by a single signal. This results in the difficulty of finding sets of spectrum compatible accelerograms over a wide range of periods if the standard MCSI spectrum is used. Similar considerations have been raised about the use of the UHS (Bommer et al., 2000; Beyer and Bommer, 2007; Katsanos et al., 2010; NIST, 2011).

A different approach could be to select the accelerograms just at one structural period of interest, for example the period of the fundamental mode of the building, defining a "Conditional" Maximum Credible Seismic Input (C-MCSI). This concept is similar to that of Conditional Mean Spectrum (CMS) proposed by Baker (2011) as a more realistic alternative to the UHS. Defining a C-MCSI response spectrum simply consists in defining the spectral accelerations just with the most hazardous source at a period of interest and selecting only the restricted range of simulations that gives the highest values of spectral acceleration at that specific period. This also means maximizing the elastic deformation energy related to a single mode.

For example, Fig. 6.25 shows the C-MCSI$_{BD}$ for a period of 1.5 s and for a period of 0.83 s (site of Trieste). Since MCSI is set equal to the value of the 95th percentile, C-MCSI is calculated by selecting the simulation that has a spectral acceleration ranging from 100th to 90th percentile at the period of interest and then choosing the median values of these simulations at each period.

The use of C-MCSI is best suited for regular structures, with a dominant vibrational mode. However, if more than one mode has an elevated mass participation it is suggested to define different C-MCSI, one for each mode, and perform dynamic analysis using those sets separately and treating the results separately to avoid underestimations due to the considerations mentioned above. This means searching for the signals that maximize, independently, the elastic deformation energies related to single modes of the structure.

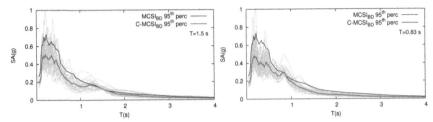

Figure 6.25. Conditional MCSI (C-MCSI) at bedrock for a vibrational period T = 1.5 s, left, and for a vibrational period T = 0.83 s (right): site of Trieste (Fasan, 2017).

Once the accelerograms are selected they need to be applied to the mathematical model of the building. This involves the choice of the ground motion axes (e.g., EW and NS components or fault-normal and fault-parallel) and how to orient the accelerograms with respect to the horizontal axes of the building (Beyer and Bommer, 2007). There are little guidance on this topic (NIST, 2011). Suggestions indicate that the choice depends on the method used to select the accelerograms (BSSC, 2015):

- if the maximum direction response spectrum is used, since it considers the maximum spectral acceleration in any direction, the selected accelerograms should be applied to the model in a random direction;
- if the site is classified as near-fault, accelerograms (representative of near source effects) should be applied in the fault-normal and fault-parallel directions as they are recorded.

The aim of MCSI is to give a conservative seismic input which envelopes uncertainties. Since MCSI accelerograms are site specific results of a detailed local analysis they could be applied to the model just as they are recorded, so in the EW and NS direction. However, this choice is not conservative. For example, a site whose hazard is dominated by one source in a determined position could be affected by less strong earthquakes caused by other sources having different positions and therefore different orientation of the ground motion components. Using just the orientation of the source that has the greater hazard may lead to an intensity of the seismic action in the other orientations smaller than that required by the other possible sources. Hence it is believed that the same set of accelerograms should be applied to the model in several directions. The same should be done if the site is classified near-fault since, obviously, it could be also far-fault respect to other sources (Kalkan and Kwong, 2013). Given that modern buildings designed to be earthquake-proof should be regular in plan and height and have a uniform distribution of masses and stiffnesses, this choice should not affect much the final design.

6.4.7.2 Selection Using the Signals

Finding signals matching a given MCSI spectrum is difficult because spectra are related to a usually huge number of accelerograms. On the other hand, asking to comply with the spectra in the range 90% to 130%, in a wide interval of periods (as done in the normally accepted practice), means introducing another variability that may somehow modify the envelope behind the concept of spectrum.

So a possible alternative should be found, that should be easy, quick and meaningful. The in depth analysis of the selected accelerograms will then provide detailed results.

Instead of using the $MCSI_{SS}$ in order to select signals compatible with it, the whole set of the signals used to define $MCSI_{SS}$ can be preliminarily processed with a quick Linear Time-History Analysis in order to extract a subset of them.

Albeit the structures may leave the elastic range during an earthquake, the evaluation of the maximum elastic deformation energy reached, related to each signal, E_{max}, is an important measure of the dynamic features of the structure in relation to the entering signal.

It must in fact be emphasized that the evaluation of the free frequencies of the structure used to enter in a response spectrum are based on the elastic properties of the system, and that those frequencies are abandoned once the elastic range is left behind. So, it is implicitly already commonly accepted to assume that the elastic behaviour of a structure is also meaningful as a predictor of what will happen when the structure leaves the elastic range.

Given a signal, a MTHA, Modal Time History Analysis, in the elastic range, runs very quickly (a few seconds). To this end if Ψ_i is the modal shape of mode i (not time dependent), in a system having n modes, and $\mathbf{K}$ is the stiffness matrix (not time dependent), one can see that for each mode exists a scalar $f(t)$ function of time that can be computed easily given the signal, the period of mode i and its damping, so that

$$\mathbf{u}(t) = \sum_{i=1}^{n} f_i(t)\Psi_i$$

where $u(t)$ is the response time history of the nodal displacement vector. So the elastic energy E is at any time

$$E(t) = \frac{1}{2}\mathbf{u}^T\mathbf{K}\mathbf{u} = \frac{1}{2}\left(\sum_{i=1}^{n} f_i(t)\Psi_i^T\right)\mathbf{K}\left(\sum_{j=1}^{n} f_j(t)\Psi_j\right)$$

Due to the orthogonality of the modes if i≠j $\Psi_i^T\mathbf{K}\Psi_j=0$, and so, ω_i is the angular frequency of mode i,

$$E(t) = \frac{1}{2}\left(\sum_{i=1}^{n} f_i^2(t)\Psi_i^T\mathbf{K}\Psi_i\right) = \frac{1}{2}\left(\sum_{i=1}^{n} f_i^2(t)\omega_i^2\Psi_i^T\mathbf{M}\Psi_i\right) = \frac{1}{2}\left(\sum_{i=1}^{n} f_i^2(t)H_i\right)$$

where $\mathbf{M}$ is the mass matrix, H_i is a scalar not time or accelerogram dependent that may be computed once-for-all before any analysis, for each mode

$$H_i = \omega_i^2\Psi_i^T\mathbf{M}\Psi_i = \Psi_i^T\mathbf{K}\Psi_i$$

This means that it is possible to predict the function $E(t)$ for a high number of accelerograms in a very short computational time. If the matrix **K** is evaluated considering only part of the structural members, the elastic energy related to those members is got.

Modal damping can be assumed to keep into account a 5% damping. At the end of this analysis the maximum deformation energy of the structure (or of selected parts of it) can be easily computed.

At each time of the MTHA three energies can be computed as a function of time t:

1. The total elastic energy $E(t)$.
2. The scaled elastic energy related to the first yield of the structure $E_E(t)$. This can be computed by a linear variation of the forces, and then of the displacements. If α is an overall reduction factor applied to forces and displacements ($0 < \alpha \leq 1$), then it can be easily seen that $E_E(t) = \alpha^2 E(t)$.
3. The "excess" of entering energy $E_P(t) = E(t)-E_E(t) = (1-\alpha(t)^2)E(t)$. Clearly, if the system remains elastic, $\alpha = 1$. It is evident that the needed scaling value α might be seen as the reciprocal of a needed global behaviour factor $R = 1/\alpha$.

A suggested criterion to select the signals would then be to choose the ones leading to the N extreme E_P energy values (and then maximizing the global needed behaviour factor R), where N is a suitably high number (20–40). In order to avoid the neglect of substructures having small deformation energy but important for the collapse prevention, each structural element deformation energy might be weighted by an "importance factor" considering the issue, e.g., secondary elements may be weighted null.

This method, named EDEM, Elastic Deformation Energy Maximization, embeds the C-MCSI. In fact C-MCSI may be got simply considering the elastic deformation energy related to a single mode. However, as the deformation energy is a scalar, it naturally keeps into account the phases and the modal summation issues. By considering only the excess of energy that the system is unable to absorb in the elastic range, a measure of the dissipative needs is got, in strict analogy with the currently in use methods of seismic engineering. More refined methods may consider, for each member, a local reduction factor α_m, instead of a global one α. This would imply to consider, element by element, only the single elements passing the yield in order to compute $E_P(t)$.

The EDEM method is currently under testing.

6.5 Case Studies

NDSHA is falsifiable. The detailed review of the traditional PSHA method revealed that the method is not adequate to describe the physical process of earthquake occurrence because of the assumption of a memoryless stochastic process – Poisson process. It is obvious that strain and stress renewal needs time and therefore the process of rebuilding the conditions for the next earthquake is time dependent.

Furthermore, the location of earthquakes even at the same fault is changing with time. Also the mechanical properties of the fault do vary, in particular after each event. Every big earthquake modifies the boundary conditions for the next one. This means that a mathematical probabilistic model has to be at least bivariate. This is outside of the scope of human knowledge due to lack of data and the shortness of human observation time in comparison with geological ages.

NDSHA is solving most of the problems posed, by an adequate description of the physical process of earthquake occurrence. It takes the largest event physically possible, usually termed Maximum Credible Earthquake (MCE), whose magnitude (cellular magnitude) can be tentatively and until proven otherwise set equal to the maximum observed cellular magnitude plus some multiple of the standard deviation (σ_M). In areas without information on faults or sparse data, historical and morphostructural data are used; those same data can be used to enrich existing parametric earthquake catalogues (Zuccolo et al., 2011; Parvez et al., 2017).

NDSHA aims to supply an envelope value, in other words a value that should not be exceeded, therefore it is immediately falsifiable/verifiable: if an earthquake occurs with a magnitude larger than that indicated by NDSHA it is necessary to measure the difference. If the difference turns out to be larger than γ_{EM} times the standard deviation (σ_M) used to define MCE, for instance (Dominique and Andre, 2000) equal to the maximum observed magnitude plus $2\sigma_M = 0.5$ (0.5 is a value which is representative of twice the standard deviation of magnitude determination at global scale, so $\gamma_{EM} = 2$), then maps are immediately falsified; similarly if the peak values (e.g., PGA) recorded at the bedrock at the occurrence of an earthquake after the compilation of NDSHA maps exceed, within error limits, those given in the same maps. On the contrary, the falsification of PSHA maps, that may be eventually possible only considering very large seismogenic zones and time intervals (Molchan et al., 1997; Panza et al., 2014), does not supply any useful information on account of the ambiguity of the interpretation of the violation: "Bad assumptions or bad luck?" (Stein et al., 2012).

To better describe from a practical point of view limits and possibilities of NDSHA, let us now consider a few examples.

6.5.1 India

The Neo-deterministic Seismic Hazard Assessment, expressed in terms of maximum displacement (PGD), velocity (PGV) and design ground acceleration (DGA), has been extracted from the synthetic seismograms and mapped on a regular grid of $0.2° \times 0.2°$ over the entire country (Parvez et al., 2017). The highest seismic hazard, expressed in terms of DGA (in the range of 0.6–1.2 g), is mainly distributed: (i) in western Himalayas and Central Himalayas along the epicentral zone of the Bihar Nepal 2015 earthquake; (ii) part of NE India and (iii) in the Gujarat (Kachchh region). A similar pattern has been found in peak velocities and peak displacements in the same areas. For the same event, using the conversion from acceleration to EMS (European Macroseismic Scale) intensity (Llboutry, 2000), the NDSHA results have been compared with the maximum observed intensities reported in EMS scale by Martin and Szeliga (2010): where observations are available, the modeled intensities are rarely exceeded (2% of the cases when $\gamma_{EM} = 0$) by the maximum observed intensities.

6.5.2 Northern Italy—The Emilia Earthquake Crisis in 2012

Currently, the PSHA map from *Gruppo di Lavoro, Redazione della mappa di pericolosità sismica, rapporto conclusivo*, 2004 (http://zonesismiche.mi.ingv.it/mappa_ps_apr04/italia.html), is the official reference seismic hazard map for Italy, and shows bedrock PGA values that have a 10% probability of being exceeded in 50 years (i.e., once in 475 years). The Emilia 20 May 2012 M = 5.9 and 29 May 2012 M = 5.8 earthquakes occurred in a zone defined at low seismic hazard by the code based on PSHA: PGA map ("return period" 475 yr) < 0.175 g; observed PGA > 0.25 g. The NDSHA map published in 2001 (Panza et al., 2001), which expresses shaking in terms of design ground acceleration, DGA, equivalent to peak ground acceleration, PGA (see Zuccolo et al. 2011), predicted values in the range 0.20–0.35 g, in good agreement with the observed motion that exceeded 0.25 g. Seismic hazard maps seek to predict the shaking that would actually occur, therefore what occurred in Northern Italy supplies a strong motivation for the use NDSHA or similar deterministic approaches, also with the aim to minimize the necessity to revise hazard maps with time. In this view, public buildings and other critical structures should be designed to resist future earthquakes. Contrary to what implicitly suggested by PSHA, when an earthquake with a given magnitude occurs, it causes a specific ground shaking that certainly does not depend on how sporadic the event is (rare or not). Hence ground motion parameters for seismic design should be independent of how sporadic (infrequent) an earthquake is, as it is done with NDSHA (Peresan and Panza, 2012 and references therein).

6.5.3 Central Italy

6.5.3.1 The L'Aquila 2009 Event

The 6 April 2009 M = 6.3 earthquake occurred in a zone defined at high seismic hazard, but the observed acceleration values exceeded those predicted by the code based on PSHA: PGA map (475 yr) 0.250–0.275 g; observed PGA > 0.35 g. The NDSHA map predicts values in the range 0.3–0.6 g and this implies that future events may cause peak ground motion values exceeding those recorded in 2009. As far as it is known such obvious caution is not explicitly, duly considered in the ongoing reconstruction.

6.5.3.2 The Earthquake Crisis Started in 2016

The 24 August M = 6.0 and 30 October M = 6.5 earthquakes occurred in a zone defined at high seismic hazard, but the observed acceleration values exceeded those predicted by the code standards, based on PSHA: PGA map (475 yr) 0.250–0.275 g; observed PGA > 0.4 g (a value that is larger than the one recorded at L'Aquila). Alternatively, the NDSHA map predicts values in the range 0.3–0.6 g. Following the earthquake crisis, starting August 24, 2016 supplies a dramatic example of a familiar English proverb "you get what you pay for", and maybe even "a little bit of knowledge is a dangerous thing", as described in the following section (*The Lesson*).

6.5.3.3 The Lesson

After these recent events many civil engineers, designers and practitioners complained about the fact that the accelerations given in the code standard hazard maps based on PSHA are *distorted downwards*. However, the due revision of these maps encounters strong and blunt resistance, particularly regarding the *methodological* approach of PSHA. Moreover, more than just these methodological aspects, it is also necessary to seriously consider the contrast between the very real benefits of risk reduction, as against the building costs increment—if more *reliable* and *robust risk coefficients* are considered (consistent with the magnitude size of *possible* events, like these recent ones). Beyond the untold human toll, *a posteriori* retrofitting costs about 30 times more than the upgrading of earthquake resistant design standards at the time of *new* construction (here to the more realistic and therefore more stringent earthquake resistance measures as identified by NDSHA).

Thus it looks very appropriate to remember the wisdom of the ancient Greek teachings and more modern everyday proverbs: "adequate prevention is better than cure"—as Hippocrates said about 2500 years ago; and the proverb "you get what you pay for". A dramatic example is provided by the city of Norcia.

Norcia had been retrofitted after the Umbria-Marche earthquake crisis started September 26, 1997. All reconstruction works used as a benchmark the PSHA map (475 yrs) on which the seismic code was based. That map proved totally misrepresentative and erroneous at the occurrence of the 30 October 2016 M = 6.5 earthquake, where in Norcia the earthquake ground motion was much larger than what had been predicted by PSHA. The resulting damage was large, corresponding to I_{MCS} = IX (http://www.6aprile.it/wp-content/uploads/2016/12/QUEST_rapporto_15nov.pdf reports I_{MCS} = VIII–IX, but it should be kept in mind that any intensity scale is discrete with unit incremental step). On the NDSHA map, the hazard value indicated is slightly above the experienced ground motion generated by the 30 October 2016 earthquake. In all likelihood, if the reconstruction and retrofitting that followed the 1997 Umbria-Marche earthquakes would have been undertaken in due account of the NDSHA estimates, the damage would have been much less (if not negligible) with respect to that *actually observed* after the (30/10/2016) event.

To have followed PSHA designated design strength and detailing requirements for new buildings, while neglecting that the Italian seismic code further provides: "*L'uso di accelerogrammi generati mediante simulazione del meccanismo di sorgente e della propagazione è ammesso a condizione che siano adeguatamente giustificate le ipotesi relative alle caratteristiche sismogenetiche della sorgente e del mezzo di propagazione"* (NTC08 Section 3.2.3.6) ["The use of accelerograms generated simulating source mechanism and wave propagation is allowed provided the hypotheses about the seismogenic characteristics of the source and the properties along the pathway are duly justified."]—certainly allowed some (marginal) cost saving during the reconstruction and retrofitting following the 1997 events, when compared to the higher earthquake resistance requirements indicated under NDSHA. Nonetheless this apparent "saving" has been unrealized and ultimately *frustrated* by the October 2016 earthquake—and now it is necessary to consider in the reconstruction and retrofitting the NDSHA values, which were unwisely *ignored* after the 1997 earthquakes.

Lastly to consider (but not least), *before* the occurrence of the $M = 6.5$ event on 30 October, when Norcia was almost completely destroyed:

a) Fasan et al. (2016) did show that the spectral accelerations for the 30 October 2016 $M = 6.5$ event, with magnitude close to the maximum ever historically observed in the area, are in very good agreement with what had earlier been predicted, based on NDSHA ground motion simulations;

b) Panza and Peresan (2016) issued the warning that the 24 August 2016 $M = 6.0$ earthquake did not necessarily generate the largest possible ground motion in the area: since the area had been previously hit by the 14 January 1703 $M = 6.9$, Valnerina earthquake. They further warned that, in the ensuing reconstruction and retrofitting activity, engineers should take into account as well that, in the future, seismic source and local soil effects may lead to ground motion values exceeding the NDSHA value of 0.6 g (predicted at the bedrock).

Therefore many now believe that it is well validated and justified to claim that NDSHA is a reliable and ready alternative to the presently widespread use of PSHA, particularly since the use of PSHA has been widely proven in the professional journals and publications to be a totally unjustified and unreliable procedure (e.g., Klügel, 2007a; 2007b; PAGEOPH Topical Volume 168, 2011; Mulargia et al., 2017; Fasan, 2017).

An implicit but important confirmation of the validity to consider NDSHA as an effective preventive tool is given in NTC18 (CSLP, 2018) that deepens and expands the concept contained in Section 3.2.3.6 of NTC08 as follows *"L'uso di storie temporali del moto del terreno generate mediante simulazione del meccanismo di sorgente e di propagazione è ammesso a condizione che siano adeguatamente giustificate le ipotesi relative alle caratteristiche sismo genetiche della sorgente e del mezzo di propagazione e che, negli intervalli di periodo sopraindicati, l'ordinata spettrale media non presenti uno scarto in difetto superiore al 20% rispetto alla corrispondente componente dello spettro elastico."* ["The use of accelerograms generated simulating source mechanism and wave propagation is allowed provided the hypotheses about the seismogenic characteristics of the source and the properties along the pathway are duly justified and that, in the considered period intervals, the average spectral ordinate is not less that 20% of the corresponding component of the elastic spectrum".]

6.6 Concluding Remarks and Further Studies

NDSHA has provided, for over two decades now, both *reliable* and *effective* earthquake hazard assessment tools for understanding, communicating and mitigating earthquake risk (Panza et al., 2001). And the NDSHA procedure for the development of seismic hazard maps at the regional scale is described in some detail at http://www.xeris.it/Hazard/index.html. Moreover, NDSHA seismic hazard assessment has been well validated by all events occurring in regions where NDSHA maps were available at the time of the quake; including observations from four recent destructive earthquakes: $M = 6$ Emilia, Italy 2012; $M = 6.3$ L'Aquila, Italy 2009; $M = 5.5–6.6$ Central Italy 2016–17 seismic crises; and $M = 7.8$ Nepal 2015. This good performance suggests that the wider adoption of NDSHA (especially in tectonically active areas—*but* with relatively prolonged quiescence, i.e., where only few major events have occurred in historical time) can better prepare civil societies for the entire suite of potential earthquakes that can... and will occur! Better to retire and then bury PSHA, which is more *concept* than it is a tested pathway to seismic safety, ... R.I.P. (Molchan et al., 1997; Kagan et al., 2012) than to experience *more* earthquake disasters and catastrophes, where erroneous hazard maps depicted only "low hazard," but the active tectonic regions acted otherwise!

PSHA, *unlike NDSHA*, has: (1) never been validated by "objective testing;" (2) actually been proven *unreliable* (Panza et al., 2014) as a forecasting method on the rates (but claimed probabilities) of earthquake occurrence; and (3) staked its hype and dominance on assumptions that both earthquake resistant design standards and societal earthquake preparedness and planning should be based on "engineering seismic risk analysis" models—models which incorporate assumptions, really fabulations (or "magical realism") now known to conflict with what we have learned scientifically regarding earthquake geology and earthquake physics over the same (almost 50-yr) time frame of PSHA's: (i) initial hype; (ii) acceptance; and (iii) eventual 40-yr rise to dominance. PSHA, because it has too often delivered not only erroneous but also too deadly results (Wyss et al., 2012; Panza et al., 2014; Bela, 2014), has been extensively debated over many years; a sample of contributions is contained in the PAGEOPH Topical Volume 168 (2011) and references therein. In evidence against PSHA: too many damaging and deadly earthquakes (like the 1988 $M = 6.8$ Spitak, Armenia; the 2011 $M = 9$ Tohoku, Japan Megathrust; and the 2012 $M = 6$ Emilia, Italy events) have all occurred in regions rated to be "low-risk" by PSHA derived seismic hazard maps.

It should therefore be widely *taken to heart* that the continued practice of PSHA for determining earthquake resistant design standards for civil protection, mitigation of heritage and existing buildings and lifelines, and community economic well-being and resilience... is in a *state-of-crisis.* And alternative methods, which are already available and ready-to-use, like NDSHA, should be applied worldwide. The results will then be twofold: (1) not only to extensively *test* these alternative methods; but (2) to *prove* that they globally actually perform more reliably and safely than PSHA.

List of Acronyms

C-MCSI	Conditional Maximum Credible Seismic Input
CMS	Conditional Mean Spectrum
CN	California-Nevada algorithm for mid-term middle-range earthquake prediction
CP	Collapse Prevention
DGA	Design Ground Acceleration
DWN	Discrete Wave Number technique
EDEM	Elastic Deformation Energy Maximization
EDP	Engineering Demand Parameter
EMS	European Macroseismic Scale
ES	Extended Source
GMPE	Ground Motion Prediction Equation
IM	Intensity Measure
IO	Immediate Occupancy
LPL	Lower Performance Level
LS	Life Safety
M8S	Algorithm for mid-term middle-range earthquake prediction
MCE	Maximum Credible Earthquake
MCE_R	Risk-Targeted Maximum Considered Earthquake
MCSI	Maximum Credible Seismic Input
$MCSI_{BD}$	Maximum Credible Seismic Input, at bedrock
$MCSI_{SS}$	Maximum Credible Seismic Input, site-specific
MS	Modal Summation
MTHA	Modal Time History Analysis
NDSHA	Neo-Deterministic Seismic Hazard Assessment
NLTHA	Non-linear Time History Analysis
OL	Operational Limit
PBSD	Performance Based Seismic Design
PGA	Peak Ground Acceleration
PGD	Peak Ground Displacement
PGV	Peak Ground Velocity
PL	Performance Level

PO	Performance Objective
PSHA	Probabilistic Seismic Hazard Assessment
R.I.P.	Requiescat In Pace (May he/she rest in peace)
RSA	Regional Scale Analysis
SA	Spectral Acceleration
SHA	Seismic Hazard Assessment
SLSS	Scaling Law for Source Spectra
STSPS	Size- and Time-Scaled Point Sources
SSA	Site Specific Analysis
TPL	Target Performance Level
UHS	Uniform Hazard Spectrum

References

Al Atik, L. and Abrahamson, N. 2010. An improved method for nonstationary spectral matching. Earthquake Spectra 26: 601–617.

Algermissen, S. and Perkins, D. 1976. A Probabilistic Estimate of Maximum Accelerations in Rock in the Contiguous United States. US Geological Survey (Report 76–416).

Ancheta, T.D., Darragh, R.B., Stewart, J.P., Seyhan, E., Silva, W.J., Chiou, B.S.-J., Wooddell, K.E., Graves, R.W., Kottke, A.R., Boore, D.M., Kishida, T. and Donahue, J.L. 2014. NGA-West2 database. Earthquake Spectra 30: 989–1005.

Archuleta, R.J. and Hartzell, S.H. 1981. Effects of fault finiteness on near-source ground motion. Bulletin of the Seismological Society of America 71: 939–957.

ASCE. 2013. Minimum Design Loads for Buildings and Other Structures (ASCE/SEI 7–10). American Society of Civil Engineers, Reston, Virginia.

ATC. 1978. Tentative Provisions for the Development of Seismic Regulations for Buildings, ATC 3-06 (NBS SP-510). Applied Technology Council.

ATC. 2012. Seismic performance assessment of buildings: Volume I—Methodology (FEMA P-58-1). Federal Emergengy Management Agency, Washington, D.C.

Baker, J.W. and Cornell, A.C. 2005. A vector-valued ground motion intensity measure consisting of spectral acceleration and epsilon. Earthquake Engineering & Structural Dynamics 34: 1193–1217.

Baker, J.W. and Cornell, C.A. 2006a. Which spectral acceleration are you using? Earthquake Spectra 22: 293–312.

Baker, J.W. and Cornell, A.C. 2006b. Spectral shape, epsilon and record selection. Earthquake Engineering & Structural Dynamics 35: 1077–1095.

Baker, J.W. 2007. Quantitative classification of near-fault ground motions using wavelet analysis. Bulletin of the Seismological Society of America 97: 1486–1501.

Baker, J.W. and Cornell, C.A. 2008. Vector-valued intensity measures incorporating spectral shape for prediction of structural response. Journal of Earthquake Engineering 12: 534–554.

Baker, J.W. 2011. Conditional mean spectrum: tool for ground motion selection. Journal of Structural Engineering 137: 322–331.

Bazzurro, P. and Luco, N. 2006. Do scaled and spectrum-matched near-source records produce biased nonlinear structural responses? Proceedings, 8th National Conference on Earthquake Engineering, San Francisco, California.

Bela, J. 2014. Too generous to a fault? Is reliable earthquake safety a lost art? Errors in expected human losses due to incorrect seismic hazard estimates. Earth's Future 2: 569–578.

Bergen, K.J., Shaw, J.H., Leon, L.A., Dolan, J.F., Pratt, T.L., Ponti, D.J., Morrow, E., Barrera, W., Rhodes, E.J., Murariand, M.K. and Owen, L.A. 2017. Accelerating slip rates on the Puente Hills blind thrust fault system beneath metropolitan Los Angeles, California, USA. Geology 45(3): 227–230.

Bertero, R.D. and Bertero, V.V. 2002. Performance-based seismic engineering: The need for a reliable conceptual comprehensive approach. Earthquake Engineering and Structural Dynamics 31: 627–652.

Beyer, K. and Bommer, J.J. 2007. Selection and scaling of real accelerograms for bi-directional loading: a review of current practice and code provisions. Journal of Earthquake Engineering 11: 13–45.

Bizzarri, A. and Crupi, P. 2013. Linking the recurrence time of earthquakes to source parameters: a dream or a real possibility? Pure and Applied Geophysics 171: 2537–2553.

Bojórquez, E. and Iervolino, I. 2011. Spectral shape proxies and nonlinear structural response. Soil Dynamics and Earthquake Engineering 31: 996–1008.

Bommer, J.J., Scott, S.G. and Sarma, S.K. 2000. Hazard-consistent earthquake scenarios. Soil Dynamics and Earthquake Engineering 19: 219–231.

Bommer, J.J. and Acevedo, A.B. 2004. The use of real earthquake accelerograms as input to dynamic analysis. Journal of Earthquake Engineering 8: 43–91.

Bommer, J.J. and Pinho, R. 2006. Adapting earthquake actions in Eurocode 8 for performance-based seismic design. Earthquake Engineering and Structural Dynamics 35: 39–55.

Boore, D.M. 1986. The effect of finite bandwidth on seismic scaling relationships. pp. 275–283. *In*: Das, S., Boatwright, J. and Scholz, C. (eds.). Earthquake Source Mechanics. Geophys. Monograph 37, American Geophysical Union, Washington, D.C.

Brandmayr, E., Raykova, R.B., Zuri, M., Romanelli, F., Doglioni, C. and Panza, G.F. 2010. The lithosphere in Italy: structure and seismicity. Journal of the Virtual Explorer 36.

Bray, J.D. and Rodriguez-Marek, A. 2004. Characterization of forward-directivity ground motions in the near-fault region. Soil Dynamics and Earthquake Engineering 24: 815–828.

Bray, J.D., Rodriguez-Marek, A. and Gillie, J.L. 2009. Design ground motions near active faults. Bulletin of the New Zealand Society for Earthquake Engineering 42: 1–8.

Brooks, E.M., Stein, S. and Spencer, B.D. 2017. Investigating the effects of smoothing on the performance of earthquake hazard maps. International Journal of Earthquake and Impact Engineering 2: 121–134.

Bruneau, M., Chang, S.E., Eguchi, R.T., Lee, G.C., O'Rourke, T.D., Reinhorn, A.M., Shinozuka, M., Tierney, K., Wallace, W.A. and von Winterfeldt, D. 2003. A framework to quantitatively assess and enhance the seismic resilience of communities. Earthquake Spectra 19: 733–752.

BSSC. 2009. NEHRP recommended seismic provisions for new buildings and other structures (FEMA P-750). Federal Emergengy Management Agency, Washington, D.C.

BSSC. 2015. NEHRP recommended seismic provisions for new buildings and other structures (FEMA P-1050-2). Federal Emergency Management Agency, Washington, D.C.

Burks, L.S. and Baker, J.W. 2016. A predictive model for fling-step in near-fault ground motions based on recordings and simulations. Soil Dynamics and Earthquake Engineering 80: 119–126.

Båth, M. 1973. Introduction to Seismology. Birkhäuser Verlag, Basel.

Cancani, A. 1904. Sur l'emploi d'une double échelle sismique des intensités, empirique et absolue. Gerlands Beitrage Geophysik 2: 281–283.

Caputo, M., Keilis-Borok, V., Kronrod, T., Molchan, G., Panza, G.F., Piva, A., Podgaezkaya, V. and Postpischl, D. 1973. Models of earthquake occurrence and isoseismals in Italy. Ann. Geofis. 26: 421–444.

Catino, M. 2014. Organizational Myopia. Problems of Rationality and Foresight in Organizations. Cambridge University Press.

CEN. 2004. Eurocode 8: Design of structures for earthquake resistance. Part 1: General rules, seismic actions and rules for building (EC8-1). European Committee for Standardization.

Cowie, P.A., Phillips, R.J., Roberts, G.P., McCaffrey, K., Zijerveld, L.J.J., Gregory, L.C., Faure Walker, J., Wedmore, L.N.J., Dunai, T.J., Binnie, S.A., Freeman, S.P.H.T., Wilcken, K., Shanks, R.P., Huismans, R.S., Papanikolaou, I., Michetti, A.M. and Wilkinson, M. 2017. Orogen-scale uplift in the central Italian Apennines drives episodic behaviour of earthquake faults. Sci. Rep. 7: 44858. doi:10.1038/srep44858.

CPTI Working Group. 2004. Catalogo Parametrico dei Terremoti Italiani, versione 2004 (CPTI04). Bologna, Italy.

CRESME. 2012. Primo Rapporto ANCE/CRESME – Lo stato del territorio italiano 2012 (in Italian).

CSLP. 2008. Italian Building Code (NTC08). Consiglio Superiore dei Lavori Pubblici.D.M. 14 gennaio 2008 – Norme tecniche per le costruzioni (in Italian), Ministero delle Infrastrutture [online] <http://www.gazzettaufficiale.it/eli/id/2008/02/04/08A00368/sg>.

CSLP. 2018. Italian Building Code (NTC18). Consiglio Superiore dei Lavori Pubblici. D.M. 17 gennaio 2018—Aggiornamento delle «Norme tecniche per le costruzioni» (in Italian), Ministero delle Infrastrutture [online] <http://www.gazzettaufficiale.it/eli/id/2018/2/20/18A00716/sg>.

D'Amico, V., Albarello, D. and Mantovani, E. 1999. A distribution-free analysis of magnitude-intensity relationships: an application to the Mediterranean region. Physics and Chemistry of the Earth, Part A: Solid Earth and Geodesy 24: 517–521.

DISS Working Group. 2015. Database of Individual Seismogenic Sources (DISS), Version 3.2.0: A compilation of potential sources for earthquakes larger than M 5.5 in Italy and surrounding areas.

Doglioni, C. 2016. Plate tectonics, earthquakes and seismic hazard. *In*: Atti dei Convegni Lincei 306. Accademia Nazionale dei Lincei. Rome, Italy.

Dolan, J.F., McAuliffe, L.J., Rhodes, E.J., McGill, S.F. and Zinke, R. 2016. Extreme multi-millennial slip rate variations on the Garlock fault, California: Strain super-cycles, potentially time-variable fault strength, and implications for system-level earthquake occurrence. Earth and Planetary Science Letters 446: 123–136.

Dominique, P. and Andre, E. 2000. Probabilistic seismic hazard map on the french national territory. Contribution n. 0632. *In*: Proceedings of the 12th World Conference on Earthquake Engineering, 2000.

Fäh, D. and Panza, G.F. 1994. Realistic modelling of observed seismic motion in complex sedimentary basins. Annals of Geophysics 37: 1771–1797.

Fasan, M., Amadio, C., Noè, S., Panza, G.F., Magrin, A., Romanelli, F. and Vaccari, F. 2015. A new design strategy based on a deterministic definition of the seismic input to overcome the limits of design procedures based on probabilistic approaches. *In*: Convegno ANIDIS 2015. L'Aquila, Italy.

Fasan, M., Magrin, A., Amadio, C., Romanelli, F., Vaccari, F. and Panza, G.F. 2016. A seismological and engineering perspective on the 2016 Central Italy earthquakes. International Journal of Earthquake and Impact Engineering 1: 395–420.

Fasan, M. 2017. Advanced Seismological and Engineering Analysis for Structural Seismic Design. University of Trieste, Italy.

Fasan, M., Magrin, A., Amadio, C., Panza, G.F., Romanelli, F. and Vaccari, F. 2017. A possible revision of the current seismic design process. *In*: World Conference on Earthquake Engineering. Santiago, Chile.

FIB. 2012. Probabilistic performance-based seismic design. Lausanne, Switzerland.

Florsch, N., Fäh, D., Suhadolc, P. and Panza, G.F. 1991. Complete synthetic seismograms for high-frequency multimode SH-waves. Pure and Applied Geophysics 136: 529–560.

Gasparini, D. and Vanmarke, E.H. 1976. Simulated earthquake motions compatible with prescribed response spectra. Cambridge, Massachusetts.

Gelfand, I.M., Guberman, S.I., Izvekova, M.L., Keilis-Borok, V.I. and Ranzman, E.J. 1972. Criteria of high seismicity, determined by pattern recognition. Tectonophysics 13: 415–422.

Geller, R.J., Mulargia, F. and Stark, P.B. 2015. Why we need a new paradigm of earthquake occurrence. pp. 183–191. *In*: Morra, G. et al. (eds.). Subduction Dynamics: From Mantle Flow to Mega Disasters, Geophysical Monograph 211. American Geophysical Union, Washington, DC, USA.

Gorshkov, A., Panza, G.F., Soloviev, A.A. and Aoudia, A. 2002. Morphostructural Zonation and preliminary recognition of seismogenic nodes around the Adria margin in Peninsular Italy and Sicily. Journal of Seismology and Earthquake Engineering 4: 1–24.

Gorshkov, A.I., Panza, G.F., Soloviev, A. and Aoudia, A. 2004. Identification of seismogenic nodes in the Alps and Dinarides. Italian Journal of Geoscience 123: 3–18.

Gorshkov, A.I., Panza, G.F., Soloviev, A.A., Aoudia, A. and Peresan, A. 2009. Delineation of the geometry of nodes in the Alps-Dinarides hinge zone and recognition of seismogenic nodes (M ≥ 6). Terra Nova 21: 257–264.

Grant, D.N. and Diaferia, R. 2013. Assessing adequacy of spectrum-matched ground motions for response history analysis. Earthquake Engineering & Structural Dynamics: 1265–1280.

Graves, R.W. and Pitarka, A. 2010. Broadband ground-motion simulation using a hybrid approach. Bulletin of the Seismological Society of America 100: 2095–2123.

Graves, R., Jordan, T.H., Callaghan, S., Deelman, E., Field, E., Juve, G., Kesselman, C., Maechling, P., Mehta, G., Milner, K., Okaya, D., Small, P. and Vahi, K. 2011. CyberShake: a physics-based seismic hazard model for Southern California. Pure and Applied Geophysics 168: 367–381.

Grigoriu, M. 2011. To scale or not to scale seismic ground-acceleration records. Journal of Engineering Mechanics 137: 284–293.

Gusev, A.A. 1983. Descriptive statistical model of earthquake source radiation and its application to an estimation of short-period strong motion. Journal of International Geophysics 74: 787–808.

Gusev, A.A. 2011. Broadband kinematic stochastic simulation of an earthquake source: a refined procedure for application in seismic hazard studies. Pure and Applied Geophysics 168: 155–200.

Hancock, J., Bommer, J.J. and Stafford, P.J. 2008. Numbers of scaled and matched accelerograms required for inelastic dynamic analyses. Earthquake Engineering and Structural Dynamics 37(14): 1585–1607.

Haselton, C.B. and Baker, J.W. 2006. Ground motion intensity measures for collapse capacity prediction: Choice of optimal spectral period and effect of spectral shape. pp. 1–10. In: 8th National Conference on Earthquake Engineering. San Francisco, California.

Haselton, C.B., Baker, J.W., Stewart, J.P., Whittaker, A.S., Luco, N., Fry, A., Hamburger, R.O., Zimmerman, R.B., Hooper, J.D., Charney, F.A. and Pekelnicky, R.G. 2017. Response history analysis for the design of new buildings in the NEHRP provisions and ASCE/SEI 7 Standard: Part I—Overview and specification of ground motions. Earthquake Spectra 33: 373–395.

Huang, Y.-N., Whittaker, A.S. and Luco, N. 2008. Maximum spectral demands in the near-fault region. Earthquake Spectra 24: 319–341.

Iervolino, I. and C.A. Cornell. 2005. Record selection for nonlinear seismic analysis of structures. Earthquake Spectra 21: 685–713.

Iervolino, I., De Luca, F. and Cosenza, E. 2010. Spectral shape-based assessment of SDOF nonlinear response to real, adjusted and artificial accelerograms. Engineering Structures 32: 2776–2792.

Iuchi, K., Johnson, L.A. and Olshansky, R.B. 2013. Securing Tohoku's future: planning for rebuilding in the first year following the Tohoku-Oki earthquake and tsunami. Earthquake Spectra 29: S479–S499.

Kagan, Y.Y., Jackson, D.D. and Geller, R.J. 2012. Characteristic earthquake model, 1884–2011, R.I.P. Seismol. Res. Lett. 83: 951–953.

Kaiser, A., Holden, C., Beavan, J., Beetham, D., Benites, R., Celentano, A., Collett, D., Cousins, J., Cubrinovski, M., Dellow, G., Denys, P., Fielding, E., Fry, B., Gerstenberger, M., Langridge, R., Massey, C., Motagh, M., Pondard, N., McVerry, G., Ristau, J., Stirling, M., Thomas, J., Uma, S. and Zhao, J. 2012. The Mw 6.2 Christchurch earthquake of February 2011: preliminary report. New Zealand Journal of Geology and Geophysics 55: 67–90.

Kalkan, E. and Kunnath, S.K. 2006. Effects of fling step and forward directivity on seismic response of buildings. Earthquake Spectra 22: 367–390.

Kalkan, E. and Kwong, N.S. 2013. Pros and cons of rotating ground motion records to fault-normal/parallel directions for response history analysis of buildings. Journal of Structural Engineering 140(3): 04013062.

Kanamori, H. 1977. The energy release in great earthquakes. Journal of Geophysical Research 82: 2981–2987.

Katsanos, E.I., Sextos, A.G. and Manolis, G.D. 2010. Selection of earthquake ground motion records: A state-of-the-art review from a structural engineering perspective. Soil Dynamics and Earthquake Engineering 30: 157–169.

Keilis-Borok, V.I. and Soloviev, A.A. 2003. Nonlinear dynamics of the lithosphere and earthquake prediction. Keilis-Borok, V.I. and Soloviev, A.A. (eds.). Springer, Berlin Heidelberg, Berlin, Germany.

Klügel, J.U. 2007a. Error inflation in probabilistic seismic hazard anaysis. Engineering Geology 90(3): 186–192.

Klügel, J.U. 2007b. Comment on "Why do modern probabilistic seismic-hazard analyses often lead to increased hazard estimates" by Bommer, J.J. and Abrahamson, N.A. Bulletin of the Seismological Society of America 97: 2198–2207.

Kossobokov, V.G. and Nekrasova, A.K. 2012. Global seismic hazard assessment program maps are erroneous. Seismic Instruments 48: 162–170.

Junbo, J. 2018. Soil Dynamics and Foundation Modeling. Springer International Publishing.

Lliboutry, L. 2000. Quantitative Geophysics and Geology. Springer-Verlag London, London, UK.

Luco, N. and Bazzurro, P. 2007. Does amplitude scaling of ground motion records result in biased nonlinear structural drift responses? Earthquake Engineering and Structural Dynamics 36(13): 1813–1835.

Luzi, L., Puglia, R., Russo, E. and ORFEUS WG5. 2016. Engineering Strong Motion Database, version 1.0. http://esm.mi.ingv.it.

Magrin, A. 2013. Multi-Scale Seismic Hazard Scenarios. PhD Thesis. University of Trieste, Italy.

Magrin, A., Gusev, A.A., Romanelli, F., Vaccari, F. and Panza, G.F. 2016. Broadband NDSHA computations and earthquake ground motion observations for the Italian territory. International Journal of Earthquake and Impact Engineering 1(4): 395–420.

Magrin, A., Peresan, A., Kronrod, T., Vaccari, F. and Panza, G.F. 2017. Neo-deterministic seismic hazard assessment and earthquake occurrence rate. Engineering Geology 229: 95–109.

Markušić, S., Suhadolc, P., Herak, M. and Vaccari, F. 2000. A contribution to seismic hazard assessment in Croatia from deterministic modeling. pp. 185–204. *In*: Seismic Hazard of the Circum-Pannonian Region. Birkhäuser Basel, Basel.

Martin, S.S. and Szeliga, W. 2010. A catalog of felt intensity data for 570 earthquakes in India from 1636 to 2009. Bull. Seismol. Soc. Am. 100: 6 2–569.

Meletti, C., Galadini, F., Valensise, G., Stucchi, M., Basili, R., Barba, S., Vannucci, G. and Boschi, E. 2008. A seismic source zone model for the seismic hazard assessment of the Italian territory. Tectonophysics 450: 85–108.

Miranda, E. and Aslani, H. 2003. Probabilistic Response Assessment for Building-Specific Loss Estimation. PEER report 2003/03.berkeley, California.

Molchan, G., Kronrod, T. and Panza, G.F. 1997. Multi-scale seismicity model for seismic risk. Bulletin of the Seismological Society of America 87: 1220–1229.

Molchan, G., Kronrod, T. and Panza, G.F. 2011. Hot/cold spots in Italian macroseismic data. Pure and Applied Geophysics 168(3-4): 739–752.

Mollaioli, F., Decanini, L., Bruno, S. and Panza, G.F. 2003. Analysis of the response behaviour of structures subjected to damaging pulse-type ground motions. pp. 109–119. *In*: NEA/CSNI/R 18.OECD/NEAWorkshop on the Relations between Seismological DATA and Seismic Engineering, Istanbul, 16–18 October 2002.

Morgenroth, J. and Armstrong, T. 2012. The impact of significant earthquakes on Christchurch, New Zealand's urban forest. Urban Forestry & Urban Greening 11: 383–389.

Mulargia F., Stark, P.B. and Geller, R.J. 2017. Why is probabilistic seismic hazard analysis (PSHA) still used? Physics of the Earth and Planetary Interiors 264: 63–75.

Nakamura, Y. 1989. A method for dynamic characteristics estimation of subsurface using microtremor on the ground surface. Railway Technical Research Institute (RTRI) 30: 25–33.

Nekrasova, A., Kossobokov, V., Peresan, A. and Magrin, A. 2014. The comparison of the NDSHA, PSHA seismic hazard maps and real seismicity for the Italian territory. Hazards 70 Natural: 629–641.

Nekrasova, A., Kossobokov, V.G., Parvez, I.A. and Tao, X. 2015. Seismic hazard and risk assessment based on the unified scaling law for earthquakes. Acta Geodaetica et Geophysica 50: 21–37.

NIST. 2011. Selecting and Scaling Earthquake Ground Motions for Performing Response-History Analyses. Washington, D.C.

PAGEOPH Topical Volume 168. 2011. Advanced Seismic Hazard Assessment, Vol. 1 and Vol. 2, Editors: Panza, G.F., Irikura, K., Kouteva-Guentcheva, M., Peresan, A., Wang, Z. and Saragoni, R., Pure Appl. Geophys., Birkhäuser, Basel, Switzerland.

Panza, G.F., Schwab, F.A. and Knopoff, L. 1972. Channel and crustal Rayleigh waves. Geophys. J. R. Astr. Soc. 30: 273–280.

Panza, G.F. 1985. Synthetic seismograms: the Rayleigh modal summation technique. Journal of Geophysics 58: 125–145.

Panza, G.F., Romanelli, F. and Vaccari, F. 2001. Seismic wave propagation in laterally heterogeneous anelastic media: Theory and applications to seismic zonation. Advances in Geophysics 43: 1–95.

Panza, G.F., La Mura, C., Peresan, A., Romanelli, F. and Vaccari, F. 2012. Seismic hazard scenarios as preventive tools for a disaster resilient society. Advances in Geophysics 53: 93–165.

Panza, G.F., Kossobovok, V., Peresan, A. and Nekrasova, A. 2014. Why are the standard probabilistic methods of estimating seismic hazard and risks too often wrong. pp. 309–357. *In*: Wyss, M. (ed.). Earthquake Hazard, Risk, and Disasters. Academic Press.

Panza, G.F., Peresan, A., Sansò, F., Crespi, M., Mazzoni, A. and Nascetti, A. 2017. How geodesy can contribute to the understanding and prediction of earthquakes. Rendiconti Lincei. DOI:10.1007/s12210-017-0626-y.

Panza, G.F. and Nunziata, C. 2018. Ground shaking. pp. 149. *In*: Bobrowsky, P.T. and Marker, B. (eds.). Encyclopedia of Engineering Geology. Springer International Publishing.

Parvez, I.A., Magrin, A., Vaccari, F., Ashish, Mir, R.R., Peresan, A. and Panza, G.F. 2017. Neo-deterministic seismic hazard scenarios for India—a preventive tool for disaster mitigation. J. Seismol. doi:10.1007/s10950-017-9682-0.

Pavlov, V.M. 2009. Matrix impedance in the problem of the calculation of synthetic seismograms for a layered-homogeneous isotropic elastic medium. Izvestiya, Physics of the Solid Earth 45: 850–860.

Peresan, A., Kossobokov, V.G., Romashkova, L. and Panza, G.F. 2005. Intermediate-term middle-range earthquake predictions in Italy: a review. Earth-Science Reviews 69: 97–132.

Peresan, A., Zuccolo, E., Vaccari, F. and Panza, G.F. 2009. Neo-deterministic seismic hazard scenarios for North-Eastern Italy. Italian Journal of Geoscience 128: 229–238.

Peresan A., Zuccolo, E., Vaccari, F., Gorshkov, A. and Panza, G.F. 2011. Neo-deterministic seismic hazard and pattern recognition techniques: time dependent scenarios for North-Eastern Italy. Pure and Applied Geophysics 168(3-4): 583–607.

Peresan, A. and Panza, G.F. 2012. Improving earthquake hazard assessment in Italy: An alternative to "Texas sharpshooting". EOS Transaction, American Geophysical Union. 93: 51, 18 December 2012.

Peresan, A., Magrin, A., Nekrasova, A., Kossobokov, V.G. and Panza, G.F. 2013. Earthquake recurrence and seismic hazard assessment: a comparative analysis over the Italian territory. *In*: ERES 2013, Transactions of Wessex Institute, http://dx.doi.org/10.2495/ERES130031.

Peresan, A., Magrin, A., Vaccari, F., Romanelli, F. and Panza, G.F. 2014. Neo-deterministic seismic hazard assessment: an operational scenario-based approach from national to local scale. *In*: Proceedings of the 5th National Conference on Earthquake Engineering and 1st National Conference on Earthquake Engineering and Seismology. Bucharest, Romania, June 19–20, 2014.

Peresan, A., Gorshkov, A., Soloviev, A. and Panza, G.F. 2015. The contribution of pattern recognition of seismic and morphostructural data to seismic hazard assessment. Boll. Geofis. Teor. Appl. 56(2): 295–328.

Peresan, A., Kossobokov, V., Romashkova, L., Magrin, A., Soloviev, A. and Panza, G.F. 2016. Time-dependent neo-deterministic seismic hazard scenarios: preliminary report on the M6.2 Central Italy earthquake, 24th August 2016. N. Concepts Glob. Tectonics J. 4(3): 487–493.

Peresan, A. 2018. Recent developments in the detection of seismicity patterns for the Italian region. pp. 149–172. *In*: Ouzounov, D., Pulinets, S., Hattori, K., Taylor, P. (eds.). Pre-Earthquake Processes: A Multi-disciplinary Approach to Earthquake Prediction Studies. Chapter 9, Volume 234, AGU Geophysical Monograph Series. Wiley and Sons. DOI: 10.1002/9781119156949.ch9.

Reyes, J.C., Riaño, A.C., Kalkan, E., Quintero, O.A. and Arango, C.M. 2014. Assessment of spectrum matching procedure for nonlinear analysis of symmetric- and asymmetric-plan buildings. Engineering Structures 72: 171–181.

Rugarli, P. 2008. Zone Griglie o…Stanze? Ingegneria Sismica. 1-2008.

Rugarli, P. 2014. Validazione Strutturale. EPC Libri, Rome, Italy.

Rugarli, P. 2015a. The role of the standards in the invention of the truth. Acta, Convegno La Resilienza delle Città d'Arte ai Terremoti, 3–4 Novembre 2015, Accademia Nazionale deiLincei, Rome.

Rugarli, P. 2015b. Primum: non nocere. Acta, Convegno La Resilienza delle Città d'Arte aiTerremoti, 3–4 Novembre 2015, Accademia Nazionale dei Lincei, Rome.

Schwab, P. and Lestuzzi, P. 2007. Assessment of the seismic non-linear behavior of ductile wall structures due to synthetic earthquakes. Bulletin of Earthquake Engineering 5: 67–84.

SEAOC. 1995. Vision 2000: Performance Based Seismic Engineering of Buildings. Structural Engineers Association of California, Sacramento, California.

Shahi, S.K. and Baker, J.W. 2011. An empirically calibrated framework for including the effects of near-fault directivity in probabilistic seismic hazard analysis. Bulletin of the Seismological Society of America 101: 742–755.

Stark, P.B. and Freedman, D.A. 2001. What is the chance of an earthquake? pp. 201–213. *In*: Mulargia, F. and Geller, R.J. (eds.). Earthquake Science and Seismic Risk Reduction, NATO Science Series IV: Earth and Environmental Sciences, v. 32, Kluwer, Dordrecht, The Netherlands.

Stein, S., Geller, R. and Liu, M. 2012. Bad assumptions or bad luck: why earthquake hazard maps need objective testing. Seismol. Res. Lett. 82: 623–626.

Stockmeyer, J.M., Shaw, J.H., Brown, N.D., Rhodes, E.J., Richardson, P.W., Wang, M., Lavin, L.C. and Guan, S. 2017. Active thrust sheet deformation over multiple rupture cycles: A quantitative basis for relating terrace folds to fault slip rates. G.S.A. Bulletin 129(9-10): 1337–1356.

Taleb, N. 2007. The Black Swan: The Impact of the Highly Improbable. Random House, New York.

Theophilou, A.I., Chryssanthopoulos, M.K. and Kappos, A.J. 2017. A vector-valued ground motion intensity measure incorporating normalized spectral area. Bulletin of Earthquake Engineering 15: 249–270.

Trifunac, M.D. 2012. Earthquake response spectra for performance based design—A critical review. Soil Dynamics and Earthquake Engineering 37: 73–83.

Vaccari, F. 2016. A web application prototype for the multiscale modelling of seismic input. pp. 563–584. *In*: Earthquakes and Their Impact on Society. Springer International Publishing, Cham.

Wang, Z. 2010. Seismic hazard assessment: issues and alternatives. Pure and Applied Geophysics 168: 11–25.

Watson-Lamprey, J.A. and Boore, D.M. 2007. Beyond SaGMRotI: conversion to SaArb, SaSN, and SaMaxRot. Bulletin of the Seismological Society of America 97: 1511–1524.

Wells, D.L. and Coppersmith, K.J. 1994. New empirical relationships among magnitude, rupture length, rupture width, rupture area, and surface displacement. Bulletin of the Seismological Society of America 84: 974–1002.

Wyss, M., Nekrasova, A. and Kossobokov, V. 2012. Errors in expected human losses due to incorrect seismic hazard estimates. Nat. Haz. 62: 927–935.

Wu, J., Li, N., Hallegatte, S., Shi, P., Hu, A. and Liu, X. 2012. Regional indirect economic impact evaluation of the 2008 Wenchuan Earthquake. Environmental Earth Sciences 65: 161–172.

Zimmerman, R.B., Baker, J.W., Hooper, J.D., Bono, S., Haselton, C.B., Engel, A., Hamburger, R.O., Celikbas, A. and Jalalian, A. 2017. Response history analysis for the design of new buildings in the NEHRP provisions and ASCE/SEI 7 standard: Part III—example applications illustrating the recommended methodology. Earthquake Spectra 33(2): 397–417.

Zinke, R., Dolan, J.F., Rhodes, E.J., Van Dissen, R. and McGuire, C.P. 2017. Highly variable latest Pleistocene-Holocene incremental slip rates on the Awatere fault at Saxton River, South Island, New Zealand, revealed by lidar mapping and luminescence dating. Geophysical Research Letters 44: 11301–11310.

Živčić, M., Suhadolc, P. and Vaccari, F. 2000. Seismic zoning of Slovenia based on deterministic hazard computations. pp. 171–184. *In*: Seismic Hazard of the Circum-Pannonian Region. Birkhäuser Basel, Basel.

Zuccolo, E., Vaccari, F., Peresan, A. and Panza, G.F. 2011. Neo-deterministic and probabilistic seismic hazard assessments: a comparison over the Italian territory. Pure and Applied Geophysics 168: 69–83.

7

Ground Motion Prediction Equations for Energy-Based Demand Parameters

Regional Application to Northwestern Turkey

Ali Sari[1],* and *Lance Manuel*[2]

7.1 Introduction

The severity of earthquake ground shaking is commonly characterized using intensity measures such as peak ground acceleration and spectral acceleration, which are force-based or strength-based parameters. Modern seismic provisions adopt such strength-based design procedures, where seismic demand is represented in the form of an elastic response spectrum. Design procedures, then, implicitly account for the ductility capacity that a structure might possess by the use of reduction factors, but they do not include in a direct way consideration of the cyclic nature of the response that the structure undergoes and the resulting cumulative damage. In contrast with such strength-based parameters, energy-based parameters may be just as easily defined but they can usefully include the effects of both the amplitude as well as the number of cycles of the response oscillations experienced by a structure. Such energy-based parameters are expected to correlate equally well or better with structural damage than the conventional strength-based parameters (Sari, 2003).

[1] Faculty of Civil Engineering, Istanbul Technical University, Istanbul, Turkey.
[2] Department of Civil, Architectural, and Environmental Engineering, University of Texas at Austin, Austin, TX 78712, USA.
    Email: lmanuel@mail.utexas.edu
* Corresponding author: asari@itu.edu.tr

Of interest here is the development of energy-based ground motion parameter prediction equations where a random effects model is employed that takes into consideration the correlation of data recorded by a single event (Abrahamson and Youngs, 1992). As a result of an increase in the database on strong motion data for Northwestern Turkey after the 1999 earthquakes and because there exists currently only a very limited number of ground motion prediction equations for any in the world for energy-based parameters, we seek to develop new region-specific energy-based ground motion parameter prediction relationships for Northwestern Turkey. A database consisting of 195 recordings from 17 seismic events is employed for this purpose. This database is the same as that employed by Özbey et al. (2004).

The empirical relationships developed are for the geometric mean of the two horizontal components of 5% damped energy parameters that will be defined in the following sections. For elastic energy demand as well as for different inelastic energy demand levels (associated with a range of ductility values), results are reported. For a subset of the cases studied that are thought to be of most interest, period-dependent model coefficients based upon a random effects model are presented in tabular form to facilitate their use.

The effect of magnitude, soil class, and distance on the predicted energy spectra is studied. The effect of ductility level is also studied. Model predictions for the energy parameters are studied in detail and compared with observed data for the different soil classes. The proposed energy-based ground motion prediction relationships are also compared with the few available Western U.S. models for energy-based parameters. There is virtually no previous work published on amplification of energy-based parameters due to site conditions except a brief study carried out by Chou and Uang (2000). A similar brief examination of the empirical prediction of short- and long-period motions based on energy considerations for different site/soil classes permits us to present soil amplification factors for energy-based ground motion parameters here.

7.2 Model Parameters and Strong Motion Database

The model parameters included to establish the empirical ground motion prediction models in this study are that same as those employed by Özbey et al. (2004). These parameters include earthquake moment magnitude, M_w; the Joyner-Boore source-to-site distance, R (a measure of the closest horizontal distance to the vertical projection of the rupture plane); and local site conditions (as specified by different site classes described in the following). It should be noted that style of faulting is not explicitly included as a model parameter. The empirical relationships developed are for normal and strike-slip earthquakes and should not be used for reverse and reverse-oblique events since the fault mechanisms for earthquakes on the North Anatolian Fault System (including most of the Kocaeli aftershocks) are predominantly strike-slip in character with normal faulting observed in some cases (Orgulu and Aktar, 2001).

A database made up of 195 ground motion records from 17 earthquakes (main shocks and aftershocks with magnitude greater than 5.0) is employed here. These records comprise a subset of a larger database of events of various magnitudes that were obtained from stations operated by Boğaziçi University's Kandilli Observatory and Earthquake Research Institute (KOERI), by Istanbul Technical University (ITU), and by the General Directorate of Disaster Affairs' Earthquake Research Department (ERD).

Table 7.1 includes a summary of the records included in the selected strong motion database (which is the same as that used by Sari (2003) and Özbey et al. (2004)) where the distribution in site classes A, B, C, and D (based on the average shear wave velocity over the top 30 meters), as defined in Table 7.2, is also shown.

The distribution of the strong motion data as a function of moment magnitude and distance is shown graphically according to site class in Fig. 7.1. Note that site classes A and B have been included together in this figure.

7.3 Background on Energy-based Ground Motion Parameters

Ground motions may be described quantitatively in different terms when one is interested in understanding their effect on structures with different natural period and damping values. Conventionally, strength-based parameters (such as spectral acceleration (S_a) or velocity (S_v) or even peak ground acceleration) have been used for this purpose. Regression-based models describing the variation in these parameters as a function of magnitude and distance (for specified site conditions and sometimes by faulting type) have been developed for various regions of the world. Not as extensively studied are energy-based ground motion parameters and models describing their variation with magnitude, distance, etc.

While energy-based measures of ground shaking have been proposed in many studies over the years [see, for example, Akiyama (1985)], it is only in a few recent studies that such studies shaking have gained interest. Since there are several different energy descriptors that have been used, it is important to define the energy-based parameters that are employed here. In order to do so, it is useful to start with the equation of motion of a single-degree-of-freedom structural (SDOF) system. This may be written as:

$$m\ddot{u}_t + c\dot{u} + f_s = 0 \tag{7.1}$$

where m, c, and f_s are the mass, viscous damping coefficient, and restoring force, respectively, of the SDOF system. Also, u_t is the absolute (total) displacement of the mass, while $u = u_t - u_g$ is the relative displacement of the mass with respect to ground, and u_g is the ground displacement.

Transformation of the equation of motion into an energy balance equation can be easily accomplished by integrating Eq. (7.1) with respect to u from the beginning of the input ground shaking (see, for example, Uang and Bertero, 1988) up to any time of interest, t. This leads to

$$\frac{m\ddot{u}_t^2}{2} + \int_0^t c\dot{u}du + \int_0^t f_s du = \int_0^t m\ddot{u}_t du_g \tag{7.2}$$

Table 7.1. Database of strong motion records used in the regression analyses.

Event No.	Event Name	Event Date	Origin Time	Lat. (deg.)	Long. (deg.)	M	H (km)	No. of Recordings			
								A	B	C	D
1	Izmit	17.08.1999	12:01:38 AM	40.76	29.97	7.4	19.6	3	5	7	7
2	Düzce-Bolu	12.11.1999	4:57:21 PM	40.74	31.21	7.2	25.0	1	3	5	18
3	Izmit	13.09.1999	11:55:29 AM	40.77	30.10	5.8	19.6	0	2	5	18
4	Hendek-Akyazi	23.08.2000	1:41:28 PM	40.68	30.71	5.8	15.3	0	1	3	8
5	Sapanca-Adapazari	11.11.1999	2:41:26 PM	40.74	30.27	5.7	22.0	0	1	4	11
6	Izmit	17.08.1999	3:14:01 AM	40.64	30.65	5.5	15.3	0	0	0	3
7	Düzce-Bolu	12.11.1999	5:18:00 PM	40.74	31.05	5.4	10.0	0	1	1	12
8	Izmit	31.08.1999	8:10:51 AM	40.75	29.92	5.2	17.7	0	1	3	13
9	Düzce-Bolu	12.11.1999	5:17:00 PM	40.75	31.10	5.2	10.0	0	2	1	11
10	Marmara Sea	20.09.1999	9:28:00 PM	40.69	27.58	5.0	16.4	0	1	4	10
11	Northeast of Bolu	14.02.2000	6:56:36 AM	40.90	31.75	5.0	15.7	0	0	0	5
12	Cinarcik-Yalova	19.08.1999	3:17:45 PM	40.59	29.08	5.0	11.5	0	0	1	5
13	Kaynasli-Bolu	12.11.1999	6:14:00 PM	40.75	31.36	5.0	10.0	0	0	0	1
14	Hendek-Adapazari	07.11.1999	4:54:42 PM	40.71	30.70	5.0	10.0	0	0	0	4
15	Izmit	19.08.1999	3:17:45 PM	40.36	29.56	5.0	9.8	0	1	1	2
16	Düzce-Bolu	19.11.1999	7:59:08 PM	40.78	30.97	5.0	9.2	0	2	0	3
17	Hendek-Adapazari	22.08.1999	2:30:59 PM	40.74	30.68	5.0	5.4	0	0	0	5
Total Number of Records:								4	20	35	136

Table 7.2. Definition of site classes used in the empirical model development.

Site class	Shear wave velocity
A	> 750 m/s
B	360 m/s to 750 m/s
C	180 m/s to 360 m/s
D	< 180 m/s

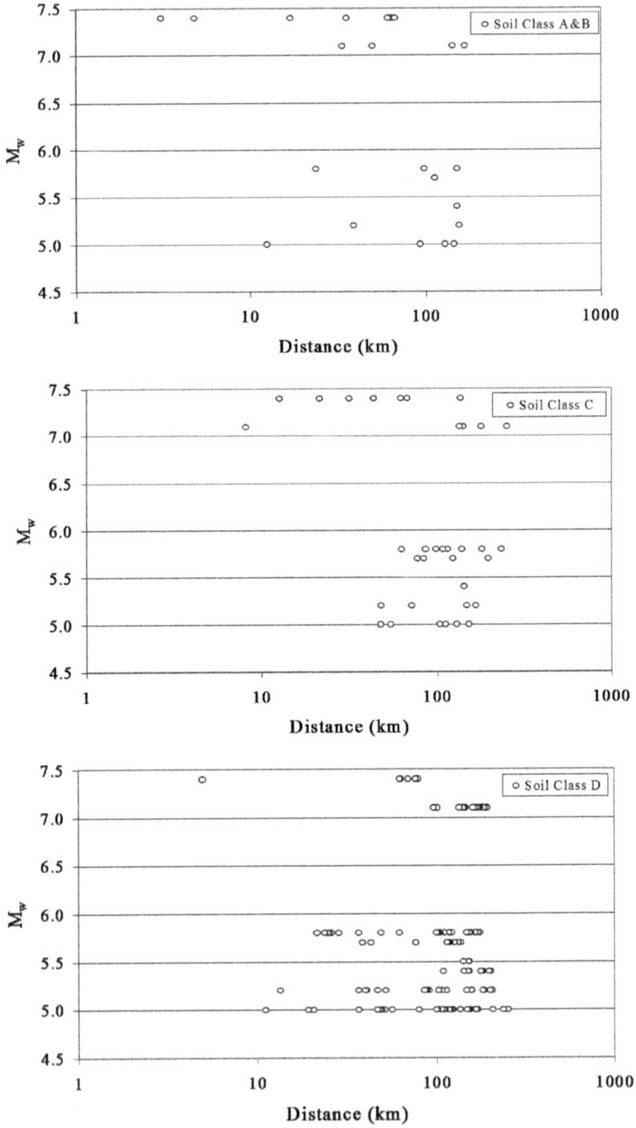

Figure 7.1. Data dstribution for soil classes (A&B), C and D.

Since the inertia force, $m\ddot{u}_t$, equals (in magnitude) the sum of the damping and restoring forces, it is also equal to the total force applied at the base of the structure. Therefore, the right-hand side of Eq. (7.2) is, by definition, the energy input to the system at any time, t. Hereinafter, following Uang and Bertero (1988), we will define Input Energy, E_i, as the *maximum* value of the energy input into the system during ground shaking. It can also be thought of as the maximum value of the work done by the total base shear on foundation/ground displacement during the ground motion. Thus, we have:

$$E_i = \max\left\{\int_0^t m\ddot{u}_t\, du_g\right\} \tag{7.3}$$

The first term on the left hand side of Eq. (7.2) is the *kinetic energy*, $E_k(t)$, while the second term is the *damping energy*, $E_d(t)$, and the last term is made up of the sum of recoverable *elastic strain energy*, $E_s(t)$, and unrecoverable *hysteretic energy*, $E_h(t)$. Thus, the input energy, E_i, can also be described as follows:

$$E_i = \max\left\{\frac{m\ddot{u}_t^2}{2} + \int_0^t c\dot{u}\,du + \int_0^t f_s\,du\right\} = \max\left\{E_k(t) + E_d(t) + \left[E_s(t) + E_h(t)\right]\right\} \tag{7.4}$$

To facilitate the reporting of various results in this study and comparisons with other studies, it is useful to define (as is done by Uang and Bertero (1988)), Absorbed Energy, E_a, as the maximum value of the sum of the recoverable strain energy and the unrecoverable hysteretic energy as follows:

$$E_a = \max\left\{\int_0^t f_s\,du\right\} = \max\left\{E_s(t) + E_h(t)\right\} \tag{7.5}$$

It is convenient to define a velocity parameter (V_i) based on input energy which we will term, "input energy-equivalent velocity"

$$V_i = \sqrt{\frac{2E_i}{m}} \tag{7.6}$$

Similarly, we define another velocity parameter (V_a) based on absorbed energy which we will term, "absorbed energy-equivalent velocity":

$$V_a = \sqrt{\frac{2E_a}{m}} = \sqrt{\frac{2(E_s + E_h)}{m}} \tag{7.7}$$

These definitions in Eqs. (7.6) and (7.7) are similar to those employed by Akiyama (1985) and Uang and Bertero (1988).

To facilitate comparison with studies that involve the more conventional design parameter, spectral acceleration, S_a, we also define an "input energy-equivalent acceleration," A_i, in terms of V_i and an "absorbed energy-equivalent acceleration," A_a, in terms of V_a, as follows:

$$A_i = \omega V_i ; \quad A_a = \omega V_a \tag{7.8}$$

Of these various parameters, input energy-equivalent velocity, V_i, was used as a ground motion parameter in studies for the Western U.S. by Lawson (1996) and Chapman (1999); the latter study also included the use of elastic input energy-based parameters in probabilistic seismic hazard analyses. Also, inelastic absorbed energy-equivalent velocity was employed as an energy demand parameter in a study by Chou and Uang (2000).

It is important to note that for linear elastic systems, the absorbed energy-equivalent velocity, V_a, is the same as spectral velocity, S_v, and the absorbed energy-equivalent acceleration, A_a, is the same as spectral acceleration, S_a. Thus, in a sense, elastic absorbed energy-equivalent velocity and acceleration represent parameters that may be defined either from strength or energy considerations.

7.4 Regression Model

A nonlinear mixed effects model is employed in the development of the empirical ground motion prediction relationships in this study. Such a model can account for both inter-event and intra-event variability. Most commonly developed models do not distinguish between these two types of variability. The mixed effects model used here has been discussed extensively by Özbey et al. (2004) and by Sari (2003). Such a model describes the relationship between a response variable, the ground motion parameter, and some covariates in the data that are grouped according to one or more classification (for example, magnitude).

The error associated with residuals (differences) between predicted and observed values of the selected ground motion parameter is comprised of an inter-event term (that represents "between-group" variability resulting from differences in the data recorded from different earthquakes) and an intra-event term (that represents "within-group" variability resulting from differences in the data recorded among the different stations for the same earthquake).

The selected functional form for the ground motion prediction relationships that are developed for elastic and in elastic energy-based ground motion parameters using the mixed effects model is as follows:

$$\log(Y_{ij}) = a + b(M_i - 6) + c(M_i - 6)^2 + d\log\sqrt{R_{ij}^2 + h^2} + eG_1 + fG_2 \tag{7.9}$$

where Y_{ij} is the geometric mean of the two horizontal components of the energy-based parameter (e.g., A_i or A_a in units of cm/s²) from the jth recording of the ith event, M_i is the moment magnitude of the ith event, and R_{ij} is the closest horizontal distance to the vertical projection of the rupture from the ith event to the location of the jth recording.

The coefficients G_1 and G_2 take on values as follows: $G_1 = 0$ and $G_2 = 0$ for site classes A and B; $G_1 = 1$ and $G_2 = 0$ for site class C; and $G_1 = 0$, $G_2 = 1$ for site class D. The coefficients to be estimated are, hence, a, b, c, d, e, f, and h.

7.5 Regression Results

To facilitate comparisons with spectral acceleration, regression results for the energy parameters are presented and discussed in terms of absorbed energy-equivalent acceleration (A_a) and input energy-equivalent acceleration (A_i) defined based on Eq. (7.8). Our interest is in predictions of both elastic and inelastic measures of energy. However, it has been reported that absorbed energy is a better measure of damage to structures than input energy (see Chou and Uang (2000)); accordingly, when we are interested in inelastic demand, we will focus on absorbed energy-equivalent acceleration, A_a. For elastic energy demand, on the other hand, we will study input energy-equivalent acceleration, A_i. It is not necessary to consider elastic A_a because it is exactly the same as spectral acceleration, S_a [empirical relationships for which have already been developed by Özbey et al. (2004)]. To summarize then, inelastic A_a and elastic A_i prediction relationships are discussed in the following.

Note that the evaluation of absorbed energy-equivalent acceleration (A_a) and input energy-equivalent acceleration (A_i) is limited to SDOF systems with 5% damping. Also, whenever inelastic structural response is of interest in the following, bilinear force-deformation characteristics with a 0% strain hardening ratio (or, equivalently, elastic perfectly-plastic behavior) are considered. The degree of strain hardening has been found to have insignificant influence on absorbed energy demands as has been reported by Chou and Uang (2000) and Seneviratna and Krawinkler (1997). This was also discussed by Sari and Manuel (2002). Hence, no other strain hardening ratios are considered.

7.5.1 Absorbed Energy

Based on the mixed effects model, Table 7.3 presents the model coefficients, a, b, c, d, e, f, and h, and the logarithmic standard error for 5%-damped inelastic absorbed energy-equivalent acceleration (A_a) for periods up to 4 seconds and for a ductility factor of 4. (A ductility factor of 4 implies that the SDOF system experiences a maximum displacement that is 4 times the yield displacement.)

Figure 7.2 shows absorbed energy-equivalent acceleration spectra based on model predictions for four different ductility levels for soil classes A and B when $M_w = 7.5$ and $R = 10$ km. Insignificant differences in absorbed energy-equivalent acceleration (A_a) are observed for ductility levels in the inelastic range from 2 to 8, except at very short periods. Recognizing this lack of sensitivity of absorbed energy demands to ductility levels, in the following while discussing inelastic A_a variation, only a ductility factor of 4 is studied.

The effect of magnitude on predicted inelastic absorbed energy-equivalent acceleration spectra is studied in Fig. 7.3 for soil classes A and B for $R = 20$ km, and for a ductility factor of 4. A systematic decrease in amplitude of the spectra with decreasing magnitude is observed at all frequencies as expected. Figure 7.4 shows the effect of soil class on inelastic absorbed energy-equivalent acceleration spectra based on the proposed model predictions for $M_w = 7.5$ and $R = 10$ km and

Table 7.3. Empirical ground motion prediction model coefficients and logarithmic standard deviation for the 5%-damped absorbed energy-equivalent acceleration, A_a (in cm/s^2) for ductility = 4.

Period(s)	a	b	c	D	h	e	f	$\sigma_{log(Y)}$
0.10	3.742	0.510	−0.085	−1.0526	13.01	0.133	0.284	0.246
0.15	3.643	0.523	−0.083	−0.9856	12.66	0.112	0.278	0.236
0.20	3.517	0.556	−0.106	−0.9176	11.92	0.078	0.304	0.228
0.25	3.320	0.582	−0.111	−0.8343	8.80	0.071	0.325	0.232
0.30	3.219	0.602	−0.116	−0.8039	8.71	0.077	0.349	0.235
0.35	3.132	0.616	−0.119	−0.7789	8.15	0.076	0.360	0.239
0.40	3.071	0.628	−0.126	−0.7681	8.26	0.078	0.377	0.245
0.45	2.996	0.639	−0.127	−0.7457	7.74	0.073	0.383	0.251
0.50	2.943	0.648	−0.129	−0.7372	7.47	0.074	0.393	0.255
0.55	2.887	0.655	−0.128	−0.7257	7.51	0.074	0.400	0.258
0.60	2.839	0.660	−0.128	−0.7178	7.57	0.076	0.404	0.258
0.65	2.810	0.665	−0.130	−0.7180	7.75	0.077	0.408	0.261
0.70	2.782	0.669	−0.131	−0.7187	7.67	0.079	0.411	0.262
0.75	2.761	0.675	−0.132	−0.7194	7.80	0.081	0.414	0.263
0.80	2.740	0.677	−0.134	−0.7212	7.85	0.083	0.415	0.264
0.85	2.715	0.683	−0.136	−0.7194	8.00	0.085	0.420	0.266
0.90	2.693	0.687	−0.137	−0.7198	8.05	0.088	0.425	0.267
0.95	2.672	0.690	−0.139	−0.7202	8.12	0.089	0.427	0.268
1.00	2.660	0.693	−0.142	−0.7231	8.30	0.090	0.429	0.268
1.10	2.630	0.696	−0.145	−0.7266	8.41	0.090	0.431	0.267
1.20	2.605	0.700	−0.148	−0.7309	8.46	0.091	0.433	0.267
1.30	2.588	0.703	−0.151	−0.7373	8.61	0.091	0.433	0.267
1.40	2.566	0.706	−0.153	−0.7403	8.70	0.091	0.433	0.268
1.50	2.548	0.708	−0.154	−0.7452	8.88	0.092	0.434	0.268
1.75	2.512	0.713	−0.159	−0.7574	9.11	0.092	0.436	0.268
2.00	2.469	0.715	−0.161	−0.7629	9.25	0.091	0.436	0.268
2.25	2.431	0.717	−0.161	−0.7670	9.33	0.085	0.432	0.267
2.75	2.356	0.720	−0.158	−0.7697	9.57	0.077	0.425	0.266
3.00	2.326	0.722	−0.157	−0.7725	9.53	0.075	0.423	0.266
3.50	2.275	0.726	−0.156	−0.7792	9.31	0.075	0.421	0.265
4.00	2.232	0.728	−0.154	−0.7846	9.44	0.072	0.417	0.265

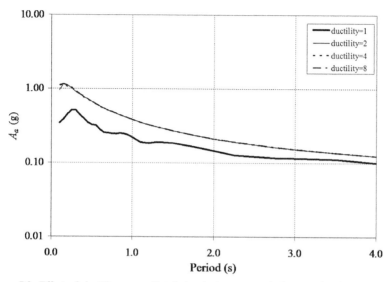

Figure 7.2. Effect of ductility on predicted absorbed energy-equivalent acceleration spectra for $M_w = 7.5$, $R = 10$ km, and for soil classes A and B.

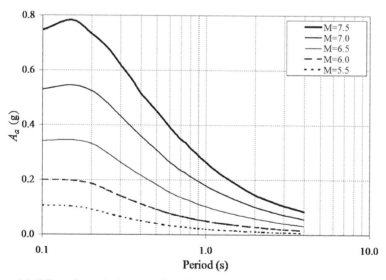

Figure 7.3. Effect of magnitude on predicted inelastic absorbed energy-equivalent acceleration spectra for soil classes A and B, $R = 20$ km, and ductility factor = 4.

for a ductility factor of 4. For all soil classes, it may be seen that the inelastic (ductility = 4) A_a has its peak value at a period lower than 0.25 seconds, which is somewhat shorter than the period of 0.3 seconds at which the elastic A_a (or spectral acceleration) has its own peak value as can be seen in Özbey et al. (2004). The effect of distance on predicted inelastic absorbed energy-equivalent acceleration spectra is studied in Fig. 7.5 for soil classes A and B for $M_w = 7.5$ and for a ductility factor of 4. An expected trend of reduced ground motion attenuation with distance is seen at longer periods.

In Fig. 7.6, model predictions of inelastic (ductility = 4) 1.0-second absorbed energy-equivalent acceleration (A_a) for mean and plus/minus one standard deviation levels are compared with data from the Kocaeli earthquake. To highlight the differences in the predicted motions for each soil class, the comparisons for soil classes (A&B), C, and D are shown separately. Reasonable fits of the model to the Kocaeli data are seen in Fig. 7.6.

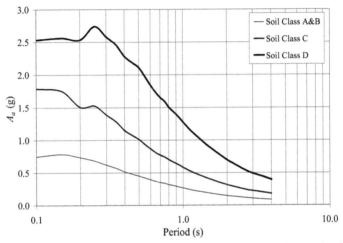

Figure 7.4. Effect of soil class on predicted inelastic absorbed energy-equivalent acceleration spectra for $M_w = 7.5$, $R = 10$ km, and ductility factor = 4.

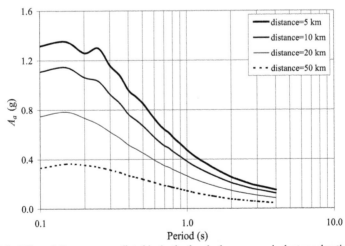

Figure 7.5. Effect of distance on predicted inelastic absorbed energy-equivalent acceleration spectra for soil classes A and B and for $M_w = 7.5$ and ductility factor = 4.

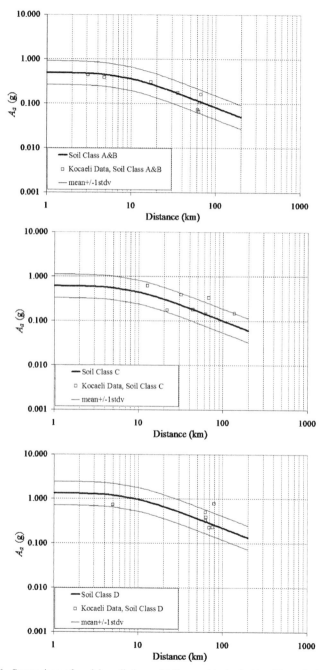

Figure 7.6. Comparison of model predictions of 1.0-second inelastic (ductility = 4) A_a at mean and mean ± 1 std. dev. levels with observed data from the Kocaeli earthquake for soil classes (A&B), C, and D.

7.5.2 Input Energy

Based on the mixed effects model, Table 7.4 presents the model coefficients, a, b, c, d, e, f, and h, and the logarithmic standard error for 5%-damped elastic input energy-equivalent acceleration (A_i) for periods up to 4 seconds.

Even though we do not study inelastic input energy demand in any detail in the following (for reasons already stated), we briefly examine the effect of ductility

Table 7.4. Empirical ground motion prediction model coefficients and logarithmic standard deviation for the 5%-damped elastic input energy-equivalent acceleration, A_i (in cm/s²).

Period(s)	a	b	c	D	H	e	f	$\sigma_{\log(Y)}$
0.10	3.877	0.614	−0.084	−1.0480	9.95	0.115	0.267	0.250
0.15	4.042	0.588	−0.096	−1.1460	16.61	0.171	0.271	0.243
0.20	3.745	0.582	−0.085	−1.0063	13.87	0.094	0.273	0.227
0.25	3.497	0.595	−0.090	−0.8887	10.21	0.039	0.278	0.231
0.30	3.376	0.602	−0.102	−0.8421	7.99	0.035	0.288	0.237
0.35	3.278	0.623	−0.124	−0.8092	6.88	0.033	0.323	0.240
0.40	3.127	0.632	−0.115	−0.7716	5.99	0.055	0.367	0.260
0.45	3.028	0.649	−0.133	−0.7335	5.43	0.062	0.388	0.268
0.50	2.943	0.687	−0.168	−0.6906	4.59	0.036	0.390	0.286
0.55	2.844	0.718	−0.186	−0.6534	3.56	0.023	0.397	0.301
0.60	2.706	0.730	−0.184	−0.6037	2.80	0.022	0.396	0.305
0.65	2.609	0.737	−0.179	−0.5784	2.30	0.024	0.405	0.300
0.70	2.563	0.744	−0.180	−0.5711	2.09	0.020	0.405	0.299
0.75	2.554	0.769	−0.187	−0.5909	2.62	0.029	0.411	0.305
0.80	2.548	0.775	−0.185	−0.6072	3.08	0.045	0.407	0.308
0.85	2.565	0.800	−0.184	−0.6374	4.71	0.066	0.416	0.318
0.90	2.586	0.813	−0.190	−0.6627	6.33	0.071	0.425	0.326
0.95	2.583	0.828	−0.203	−0.6696	6.96	0.071	0.431	0.334
1.00	2.592	0.842	−0.221	−0.6807	6.69	0.069	0.433	0.334
1.10	2.585	0.857	−0.247	−0.6914	6.23	0.071	0.435	0.335
1.20	2.590	0.873	−0.273	−0.7117	6.13	0.086	0.444	0.336
1.30	2.626	0.893	−0.285	−0.7452	6.58	0.070	0.416	0.332
1.40	2.665	0.917	−0.294	−0.7821	7.94	0.050	0.404	0.339
1.50	2.670	0.933	−0.297	−0.7992	9.23	0.032	0.376	0.350
1.75	2.746	0.952	−0.320	−0.8832	10.04	0.031	0.343	0.351
2.00	2.710	0.956	−0.322	−0.9258	9.43	0.050	0.350	0.337
2.25	2.782	0.962	−0.320	−1.0119	12.42	0.049	0.355	0.329
2.75	2.663	0.956	−0.268	−1.0467	12.89	0.032	0.331	0.314
3.00	2.623	0.968	−0.260	−1.0540	13.01	0.034	0.324	0.316
3.50	2.545	0.972	−0.256	−1.0674	10.36	0.023	0.311	0.318
4.00	2.438	0.986	−0.244	−1.0699	10.41	0.014	0.315	0.309

level on input energy-equivalent acceleration (A_i) spectra first. Figure 7.7 shows these A_i spectra based on model predictions for four different ductility levels for soil classes A and B when $M_w = 7.5$ and $R = 10$ km. Again, as was the case when studying A_a, insignificant differences in input energy-equivalent acceleration (A_i) are observed for all ductility levels ranging from 1 (elastic) to 8.

The effect of magnitude on predicted elastic input energy-equivalent acceleration spectra is studied in Fig. 7.8 for soil classes A and B for $R = 20$ km. A systematic decrease in amplitude of the spectra with decrease in magnitude is

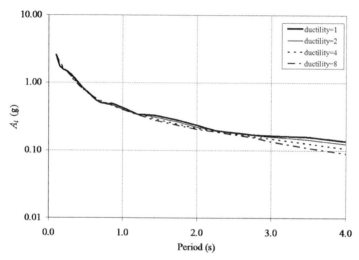

Figure 7.7. Effect of ductility on predicted input energy-equivalent acceleration spectra for $M_w = 7.5$, $R = 10$ km, and for soil classes A and B.

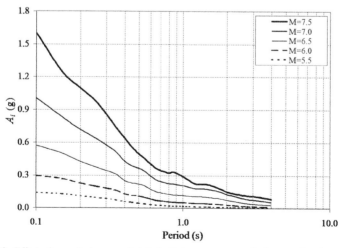

Figure 7.8. Effect of magnitude on predicted elastic input energy-equivalent acceleration spectra for soil classes A and B and for $R = 20$ km.

observed at all frequencies as expected. The shapes of the spectra are such that peak values of elastic A_i occur at the shortest periods unlike spectra for spectral acceleration which have peaks at intermediate periods generally. Figure 7.9 shows the effect of soil class on elastic input energy-equivalent acceleration spectra based on the proposed model predictions for $M_w = 7.5$ and $R = 10$ km. Again, it may be seen that the elastic A_i is largest at very short periods for all soil classes. The effect of distance on predicted elastic input energy-equivalent acceleration spectra is studied in Fig. 10 for soil classes A and B for $M_w = 7.5$. The expected trend of reduced ground motion attenuation with distance is seen at longer periods as was also observed for A_a spectra.

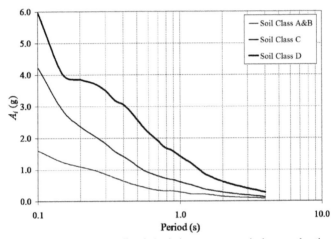

Figure 7.9. Effect of soil class on predicted elastic input energy-equivalent acceleration spectra for $M_w = 7.5$ and $R = 10$ km.

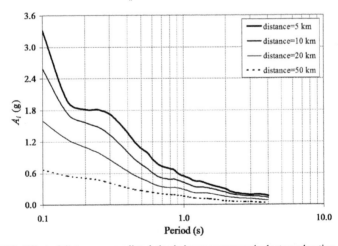

Figure 7.10. Effect of distance on predicted elastic input energy-equivalent acceleration spectra for soil classes A and B and for $M_w = 7.5$.

In Fig. 7.11, model predictions of elastic 1.0-second input energy-equivalent acceleration (A_i), for mean and plus/minus one standard deviation levels are compared with data from the Kocaeli earthquake for soil classes (A&B), C, and D.

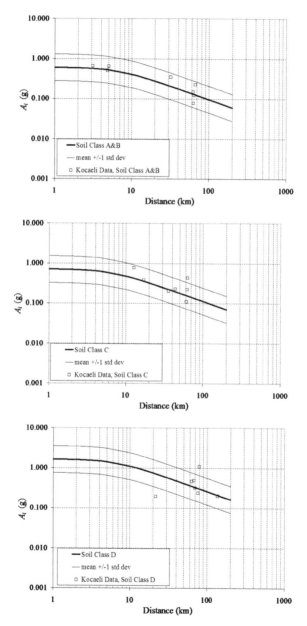

Figure 7.11. Comparison of model predictions of 1.0-second elastic A_i at mean and mean ± 1 std. dev. levels with observed data from the Kocaeli earthquake for soil classes (A&B), C, and D.

Reasonable fits of the model to the Kocaeli data are seen in Fig. 7.11. The elastic A_i levels for the various soil classes are comparable with each other (highest levels are seen for soil class D) and somewhat higher than the levels of inelastic A_a (for a ductility of 4) that were seen in Fig. 7.6.

7.6 Discussion

We discuss below three different issues next related to the variation of energy-based parameters studied. First, we compare the variation of the energy-based acceleration parameters, A_a and A_i, discussed in the preceding with that of spectral acceleration [from Özbey et al. (2004)] based on the same ground motion database. Second, we discuss how predictions of the variation of energy-based parameters according to the empirical relationships developed here for Northwestern Turkey compare with similar relations developed using a Western U.S. strong motion database. Finally, we discuss how the empirical relations developed may be employed to gain an understanding of the amplification of energy demands based on site conditions.

7.6.1 Comparison of the Predicted Variation of Strength- and Energy-Based Parameters

Predicted spectral acceleration levels for Northwestern Turkey, based on the study byÖzbey et al. (2004), and levels of absorbed energy-equivalent acceleration, A_a (for a ductility of 4) as well as elastic input energy-equivalent acceleration (A_i), based on the proposed model are compared in Figs. 7.12 and 7.13 for natural periods of

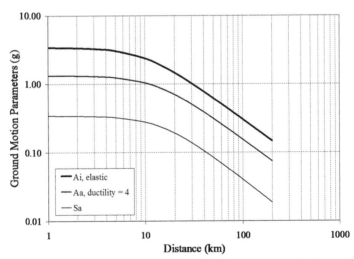

Figure 7.12. Comparison of 0.1-second elastic input energy-equivalent acceleration and inelastic (ductility = 4) absorbed energy-equivalent acceleration predictions from the proposed models with 0.1-second spectral acceleration predictions for soil classes A and B and for $M_w = 7.4$.

0.1 and 1.0 second, respectively. These comparisons are for an event with moment magnitude, M_w, equal to 7.4 and for soil classes A and B. The attenuation of motions with distance is shown in the figures.

It may be observed that the ratio of A_a (for a ductility of 4) to S_a is higher at short periods (see Fig. 7.12 for the 0.1 second period) compared to intermediate or long periods (see Fig. 7.13). When the response is elastic (ductility equal to 1), A_a is the same as S_a and the A_a/S_a ratio is unity; with increasing ductility this ratio increases. For a ductility factor is 4, the A_a/S_a ratio is about 3.3 at a distance of 10 km for a 0.1 second natural period, while the ratio is only about 1.5 at the same distance when the natural period is increased to 1 second. It is clear that predicted A_a levels should increase with ductility because hysteretic energy (E_h) contributes a relatively greater proportion to the total absorbed energy as ductility levels increase (whereas in the elastic case, E_h is equal to 0). Predicted A_i levels are higher than both A_a and S_a at the two periods, 0.1 and 1.0 seconds. However, the predicted levels for A_a and A_i are more similar to each other at the 1.0 second period than at the 0.1 second period.

When comparing attenuation of the various parameters as a function of distance, the A_a/S_a ratio is expected to increase with distance according to Chou and Uang (2000). However, in the present study, this ratio is seen in Figs. 7.12 and 7.13 to be fairly uniform at all distances greater than around 10 km. Predicted A_i levels are higher than both A_a and S_a at all distances but the differences between A_a and A_i decrease at longer periods.

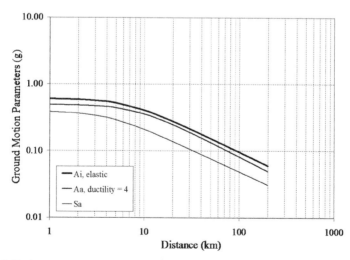

Figure 7.13. Comparison of 1.0-second elastic input energy-equivalent acceleration and inelastic (ductility = 4) absorbed energy-equivalent acceleration predictions from the proposed models with 1.0-second spectral acceleration predictions for soil classes A and B and for $M_w = 7.4$.

7.6.2 Comparison of the Proposed Model with Western U.S. Models

7.6.2.1 Absorbed Energy

Chou and Uang (2000) established empirical relationships for absorbed energy-equivalent velocity (V_a) for the Western U.S. using a two-stage regression analysis procedure. The geometric mean of the two horizontal components of each ground motion recording was used in the study by Chou and Uang (2000) and the functional form of their predictive model was similar to the one used here. A comparison of the proposed model predictions for a ductility factor of 4 with those of Chou and Uang (2000) is presented in Figs. 7.14 and 7.15 for an event with moment magnitude, M_w,

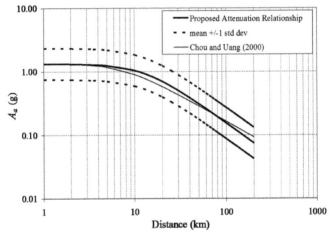

Figure 7.14. Comparison of predictions from the proposed model with those from a Western U.S. model for 0.1-second inelastic A_a (ductility = 4) for soil classes A and B and for M_w = 7.4.

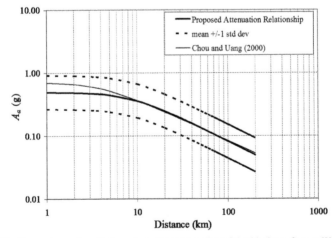

Figure 7.15. Comparison of predictions from the proposed model with those from a Western U.S. model for 1.0-second inelastic A_a (ductility = 4) for soil classes A and B and for M_w = 7.4.

of 7.4 and for soil classes A and B. The proposed model predicts generally similar levels of absorbed energy-equivalent acceleration (A_a) as the model by Chou and Uang (2000) at both periods, 0.1 and 1.0 seconds, as is seen in the figures. At short distances (say $R < 10$ km), the model by Chou and Uang (2000) predicts slightly higher 1.0-second A_a mean levels compared to the proposed model. In general, though, the Western U.S. model predictions are within one standard deviation of the mean predictions from the proposed model.

   In Fig. 7.16, predicted A_a spectra based on the proposed model are compared with the Western U.S. model of Chou and Uang (2000) at distances of 20, 60, and 150 km. Predictions of the mean absorbed energy-equivalent acceleration (A_a) spectra from the two models are very similar deviating only very slightly at longer distances than around 60 km. Again, though, the Western U.S. model predictions of mean A_a spectra lie well within one standard deviation of the predicted mean spectra based on the proposed model.

7.6.2.2 Input Energy

Lawson (1996) established empirical relationships for input energy-equivalent velocity (V_i) for the Western U.S. using a similar regression model to that used here. That study was based on 126 ground motion records. The V_i value used there was based on the larger of the two horizontal components rather than the geometric mean of the two components that is used here. Chapman (1999) also proposed empirical relationships for V_i for the Western U.S. based on 303 ground motion records. The geometric mean of two horizontal components was used in that study. A comparison of the proposed model predictions for elastic A_i with those of Lawson (1996) and Chapman (1999) is presented in Figs. 7.17 and 7.18 for an event with moment magnitude, M_w, of 7.4 and for soil classes A and B. For the 1.0-second predictions, both of the models based on Western U.S. data generally predict somewhat higher levels of input energy-equivalent acceleration, A_i, at all distances compared with the proposed model. At the shorter period (0.1 seconds), the Western U.S. models predict higher A_i levels only over very large distances.

   In Fig. 7.19, predicted A_i spectra based on the proposed model are compared with the two Western U.S. models of Lawson (1996) and Chapman (1999) at distances equal to 20, 60, and 150 km. Predictions of the mean input energy-equivalent acceleration (A_i) spectra from the three models are generally quite similar with largest deviations at the 150 km distance. The Western U.S. model predictions of mean A_i spectra lie well within one standard deviation of the predicted mean spectra based on the proposed model, except at the 150 km distance where especially at short periods, the Western U.S. model predictions are almost one standard deviation above the mean spectra from the proposed model.

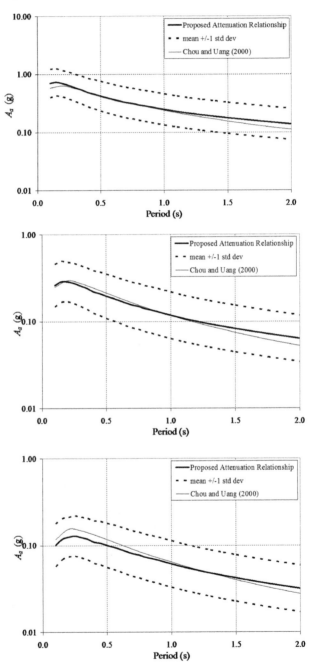

Figure 7.16. Comparison of predicted inelastic A_a spectra (ductility = 4) based on the proposed model with those based on a Western U.S. model for distances (20, 60, and 150 km) for soil classes A and B and for $M_w = 7.4$.

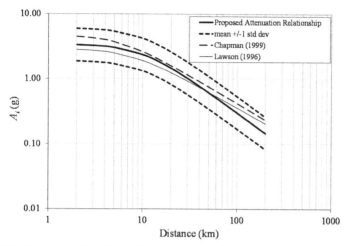

Figure 7.17. Comparison of predictions from the proposed model with those from two Western U.S. models for 0.1-second elastic A_i for soil classes A and B and for $M_w = 7.4$.

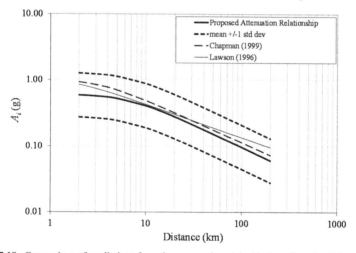

Figure 7.18. Comparison of predictions from the proposed model with those from two Western U.S. models for 1-second elastic A_i for soil classes A and B and for $M_w = 7.4$.

7.6.3 Amplification of Energy-based Parameters Due to Site Conditions

There are almost no previous studies except that of Chou and Uang (2000) that address amplification of energy-based parameters due to site/soil conditions. Here, two factors, F_a' and F_v', for short-period and mid-period amplification of energy-based acceleration are defined in a manner quite similar to the amplification factors, F_a and F_v, employed in the NEHRP Provisions (NEHRP, 2009) for amplification of spectral acceleration for different site classes (relative to a reference site class). We consider amplification factors for A_a for site classes C and D relative to site classes

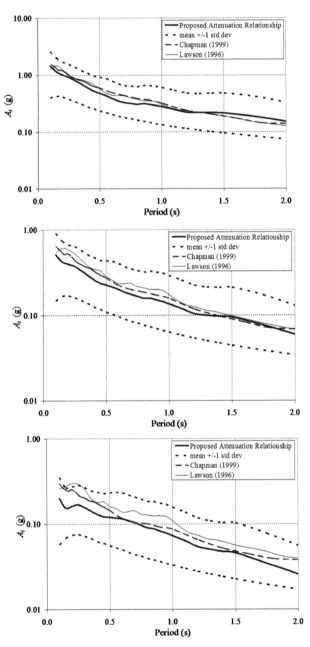

Figure 7.19. Comparison of predicted elastic A_i spectra based on the proposed model with those based on two Western U.S. models for distances (20, 60, and 150 km) for soil classes A and B and for $M_w = 7.4$.

A and B taken together as the reference site class. The amplification factor, F_a', is defined at a period of 0.2 seconds while F_v' is defined at a period of 1.0 second. Based on predicted levels of the selected energy-based acceleration parameter for the different site classes, these two soil amplification factors are computed as follows:

$$F'_a = \frac{A_{a,soil} \ at \ T = 0.2\,\text{sec}}{A_{a,(A+B)} \ at \ T = 0.2\,\text{sec}}, \ F'_v = \frac{A_{a,soil} \ at \ T = 1.0 \ \text{sec}}{A_{a,(A+B)} \ at \ T = 1.0 \ \text{sec}} \tag{7.10}$$

where the numerator in the two factors refers to predicted level of A_a (for a ductility of 4) for the site class (C or D) under consideration. Similarly, amplification factors for the elastic input energy-equivalent acceleration (A_i) were defined in a similar manner to that given by Eq. (7.9).

Table 7.5 shows amplification factors for the two energy-based parameters for site classes C and D at periods of 0.2 and 1.0 seconds. It should be noted that the soil classification used in this study is not the same as that used in the NEHRP (2009) Provisions. The soil classification used here is as given in Table 7.2 where the site class A here is similar to NEHRP's Rock Sites, A and B.

Despite the difference in site class definitions in the present study, similar patterns related to the amplification are observed as with spectral acceleration in NEHRP (2009). Greater amplification levels for A_a and A_i are generally noted at the longer periods and for the softer soil classes.

The establishment of amplification factors in this manner makes it relatively easy to account for site effects in studies involving energy-based parameters (such as in energy-based probabilistic seismic hazard analysis). However, additional studies are recommended on amplification of energy demands that consider various levels of input motion and examine other natural periods a well.

Table 7.5. Amplification factors, F_a' and F_v', for energy-based parameters for site classes C and D at periods of 0.2 and 1.0 seconds, respectively.

Site Class	A_a Amplification factors		A_i Amplification factors	
	F_a'	F_v'	F_a'	F_v'
C	1.2	1.3	1.3	1.2
D	2.0	2.7	2.1	2.8

7.7 Conclusion

The severity of earthquake ground shaking can be characterized using energy-based intensity measures such as absorbed energy and input energy. Such energy-based parameters can serve as alternatives to more conventional strength-based parameters such as spectral acceleration. New ground motion prediction model relationships have been developed for 5%-damped absorbed energy-equivalent acceleration (A_a) and input energy-equivalent acceleration (A_i) for periods up to 4 seconds. A mixed effects model has been employed here to establish these relationships that accounts for inter- and intra-event variability. Coefficients for use with the proposed models are presented for inelastic A_a predictions for a ductility factor of 4 and for elastic A_i predictions.

This study is region-specific as only recordings from earthquakes that have occurred in Northwestern Turkey have been employed in developing the predictive relationships. Comparing predictions of energy-based acceleration parameters for Northwestern Turkey with a strength-based parameter (spectral acceleration, S_a), it was found that the energy demand parameters were generally greater with elastic A_i demands highest. Also, inelastic A_a demands were higher than S_a in part because inelastic absorbed energy includes contributions from hysteretic energy. In contrast with the findings of Chou and Uang (2000), the A_a/S_a ratio was not found to increase with distance in the present study; this ratio was almost the same at all distances.

When the proposed model predictions were compared with those based on Western U.S. empirical models for shallow crustal zones not very different from that for Northwestern Turkey, it was found that the Western U.S. models generally predicted similar energy demands to those with the proposed model. The proposed model predicts generally similar levels of inelastic absorbed energy-equivalent acceleration (A_a) to the Western U.S. models at both periods, 0.1 and 1.0 seconds. The Western U.S. models generally predicted slightly higher levels of input energy-equivalent acceleration (A_i) compared to the proposed model.

For the two energy-based parameters studied, amplification factors, F_a' and F_v', were defined and proposed for site classes C and D to account for amplification of motions due to site conditions. These factors, F_a' and F_v', correspond to soil amplification factors at 0.2 and 1.0 second periods, respectively. As expected, greater amplification levels for A_a and A_i were generally noted at the longer periods and for the softer soil class (site class D). Both the energy-based acceleration parameters, A_a and A_i, were found to experience similar levels of soil amplification at the two periods studied.

Acknowledgements

The authors gratefully acknowledge the financial assistance provided by a grant awarded by the National Science Foundation.

References

Abrahamson, N.A. and Youngs, R.R. 1992. A stable algorithm for regression analyses using the random effect on the use of the elastic input energy for seismic hazard analysis. Earthquake Spectra 15: 607–637.

Chou, C.C. and Uang, C.M. 2000. An Evaluation of Seismic Energy Demand: An Attenuation Approach Pacific Earthquake Engineering Research Center Report 2000/04, College of Engineering University of California, Berkeley, USA.

Lawson, R.S. 1996. Site-dependent Inelastic Seismic Demands. Ph.D. Dissertation, Stanford University, Stanford, USA.

NEHRP. 2009. NEHRP Recommended Seismic Provisions for New Buildings and Other Structures, Federal Emergency Management Agency, FEMA P-750.

Orgulu, G. and Aktar, M. 2001. Regional moment tensor inversion for strong aftershocks of the august 17, 1999 Izmit earthquake (M_w = 7.4). Geophysical Research Letters 28(2): 371–374.

Özbey, C., Sari, A., Manuel, L., Erdik, M. and Fahjan, Y. 2004. An empirical attenuation relationship for Northwestern Turkey ground motion using a random effects approach. Soil Dynamics and Earthquake Engineering 24(2): 115–12.

Sari, A. 2003. Energy Considerations in Ground Motion Attenuation and Probabilistic Seismic Hazard Studies Site. Ph.D. Dissertation, University of Texas, Austin, TX.

Sari, A. and Manuel, L. 2002. Strength and Energy Demands from the August 1999 Kocaeli earthquake ground motions, Paper 661, proceedings of the 7th U.S. National Conference on Earthquake Engineering, Boston, USA.

Seneviratna, G.D.P.K. and Krawinkler, H. 1997. Evaluation of Inelastic MDOF Effects for Seismic Design. Department of Civil Engineering, Stanford University, USA.

Uang, C.M. and Bertero, V.V. 1988. Use of Energy as a Design Criterion in Earthquake-Resistant Design, Report No: UCB/EERC-88/18, Earthquake Engineering Research Center. College of Engineering, University of California at Berkeley.

8

Nonlinear Seismic Analysis of Framed Structures

Stelios Antoniou[1],* and *Rui Pinho*[2]

8.1 Introduction

From the onset of structural analysis, analysts have employed linear elastic methods, implicitly assuming small deformations and limited damage of the structural members and an approximately elastic performance of all the structural components. Even in today's engineering practice, elastic methods are still vastly employed for the design of new structures. This is reasonable, considering the fact that in a new structure, engineers are able to choose the strength and stiffness characteristics of the structural components so as to have a reasonable distribution of inelasticity along the different structural members, without large concentrations of inelastic deformations at particular, more vulnerable locations of the building. This, together with careful ductility detailing of the structural members (e.g., closely spaced stirrups in RC members, or diagonal reinforcement, where needed) provides an efficient, and reasonably accurate framework for the design of new structures with a high level of reliability.

However, buildings that have been designed and constructed before the introduction of earthquake resistance Codes still comprise a large percentage of the total building stock. These structures have been designed mainly for gravity loads,

---

[1] Seismosoft Ltd., 21 Perikleous Stavrou str., 34132 Chalkida, Greece.
[2] Seismosoft Ltd., Piazza Castello 19, 27100 Pavia, Italy.
  Email: rui.pinho@seismosoft.com
* Corresponding author: s.antoniou@seismosoft.com

without special considerations to withstand seismic actions in a manner similar to today's practice. As a result, very frequently they exhibit irregular arrangement of their structural members, with uneven distribution in plan or elevation of the strength, stiffness and mass, which adversely affects their behaviour under earthquake loading (e.g., soft ground storeys, short columns, coupling beams between large shear walls, indirect supports on beams, etc.).

Because of this, the use of elastic procedures for the analysis of existing buildings may lead to non-negligible inaccuracies in the estimation of both the force and the deformation demand of the structural components. What is more, in the majority of cases this approximation leads to the underestimation of the displacement demand on those members where inelastic deformations are concentrated, and which are thus the most vulnerable under seismic loading.

8.2 Current Use of Nonlinear Structural Analysis

As stated above, still nowadays the design of buildings under seismic loading is performed mainly assuming linear elastic analysis, despite the fact that it is generally acknowledged that it significantly lacks accuracy and can lead to non-negligible underestimation of the force and deformation demand, with respect to its inelastic counterpart. Linear analysis requires significantly less computational effort and resources than inelastic analysis. Until relatively recently (mid-90s) the computational power of computers was not sufficiently large to allow for the general introduction and use of inelastic analysis for structural design and assessment. Further, all the design methodologies that are being employed today have been developed several years ago, when the use of nonlinear procedures was not widespread, due to the lack of computational resources, but also the lack of the necessary corresponding knowledge and experience within the engineering community.

All the same, true structural behaviour is inevitably and inherently nonlinear, characterized by non-proportional variation of displacements with loading, particularly in the presence of large deformation demands and/or material nonlinearities. This is particularly true in the seismic response case, since most structures are not designed to respond elastically when subjected to the design earthquake, because of economic constraints and the uncertainties in predicting seismic demands, and are thus expected to instead experience significant inelastic deformations under large events. Hence, all analyses should be treated as potentially nonlinear, implying the use of an incremental-iterative solution procedure whereby loads should be applied in pre-defined increments, equilibrated through an iterative procedure within the context of nonlinear solution procedures. The objective is to take advantage of ductility and post-elastic strength to meet the established performance criteria with a minimum cost.

Such analysis methods allow for the estimation of the structural response beyond the elastic range, including strength and stiffness deterioration associated with inelastic material behaviour and large displacements, thus providing a highly efficient and accurate framework, and allowing for the determination of the realistic structural behaviour, with less simplifying assumptions and more direct (high-level) design criteria.

Supplemented by recent advancements in computing technologies, as well as the increase in the available experimental data for model calibration, the use of nonlinear structural analysis is becoming increasingly widespread. As such, nonlinear analysis already plays an important role in the assessment of existing buildings, as well as in the design of new structures, either directly, through performance based design methodologies, or indirectly by assessing the integrity of structures designed with elastic methods.

As discussed in NIST (2010), the first guidelines on the application of nonlinear analysis were published in the mid-90s, *FEMA-273: NEHRP Guidelines for the Seismic Rehabilitation of Buildings* (FEMA, 1997) and *ATC-40: Seismic Evaluation and Retrofit* of Concrete Buildings (ATC, 1996). Since then, improvements have been proposed in *FEMA-440: Improvement of Nonlinear Static Seismic Analysis Procedures* (FEMA, 2005) and *FEMA-P440A: Effects of Strength and Stiffness Degradation on Seismic Response* (FEMA, 2009). Nonlinear analysis methodologies have already been introduced into all modern assessment methodologies, such as *ASCE-41 Seismic Rehabilitation of Existing Buildings* (ASCE, 2014), *Eurocode 8, Part-3* (EN 1998-3, 2004), as well as new Codes in several other countries, such as, e.g., Italy, Greece, and Turkey. Moreover, nonlinear analysis concepts are also being employed in methods for seismic risk assessment, such as, e.g., HAZUS (Kircher et al., 1997a; Kircher et al., 1997b) and OpenQuake (Silva et al., 2014).

Typical instances where nonlinear analysis is applied in structural earthquake engineering practice are the following (NIST, 2013):

1) *Evaluation and retrofit of existing buildings.* The majority of existing buildings falls short of meeting prescriptive detailing requirements and provisions of the relevant standards for new buildings, which presents a challenge for the evaluation and the retrofit with the use of elastic analysis methods. As a result, seismic evaluation and retrofitting of existing buildings has been one of the primary drivers for the use of nonlinear analysis in engineering practice, since more accurate analysis may permit less conservative strengthening approaches and reduced intervention costs.

2) *Verification of the design for new buildings.* More recently, the role of nonlinear analysis is being expanded beyond the analysis of existing structures to quantify building performance more completely. Nonlinear analysis is thus now also used to verify structural designs based on elastic methods, so as to allow for the better estimation of the true structural response of the building. For instance, the *ATC 58: Guidelines for Seismic Performance Assessment of Buildings* (ATC, 2009) employs nonlinear dynamic analyses for seismic performance assessment of new and existing buildings alike, including fragility models that relate structural demand parameters to explicit damage and loss metrics.

3) *Design of new buildings that employ structural systems, materials or other features that do not conform to current building code requirements.* Although most new buildings are designed using elastic analysis methods and prescriptive code provisions, the use of nonlinear analysis in the design of new buildings is gradually becoming increasingly common. The significant advance in engineering technology that took place in recent decades with the introduction of new materials, such as Fiber-reinforced polymers (FRP) or Shape-memory alloys (SMA), or new systems, such as damping and base isolation devices, provides engineers with a large variety of technologies and solutions that may provide enhanced structural safety with the same or reduced financial cost. However, the use of conventional analytical methodologies is often not appropriate for such scenarios, with more advanced analysis being thus often required. In these cases, nonlinear (static, and very often dynamic) analysis is hence typically employed, so as to allow for the modelling of the full hysteretic behaviour of these new materials or systems.

4) *Design of tall buildings in high seismicity regions.* It is very often needed to estimate and assess a building performance that is at or above the level expected of standard code designs. The design of tall buildings with seismic-force-resisting systems that are not permitted by the code is a common example of the use of nonlinear analysis in design. Towards this end, several engineering resource documents outline explicit requirements for the use of nonlinear dynamic analysis to assess the performance of tall buildings, e.g., *Pacific Earthquake Engineering Research Center Guidelines for Performance-Based Seismic Design of Tall Buildings* (PEER, 2010), *Recommendations for the Seismic Design of High-rise Buildings* (Willford et al., 2008), and *PEER/ATC 72-1 Modeling and Acceptance Criteria for Seismic Design and Analysis of Tall Buildings* (PEER/ATC, 2010).

5) *Performance-based design of new buildings with specific owner requirements.* The introduction of performance-based engineering allows for the selection of particular performance levels and objectives by the owner according to his/her specific needs. These objectives can be quantified, the performance can be predicted analytically, and the cost of improved performance can be evaluated to allow rational trade-offs based on life-cycle considerations, rather than construction costs alone. Within this framework the use of nonlinear procedures that provide enhanced accuracy and significantly better performance predictions is in many ways advantageous and can easier accommodate for the specific owner/stakeholder requirements.

6) *Seismic risk assessment.* Since the introduction of seismic risk assessment methods, such as HAZUS in the late 1990s, and OpenQuake in the more recent years, the analytical procedures have been continually refined, including methods that employ building-specific analyses to improve building fragility models.

Although nonlinear analysis is undoubtedly significantly superior in terms of the accuracy in structural response predictions, this does not come without a cost. The computational resources required for nonlinear analysis are far from negligible, especially in the case of large models subjected to dynamic loading (i.e., acceleration time-histories applied at their base).

Further, whereas linear analysis takes into account only the mass and stiffness distribution of the structural members, nonlinear analysis involves the consideration of both the stiffness and the strength of the members, as well as their overall inelastic behaviour and limit states that depend on deformations and forces. It requires the definition of component models that capture the force-deformation response of components and systems based on expected strength and stiffness properties, taking also into account large deformations. These data require a deeper knowledge of the structural configuration and of the modelling procedures followed in the analytical process, better trained and more experienced analysts, as well as significant effort and additional cost to acquire a good level of knowledge of the structure.

What is more, the results of nonlinear analyses can be very sensitive to assumed input parameters and the types of models used. Without good experience in the analytical and assessment methods that are followed, engineers can be easily misled and extract conclusions about the structural performance that are far from the true response. Generally, it is advisable to have clear expectations about those portions of the structure that are expected to undergo inelastic deformations, so as to use the analyses to confirm the locations of inelastic deformations.

Finally, in contrast to linear elastic procedures and design methods that are well established, used and tested extensively for decades, nonlinear inelastic analysis techniques and their application to design/assessment are relatively recent and are still evolving. In order to keep up with the latter, engineers do need to be open to seeking regular updating and training, acquiring innovative knowledge, developing new skills and gaining confidence with continuously evolving tools.

8.3 Some Theoretical Background

8.3.1 Sources of Nonlinearity

The modelling of the mechanical properties of members is a complex and wide-ranging subject. In linear analysis, it is sufficient to assume that the material remains linear and elastic, i.e., that the deformation process is fully reversible and the stress is a unique function of strain. However, such a simplified assumption is appropriate only within a limited range, and is gradually being replaced by more realistic approaches.

The primary source of nonlinearities in low and medium-rise building structures is material inelasticity and plastic yielding in the locations of damage. In larger high-rise buildings, while material inelasticity still plays an important role, large deformations relative to the frame element's chord (known as P-delta effects), and geometric nonlinearities become equally important and should be taken into account.

These two aspects (material inelasticity and geometrical nonlinearity) are expanded upon in the subsequent Sections, taking account of the fact that seismic assessment of buildings is typically performed through the use of line finite elements (e.g., beams or rods), whilst two- and three-dimensional FE are very rarely utilized, due to the very heavy computational burden and also the lack of reliable and numerically stable 2D/3D nonlinear material constitutive models.

8.3.1.1 Material Inelasticity

Material nonlinearities occur when the stress-strain or force-displacement law is not linear, or when material properties change with the applied loads. Contrary to linear analysis procedures, where the material stresses are always proportional to the corresponding strains and a fully elastic behaviour is assumed, in nonlinear analysis the material behaviour depends on current deformation state and possibly past history of the deformation. In order to estimate the stress caused by the strain at a particular location of the structure, complete expressions for the uniaxial stress-strain relationship of the material should be provided, including hysteretic rules for unloading and reloading.

To better illustrate this concept, reference is made here to the fibre modelling approach as applied to a reinforced concrete beam, even if, as discussed subsequently in this chapter, there are alternative modelling approaches to represent material inelasticity. Thus, the source of such inelasticity can be defined at the sectional level through the creation of a fibre model for the section. A fibre section consists in the subdivision of the area in *n* smaller areas, to each of which is attributed a uniaxial material stress-strain relationship, i.e., reinforcing steel and concrete, confined and unconfined. After defining the material constitutive relationships of every material of the section, and calculating the stresses at the fibres, the sectional moment-curvature state of beam-column elements is then obtained through the integration of the nonlinear uniaxial stress-strain response of the individual fibres. The discretization of a typical reinforced concrete cross-section is depicted in Fig. 8.1.

Estimating the inelastic response of the structural member requires the integration of the stresses calculated at appropriately selected integration cross-sections along the member (called Gauss Sections a and b, in the example shown in Fig. 8.1). Finally, the global material nonlinearity of the frame is then obtained by the assembly of the contributions in stiffness and strength of the structural components.

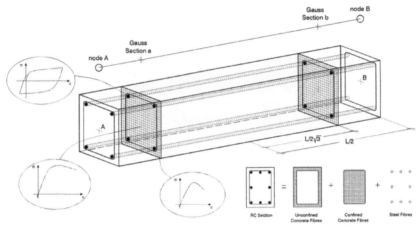

Figure 8.1. Discretization of a typical reinforced concrete cross-section.

8.3.1.2 Geometric Nonlinearities

Geometric nonlinearities involve nonlinearities in kinematic quantities, and occur due to large displacements, large rotations and large independent deformations relative to the frame element's chord (also known as P-delta effects).

The effect of geometric nonlinearities on the response of structures can range from negligible, in cases where large deformations are not expected, to extreme, in large and slender structures. In the general case, geometric nonlinearities must be modelled as they can ultimately lead to loss of lateral resistance, ratcheting (a gradual build up of residual deformations under cyclic loading), and dynamic instability (NIST, 2010, 2013). Large lateral deflections magnify the internal force and moment demands, causing a decrease in the effective lateral stiffness. With the increase of internal forces, a smaller proportion of the structure's capacity remains available to sustain lateral loads, leading to a reduction in the effective lateral strength.

For the numerical simulation of geometric nonlinearities and the inclusion of its effects in the analysis, the most advanced formulation is a total co-rotational formulation [Correia and Virtuoso, 2006], which is based on an exact description of the kinematic transformations associated with large displacements and three-dimensional rotations of the beam-column member. This leads to the correct definition of the element's independent deformations and forces, as well as to the natural definition of the effects of geometrical nonlinearities on the stiffness matrix.

This formulation considers, without losing its generality, small deformations relative to the element's chord, notwithstanding the presence of large nodal displacements and rotations. In the local chord system of the beam-column element, six basic displacement degrees-of-freedom $(\theta_{2(A)}, \theta_{3(A)}, \theta_{2(B)}, \theta_{3(B)}, \Delta, \theta_{T})$ and corresponding element internal forces $(M_{2(A)}, M_{3(A)}, M_{2(B)}, M_{3(B)}, F, M_{T})$ are defined, as shown in Fig. 8.2.

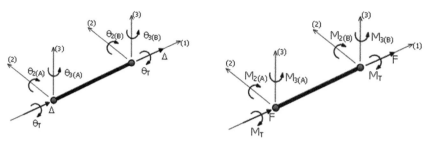

Figure 8.2. Local chord system of a beam-column element.

8.3.2 Modelling of Structural Members

Since the introduction of nonlinear analysis, various types of structural component models have been developed to simulate the beam-column behaviour. The models vary in sophistication, computational resources required and accuracy in the representation of the nonlinear response. The main ways to model the nonlinear response of a structural member are depicted in Fig. 8.3 (NIST, 2010).

As discussed in NIST (2013), the most basic models are shown at the left, and are the so-called plastic hinge models. They are concentrated plasticity models in which all of the nonlinear effects are lumped into an inelastic spring, whereas the rest of the element remains linear and elastic. At the right of Fig. 8.3 is a detailed continuum finite element model with explicit nonlinear rules for the representation of the response of its three-dimensional components. Between the two extremes, there are various types of distributed plasticity fibre elements that provide hybrid representations of the structural behaviour.

Quoting again NIST (2013); "to some extent all of the models have a phenomenological basis because they all ultimately rely on mathematical models that are calibrated to simulate nonlinear phenomena observed in tests. However, the concentrated plasticity models rely almost exclusively on phenomenological representation of the overall component behaviour, whilst the continuum finite element models include a more fundamental representation of the response,

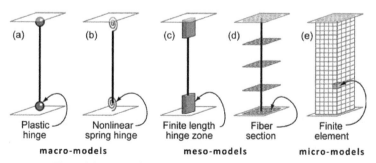

Figure 8.3. Types of structural component models (NIST, 2010).

where only the most basic aspects (e.g., material constitutive relationships) rely on empirical data."

Noting that the use of the more refined and complex FE models cannot be justified but for the modelling of single members or very small buildings, due to the considerable computational resources that are required, only two main different strategies are generally employed in the reproduction of the inelastic response of structures, the '*concentrated plasticity*' and the '*distributed inelasticity*' models.

In concentrated (or lumped) plasticity, the plastic deformations are 'lumped' at the ends of a linear elastic element and are based on the moment-rotation relationships of the end sections for a given axial force. On the other hand, distributed plasticity beam-column elements, referred to already in Section 8.3.1.1 above, allow for the formation of plastic hinges at any location along the member length, whilst inelasticity is represented in terms of stresses and strains at the fibres of the integration sections, thus accounting for the axial-moment interaction. There are different variations and implementations of these two modelling philosophies, as well as hybrid approaches that employ features of both strategies. Each FE software package follows its own formulations, although the main principles for the modelling of frame members remain.

8.3.2.1 Concentrated Plasticity Elements

Historically, concentrated plasticity elements were the first elements developed to include nonlinearities into beam-column members. In the concentrated plasticity philosophy, nonlinear behaviour is assumed at the extremities of the structural element, whilst the body is modelled as an elastic part, and the nonlinear behaviour is represented, through nonlinear rotational springs or plastic hinges. The first concentrated plasticity elements were presented by Clough and Johnston (1966), a model that consisted of two elements in parallel, and by Giberson (1967), a slightly different model consisting of two elements in series.

The concentrated plasticity modelling assumptions lead to simple models with reduced computational cost, but the simplification of the concentrated plasticity may fail to describe correctly the hysteretic behaviour of some reinforced concrete members, hence it cannot be applied to all elements without consideration. In addition, this type of element requires adequate knowledge of the user on the calibration of their inelastic parameters, in order to obtain accurate results (e.g., the location of the plastic hinges in the structure, which is difficult to know before running the analysis, the length of the plastic hinges, or the hysteretic rules of the stress-strain relationship of the nonlinear parts).

Advantages and Disadvantages of Concentrated Plasticity Models

The main advantage of concentrated plasticity models is the significant reduction in computational cost and requirements for data storage in three dimensional finite element models. Further, and despite their simplicity, concentrated plasticity models can generally capture most of the important behavioural effects of the response of beam or columns in steel or reinforced concrete moment frames, from the onset of yielding through to the highly inelastic range. Strength and stiffness degradation, associated with concrete spalling, steel yielding, concrete crushing and reinforcing bar buckling (concrete members) or local flange and web buckling (steel members) can be modelled effectively. Under certain conditions, these nonlinear effects can be captured with simple hinge models, equally or even more reliably than with other finite element or distributed plasticity models.

However, effects such as the interaction between axial, and flexural failure in concrete members, or the interaction of local and torsional-flexural buckling in steel members, are difficult to capture using concentrated plasticity models. Further, concentrated plasticity models fail to approximate adequately the behaviour of members where large bending moments are not concentrated at the two ends of the members. This inadequacy can be very significant in certain cases, such as large shear walls at the lower floors of buildings that feature large moments throughout their height. The limitations of the concentrated plasticity modelling strategy in such cases has serious implications in the accuracy of the predictions, considering that, owing to their size, such members usually dominate the structural response.

What is more, significant knowledge in advanced nonlinear modelling is usually required from the engineer, since some type of calibration of the moment-rotation curves is typically required. An accurate prediction of the structural response cannot be obtained when users possess limited knowledge on issues related to the calibration of the inelastic element parameters. As a result, the plastic hinge length and the characterization of the inelastic section parameters should be given additional attention (and time), during the structural modelling process.

8.3.2.2 Distributed Plasticity Elements—Fibre Modelling

Contrary to the concentrated plasticity models, distributed inelasticity elements allow inelastic deformations to be developed anywhere within the member. Lately, they have gained increased popularity and are becoming widely employed in earthquake engineering applications, either for research or professional engineering purposes. The fibre approach is used to represent the cross-section behaviour, where the sectional stress-strain state is obtained through the integration of the nonlinear uniaxial stress-strain response of the individual fibres, and the inelastic response of the structural member is calculated from the integration of the forces at the integration cross-sections along the member, as in Fig. 8.4 (note that the latter refers to a different fibre modelling formulation from that showed in Fig. 8.1, making use of a larger number of integration cross-sections).

In fibre modelling users are asked to define the number of section fibres employed. The ideal number of section fibres, sufficient to guarantee an adequate reproduction of the stress-strain distribution across the element's cross-section, varies with the shape and material characteristics of the latter, depending also on the degree of inelasticity to which the element will be forced to.

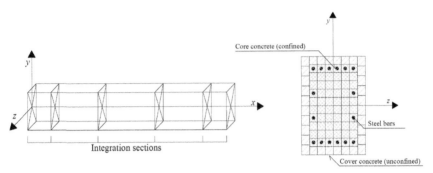

Figure 8.4. Fibre modelling and integration sections in distributed plasticity modelling.

One of the main advantages of the distributed plasticity models, with respect to the concentrated plasticity models, is the non-existence of a predetermined location or length where inelasticity occurs, which allows for a much closer approximation of the actual response of the structural member. On the other hand, however, distributed plasticity modelling requires additional computational capacity, in terms of increased analysis time, but also of memory and CPU consumption.

Types of Distributed Plasticity Elements

There are two different formulations for distributed inelasticity frame elements: the classical displacement-based (DB) formulation (e.g., Hellesland and Scordelis, 1981; Mari and Scordelis, 1984), and the more recent force-based (FB) formulation (e.g., Spacone et al., 1996; Neuenhofer and Filippou, 1997).

The *displacement-based formulation* follows a standard FE approach, where the element deformations are interpolated from an approximate displacement field. In order to approximate nonlinear element response, constant axial deformation and linear curvature distribution are enforced along the element length, which is exact only for prismatic linear elastic elements. Consequently, displacement-based formulation should be employed with members of small length, leading to the need for a mesh refinement, in order to achieve good accuracy in the case of higher order distributions of deformations.

The *force-based formulation* is the most accurate, since it is capable of capturing the inelastic behaviour along the entire length of a structural member, even when employing a single element per member. Hence, its use allows for high accuracy in the analytical results, whilst giving also users the possibility of readily employing element chord-rotations output for seismic code checks.

For linear elastic material behaviour, the two approaches obviously produce the same results, provided that only nodal forces act on the element. On the contrary, in onset of material inelasticity, imposing a displacement field does not allow capturing the real deformed shape, given that the curvature field can be highly nonlinear. In this situation, with a DB formulation a refined discretization (meshing) of the structural element (typically 4–5 elements per structural member) is required for the computation of nodal forces/displacements, in order to accept the assumption of a linear curvature field inside each of the sub-domains.

Instead, an FB formulation is always exact, since it does not depend on the assumed sectional constitutive behaviour. In fact, it does not restrain in any way the displacement field of the element. In this sense this formulation can be regarded as always 'exact', the only approximation being introduced by the discrete number of the controlling sections along the element that are used for the numerical integration. A minimum number of 3 Gauss-Lobatto integration sections are required to avoid under-integration, however such option will in general not simulate the spread of inelasticity in an acceptable way. Consequently, the suggested minimum number of integration points is 4, although 5–7 integration sections are typically used. Such feature enables to model each structural member with a single FE element, therefore allowing a one-to-one correspondence between structural members (beams and columns) and model elements, thus leading to considerably smaller models, with respect to when DB elements are used, and much faster analyses, notwithstanding the heavier element equilibrium calculations. An exception to this non-discretization rule arises when localization issues are expected, in which case special caution is needed, as discussed in Calabrese et al. (2010).

As mentioned above, the use of a single element per structural element gives users the possibility of readily employing element chord-rotations output for seismic code verifications. Instead, when the structural member has had to be discretized in two or more frame elements (necessarily the case for DB elements), then users need to post-process nodal displacements/rotation in order to estimate the members chord-rotations (e.g., Mpampatsikos et al., 2008).

Advantages and Disadvantages of Distributed Plasticity Models

The main advantage of distributed inelasticity elements is that they do not require any calibration of the inelastic response parameters, as is instead needed for concentrated-plasticity phenomenological models. Consequently, they do not require advanced modelling knowledge from the analyst, since all that is required is the introduction of the geometrical and material characteristics of structural members. Further, there is no requirement for a-priori moment-curvature analysis of members, and there is no need to introduce any element hysteretic response, since this is implicitly defined by the material constitutive models implemented in the structural analysis tool.

In addition, because of their ability to allow for the development of nonlinearities anywhere along the member length, distributed plasticity models allow yielding to occur at any location along the element, which is especially important in the case of large shear walls, or in the presence of distributed element loads (girders with high gravity loads). They possess the ability to track gradual inelasticity (e.g., steel yielding and concrete cracking) over the cross section and along the member length, to directly model the axial load-bending moment interaction, both on strength and stiffness terms, and to allow the straightforward representation of biaxial loading, and interaction between flexural strength in orthogonal directions.

Because of these significant advantages of distributed plasticity models, in general their predictions are more accurate and closer to reality with respect to their concentrated plasticity counterparts. However, these advantages come with the cost of the increased required computational resources.

Further, the ability of these models may be limited in the capture of degradation associated with bond slip in concrete joints, local buckling and fracture of steel reinforcing bars and steel members, though specialized material models have recently been developed to represent such phenomena.

Finally, the calculation of curvatures (and stresses and strains) along the member length can be sensitive to the specified hardening (or softening) modulus of the materials, the assumed displacement (or force) interpolation functions along the member length, and the type of numerical integration and discretization of integration points along the member, which can lead to considerable errors and inconsistencies in the curvature and strain demands calculated in the analysis, along with the associated stress resultants (i.e., forces) and members' stiffness (see, e.g., Calabrese et al., 2010).

8.3.2.3 Hybrid Concentrated Plasticity Fibre Elements

A combination of distributed inelasticity models with concentrated plasticity elements, featuring a similar distributed inelasticity formulation, but concentrating such inelasticity within a fixed length of the element, was proposed by Scott and Fenves (2006). The advantages of such formulation are not only reduced analysis time (since fibre integration is carried out for the two member end sections only) and increased stability of the nonlinear solutions, but also full control/calibration of the plastic hinge length (or spread of inelasticity), which allows the overcoming of localization issues, as discussed, e.g., in Calabrese et al. (2010).

8.3.3 Solving Nonlinear Problems in Structural Analysis

Once the governing equations of geometrically nonlinear structural analysis and the discretization of those equations by finite element methods is completed, a procedure is required for the solution of these equations. Nonlinear problems in structural mechanics are solved with incremental algorithms through a process that is considerably more articulated than common linear elastic analysis solvers.

All solution procedures of practical importance are strongly rooted on the concept of gradually advancing the solution by "continuation", that is to follow the equilibrium response of the structure as the control and state parameters vary by small amounts. Various algorithms exist for handling such problems, but a common feature is that continuation is a multilevel process that involves a hierarchical breakdown of the entire loading stage into incremental steps, with iterative steps within. The incremental solution methods are then divided into two broad categories: (1) Purely incremental methods, also called *predictor-only methods*, and (2) Corrective methods, also called *predictor-corrector* or *incremental-iterative methods*.

In purely incremental methods the iteration level is missing. In corrective methods a predictor step is followed by one or more iteration steps. The set of iterations is called the corrective phase, and its purpose is to eliminate or reduce the so-called drifting error, which is a problem in purely incremental methods. For this reason the corrective methods have become the standard for the solution of the nonlinear equations in all modern finite element packages.

Solutions accepted after each increment and a corrective phase are often of interest to users because they represent approximations to equilibrium states until the final loading state. They are therefore saved as they are computed. On the other hand, intermediate results of the iterative process are rarely of interest, since the solutions are not equilibrated and constitute "merely" an intermediate step until the next equilibrated solution, and hence most nonlinear structural analysis programs discard them.

The use of increments may seem at first sight unnecessary if one is interested primarily in the final solution. But breaking up a stage into increments serves different purposes:

i) The presence of path-dependent effects in nonlinear analysis problems severely restricts increment sizes because of history-tracing constraints—for example, in plasticity analysis stress states must not be allowed to stray too far outside the yield surface.

ii) The engineer can acquire a better insight into structural behaviour by studying the response plot toward the final solution, which in many cases can provide more useful information than simply the structural state at the end. It is noted that in several cases failures and critical points occur before the stage end.

iii) The breakdown of the entire loading stage can lead to numerically more stable solutions and avoid convergence problems.

8.3.3.1 Incremental-Iterative Algorithm

The basic method for the solution of nonlinear equations in the majority of finite element programs is the load-control Newton-Raphson (NR) algorithm, and variations of it. The Newton-Raphson method, in its simplest form, is a numerical method for finding the roots of a function $f(x)$. Since the method is iterative, a trial guess is made at $x = x_n$. Evaluating the function at x_n, we find that $f(x_n) \neq 0$, i.e., it is not a root. If f'(x_n) is the tangent of the function at x_n, the equation of the tangent passing from x_n is:

$$f(x_{n+1}) - f(x_n) = f'(x_n) \cdot (x_{n+1} - x_n) \qquad (8.1.)$$

With the aid of Eq. (8.1), we can obtain a second trial solution at x_{n+1}:

$$x_{n+1} = x_n - \frac{f'(x_n)}{f(x_n)} \qquad (8.2)$$

where x_{n+1} is the point where the tangent intersects axis X. If $f(x_{n+1}) \approx 0$, then we have located the root, otherwise we proceed finding a new trial solution at x_{n+2}, until convergence to the correct solution has been reached within an acceptable convergence limit. The method is schematically shown in Fig. 8.5.

In structural mechanics, the Newton-Raphson (NR) method is extended so as to accommodate the solution of a series of nonlinear equations, arising from the general equation of (nonlinear) structural equilibrium:

$$P = k \times u \qquad (8.3)$$

Because of the nonlinear nature of the problem at hand, the stiffness matrix **K** is a function of the deformation vector **u** and is constantly updated at each iteration, and hence Eq. (8.3) cannot be solved directly (i.e., as discussed above, one needs to follow an incremental solution procedure such as, e.g., the Newton-Raphson method).

The incremental-iterative Newton-Raphson method is schematically shown in Fig. 8.6. The iterative procedure follows the conventional scheme, whereby the internal forces corresponding to a displacement increment are computed and convergence is checked. If no convergence is achieved, then the out-of-balance forces (difference between applied load vector and equilibrated internal forces) are applied to the structure, and the new displacement increment is computed. Such loop proceeds until convergence has been achieved or the maximum number of iterations has been reached. Load-displacement plots, such as those of Fig. 8.6 are exact for one SDOF systems. For larger MDOF systems, they describe only schematically the structural response and the gradual convergence to the solution of the system of equations.

The full Newton-Raphson method provides a quadratic rate of convergence, meaning that it requires a small number of iterations to reach the solution. However, recalculating and inverting the stiffness matrix at every iteration requires increased computing resources. Hence, a common alternative is to recalculate and invert the stiffness matrix only at the first iteration and use it for all the corrective iterations. This approach is known as modified Newton-Raphson and is shown in Fig. 8.7.

The employment of Newton-Raphson (NR), modified Newton-Raphson (mNR) or NR-mNR hybrid solution procedures may lead to a fairly flexible solution algorithm. It is clear that the computational savings in the formation, assembly and reduction of the stiffness matrix during the iterative process can be significant

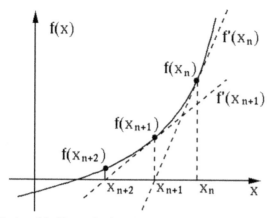

Figure 8.5. Application of the Newton-Raphson (NR) method for finding the roots of a function $f(x)$.

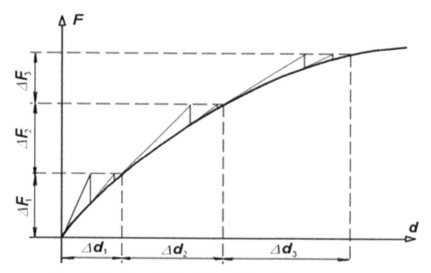

Figure 8.6. The Newton-Raphson (NR) method in nonlinear structural analysis.

when using the mNR instead of the NR procedures. However, more iterations are often required with the mNR, thus leading in some cases to an excessive computational effort. For this reason, the hybrid approach, whereby the stiffness matrix is updated only in the first few iterations of a load increment, does usually lead to an optimum scenario.

For further discussion and clarifications on the algorithms described above, readers are strongly advised to refer to available literature, such as the work by Cook et al. (1988), Crisfield (1991), Zienkiewicz and Taylor (1991), Bathe (1996) and Felippa (2002), to name but a few.

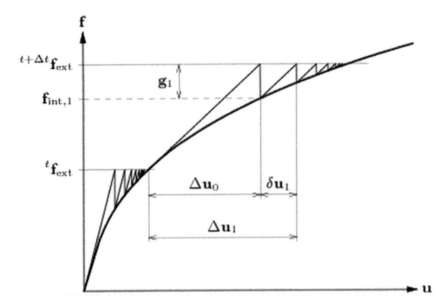

Figure 8.7. The Modified Newton-Raphson (mNR) method in nonlinear structural analysis.

8.3.3.2 Convergence Criteria

As mentioned above, in iterative-incremental solution algorithms, the iterative process at each step is continued until *'convergence is achieved'*, that is the values of a norm of the out-of-balanced forces or the unbalanced deformations of the structure become smaller than given convergence criteria that have been specified by the user at the beginning of the analysis.

There are two main distinct categories of convergence criteria in nonlinear analysis:

- i) Displacement/rotation-based criteria.
- ii) Force/moment-based criteria.

Two additional convergence check schemes may arise from the combination of the distinct criteria above:

- iii) Displacement/Rotation AND Force/Moment based scheme, where it is considered that the solution has been reached when both the deformation and the force based criteria have been achieved.
- iv) Displacement/Rotation OR Force/Moment based scheme, where convergence is achieved when either the deformation or the force based criteria has been achieved.

Usually, the *displacement/rotation* criterion consists in verifying, for each individual degree-of-freedom of the structure, that the current iterative displacement/ rotation is less or equal than a user-specified tolerance. In other words, if and when all values of displacement or rotation that result from the application of the iterative (out-of-balance) load vector are less or equal to the pre-defined displacement/ rotation tolerance factors, then the solution is deemed as having converged. This concept can be mathematically expressed in the following manner:

$$\max\left[\left|\frac{\delta d_i}{d_{tol}}\right|^{n_d}_{i=1}, \left|\frac{\delta\theta_j}{\theta_{tol}}\right|^{n_\theta}_{j=1}\right] \leq 1 \rightarrow convergence \tag{8.4}$$

where,

δd_i is the iterative displacement at translational degree of freedom i

$\delta\theta_j$ is the iterative rotation at rotational degree of freedom j

n_d is the number of translational degrees of freedom

n_θ is the number of rotational degrees of freedom

d_{tol} is the displacement tolerance, in the employed by the analysis length unit

θ_{tol} is the rotation tolerance, in rad (dimensionless)

The *force/moment* criterion, on the other hand, comprises the calculation of the Euclidean norm of the iterative out-of-balance load vector, and subsequent comparison to a user-defined tolerance factor. It is therefore a global convergence check (convergence is not checked for every individual degree-of-freedom as is done for the displacement/rotation case) that provides an image of the overall state of convergence of the solution, and which can be mathematically described in the following manner:

$$G_{norm} = \sqrt{\frac{\sum_{i=1..n}\left[\frac{G_i}{V_{REF}}\right]^2}{n}} \leq 1 \rightarrow convergence \tag{8.5}$$

where,

G_{norm} is the Euclidean norm of iterative out-of-balance load vector

G_i is the iterative out-of-balance load at degree-of-freedom i

V_{REF} is the reference 'tolerance' value for forces (translational DOFs) and moments (rotational DOFs). Typically, different values are assumed for V_{REF} for the translational and the rotational DOFs

n is the number of the degrees-of-freedom

8.3.3.3 Numerical Instability, Divergence and Iteration Prediction

It is noted that, contrary to linear analysis, where a solution, even if it is unreasonable and unrealistic, can always be found, in the case of nonlinear analysis the achievement of convergence and the calculation of the solution is not always guaranteed. This may be due to purely structural equilibrium reasons (e.g., a beam cannot withstand the applied vertical loads, fails and a solution cannot be found), but also due to numerical reasons (e.g., the solution procedure cannot accommodate large load increments, or sudden redistribution of forces, after the failure of a load-carrying member).

For this reason, in nonlinear analysis the maximum number of iterations is always specified, so as to avoid a situation whereby an unattainable solution is pursued infinitely. If the maximum number of iterations is reached and convergence has not been achieved, the analysis is either stopped or, in more advanced software packages, the load step is subdivided into smaller increments, in order to try to achieve better convergence conditions.

In addition to the convergence verification schemes described above, at the end of an iterative step three other solution checks may be carried out; numerical instability, solution divergence and iteration prediction. These criteria serve the purpose of avoiding the computation of needless equilibrium iterations in cases where it is already apparent that convergence will not be reached, thus minimising the duration of the analysis.

Numerical instability. The possibility of the solution becoming numerically unstable is checked at every iteration by comparing the Euclidean norm of out-of-balance loads, G_{norm}, with a pre-defined maximum tolerance several orders of magnitude larger than the applied load vector (e.g., 1.0E+20). If G_{norm} exceeds this tolerance, then the solution is assumed as being numerically unstable and iterations within the current increment are interrupted.

Solution divergence. Divergence of the solution is checked by comparing the value of G_{norm} obtained in the current iteration with that obtained in the previous one. If G_{norm} has increased, then it is assumed that the solution is diverging from a possible solution and iterations within the current increment are interrupted.

Iteration prediction. Finally, a logarithmic convergence rate check can also be carried out, so as to try to predict the number of iterations required for convergence to be achieved. If this estimated number of iterations is larger than the maximum number of iterations specified by the user, then it is assumed that the solution will not achieve convergence and iterations within the current increment are interrupted.

It is noted that these three additional checks described above are usually reliable and effective within the scope of applicability of nonlinear analysis, for as long as the divergence and iteration prediction check is not carried out during the first iterations of an increment, when the solution might not yet be stable enough. Hence, it is advisable that these checks are carried out after an initial number of iterations.

8.3.4 Types of Nonlinear Analysis

Within the context of modern seismic design and assessment regulations, two types of nonlinear analysis procedures are typically proposed: (i) the Nonlinear Static Procedure (NSP), which makes use of the so-called *pushover analysis*, (ii) and the Nonlinear Dynamic Procedure (NDP), which consists of the also often termed *nonlinear time-history analysis* or *nonlinear response analysis*.

8.3.4.1 Nonlinear Static Procedure (NSP)

As stated in ASCE 41-13 and other codes, guidelines, and research publications, in the Nonlinear Static Procedure a mathematical model directly incorporating the nonlinear load-deformation characteristics of individual components of the building is subjected to monotonically increasing lateral loads representing inertia forces in an earthquake, until a target displacement is exceeded. The main objective of the method is to assess the capacity of the structure, considering both the deformability and strength of all structural members.

The lateral loads are gradually applied until the displacement of a selected 'Control Node', typically located at the centre of mass of the top storey of the building, reaches the so-called 'Target Displacement', which represents an approximation of the displacement demand under earthquake ground motion. The demand parameters for the structural components at the target displacement are then compared against the respective acceptance criteria for the desired performance state. System level demand parameters, such as story drifts and base shear forces, may also be checked.

Although the nonlinear static procedure is generally a much more reliable approach for characterizing the performance of a structure than linear procedures, it is still not exact and cannot accurately account for changes in the dynamic response as the structure degrades in stiffness, nor can it account for higher mode effects in multi-degree-of-freedom (MDOF) systems, though the use of Displacement-based Adaptive Pushover (Antoniou and Pinho, 2004), amongst other non conventional pushover algorithms, has been shown to significantly limit such potential shortcomings.

8.3.4.2 Nonlinear Dynamic Procedure (NDP)

As stated in ASCE 41-13 and other codes, guidelines, and research publications, in the Nonlinear Dynamic Procedure a mathematical model directly incorporating the nonlinear load-deformation characteristics of individual components of the building is subjected to earthquake shaking represented by ground motion acceleration histories to obtain forces and displacements. The objective of the method is to assess the capacity of the structure, considering the deformability, the strength and the hysteretic behaviour of all structural members that are subjected to the specified earthquake ground motion (ASCE 2014).

The direct integration of the equations of motion is accomplished using appropriate integration algorithms, such as the numerically dissipative α-integration algorithm (Hilber et al., 1977) or a special case of the former, the well-known Newmark scheme (Newmark, 1959). The modelling of seismic action is achieved by introducing acceleration time-histories (i.e., accelerograms) at the supports of the structure. In addition, dynamic analysis may also be employed for modelling of pulse loading cases (e.g., blast, impact, etc.), in which case instead of acceleration time-histories at the foundation, force pulse functions of any given shape (rectangular, triangular, parabolic, and so on), can be employed to describe the transient loading applied to the appropriate nodes.

The NDP constitutes a sophisticated approach for examining the inelastic demands produced on a structure by a specific suite of ground motion acceleration time-histories. As with the NSP, the results of the NDP can be directly compared with test data on the behaviour of representative structural components to identify the structure's probable performance, when subjected to a specific ground motion.

As nonlinear dynamic analysis involves fewer assumptions than the nonlinear static procedure, it is subject to fewer limitations than nonlinear static procedure. It automatically accounts for higher-mode effects and shifts in inertial load patterns as structural softening occurs. In addition, for a given earthquake record, this approach directly solves for the maximum global displacement demand produced by the earthquake on the structure, eliminating the need to estimate this demand based on general relationships.

Despite these advantages, the NDP requires non-negligible judgment and experience to be carried out, and should only be used when the engineer is thoroughly familiar with nonlinear dynamic analysis techniques and limitations. The analyses can be highly sensitive to small changes in assumptions with regard to either the character of the ground motion record used or the nonlinear stiffness behaviour of the elements. For instance, two ground motion records enveloped by the same response spectrum can produce radically different results with regards to the distribution and amount of inelasticity predicted in the structure. Further, due to the inherent variability in earthquake ground motions, dynamic analyses for multiple ground motions are necessary to calculate an upper bound for the values of the demand parameters for a given earthquake scenario.

8.4 How Reliable are Numerical Predictions from Nonlinear Analysis Methods?

Through realistic modelling of the underlying mechanics, nonlinear static analysis and especially nonlinear dynamic analysis, reduce uncertainty in demand predictions, as compared to the linear methods of analysis. However, even with nonlinear dynamic analyses, it is practically impossible to always calculate accurately the demand parameters for the different structural members. Hence, very often there are discrepancies between the analytical predictions of the response parameters and their actual values during a seismic event. These discrepancies are usually largest for structural deformations and accelerations and lower in force-controlled components of capacity-designed structures where the forces are limited by the strength of yielding members.

Since no numerical representation of the components response is perfect, one of the sources of inaccuracy is the limitations related to the analytical capabilities of the selected software package with which the analysis is to be carried out. Naturally, any analytical formulation that describes the hysteretic behaviour of materials, sections or members has limitations that restrain its ability to represent the structural response in a very precise manner, especially in the highly inelastic range, where the lateral stiffness is significantly reduced and the deformation response is thus very sensitive to small changes of the loading. Analysts should be well aware of these limitations prior to the execution of the analyses, so as to avoid modelling strategies that magnify the possible errors and affect considerably the analytical predictions.

All the same, with the continuous development of new, enhanced models to represent structural behaviour, inaccuracies related to the software itself and its analytical formulations and capabilities tend to become smaller, especially in the cases of analysis of ordinary structures with components featuring relatively predictable behaviour that can be approximated reasonably well. In such cases the two main sources of inaccuracy are mostly related to human-related parameters, that is (i) uncertainties arising from the variability in the measured physical attributes of the structure such as material properties, geometry and structural details, and (ii) incomplete mathematical model representation of the actual structural behaviour by the analyst. The problems described above, coupled with hazard uncertainties in the ground motion intensity, and ground motion uncertainty arising from frequency content and duration of ground motions for a given intensity, may lead to large divergence of the analytical predictions from the true structural response during a large earthquake event.

Considering the discussion above, analysts should always bear in mind the different sources of uncertainty in the structural evaluation process, and how these may affect the predictions. The uncertainties should always be accounted for and, if possible, they should be quantified and represented in the acceptance criteria checks through the selection of appropriate values of the corresponding safety factors. Obviously, the selected values of the safety factors should reflect the level of uncertainties in the knowledge of the structure and in the analytical model that has been employed. For instance, depending on the level of knowledge of a structure, different values of the knowledge factor (ASCE 41-13) or the confidence factor (EC8, Part-3) are assigned, with more conservative values in the cases where there is lack of good knowledge of the structural configuration.

8.4.1 Blind Prediction Exercises

One measure of the accuracy that can be achieved by the analytical procedures is blind prediction. In a blind prediction exercise, engineers, who do not know the real behaviour of a structure that is to be measured in an experimental test, are asked to predict its response employing state-of-the-art analytical tools. This process provides an unbiased assessment of prediction capabilities, because prior knowledge of the structural response cannot be used to improve the simulation results.

In the recent past, a significant number of blind prediction exercises have been performed, from single components or subassemblies to large, full-scale structural systems. Teams of analysts have been invited to predict the outcome of shake-table tests, given knowledge of the building design and the nature of the applied ground motions.

It has been observed that the scatter among predictive results in blind prediction exercises is large, indicating that the comparison with measured response and observed behaviour (e.g., nonlinear mechanisms and failure modes) is not always favourable. This suggests that, in practice, many predictions are unrealistic and their reliability is low. Further, it is difficult to draw meaningful conclusions from the limited sets of analyses carried out, because differences could arise from many possible sources, including (i) whether or not the input parameters are characterized consistently, (ii) there is an error in the material model, (iii) the mesh is too coarse, (iv) or the nonlinear model is incapable of representing a particular behaviour.

However, blind prediction exercises can be used to spot and evaluate the sources of variation between analytical predictions and actual data, and to identify analytical procedures that appear to be consistently better (or worse) than others. Careful and detailed review of previous blind prediction exercises can yield important insight and information regarding the accuracy and reliability of nonlinear analysis methods. Consequently, blind prediction contests can be used as a source for general improvement of the reliability of our analytical tools.

8.4.2 Validation Examples

Experimental test data available in non-blind prediction contexts can also be used to demonstrate or test the capability of a given modelling approach, software tool, or analyst to reproduce the seismic response of structures. For exemplificative purposes, some examples of such validation exercises are included here, with a view to support also the aforementioned adequacy of nonlinear dynamic analysis.

In all cases, the SeismoStruct 2016 software package (Seismosoft, 2016) is employed for the execution of the nonlinear analyses, not only because such nonlinear structural analysis tool is freely available for non-commercial purposes, but also because it has been employed by the winning participants in recent blind predictions contests, e.g., at the NEES Large High-Performance Outdoor Shake Table at UCSD's Englekirk Structural Engineering Center, at the 3D Shake-Table of the National Laboratory for Civil Engineering (LNEC) in Lisbon (Portugal) as part of a Bind Prediction Contest organized at the 15th Word Conference on Earthquake Engineering (15WCEE), etc.

All the structural models employed feature nonlinear characteristics that enable them to correctly identify the structural behaviour and failure mechanisms in the highly-inelastic range. Material inelasticity is represented through fibre modelling with the distributed plasticity approach. Different types of frame elements were employed, e.g., force-based, force-based plastic hinge or displacement-based element types, and geometric nonlinearities are automatically incorporated in the model by the program.

Since full details of the models, and the model input files themselves, can be readily obtained from Seismosoft's website, herein only very brief description of the test specimens and of selected results are given.

8.4.2.1 Four-storey 2D RC frame

This is a full-scale, four-storey, 2D bare frame, which was designed for gravity loads and for a nominal lateral load of 8% of the structure's weight (see Fig. 8.8). The reinforcement details attempted to reproduce the construction practices used in southern European countries in the 1950's and 1960's. The frame was tested at the ELSA laboratory (Joint Research Centre, Ispra) in pseudo-dynamic fashion, considering two artificial records (aiming at representing return periods of 475 and 975 years). Further information about the frame (known as ICONS frame) and the tests conducted in ELSA can be found in Pinto et al. (1999), Carvalho et al. (1999), Pinho and Elnashai (2000) and Varum (2003).

The comparison between experimental and analytical results, in terms of top displacement against time is shown in Fig. 8.9 for one of the records, where it can be observed that the analysis provides a good estimate of the structural response throughout the duration of the test run.

Figure 8.8. ICONS frame tested at the ELSA laboratory of Ispra (Pinto et al., 1999).

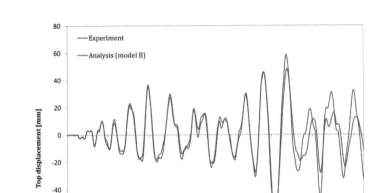

Figure 8.9. Experimental vs. Analytical results—top displacement vs. time (475 yrp).

8.4.2.2 *Seven-storey RC Shear Wall Building*

This is a slice of a seven-storey, full-scale RC shear wall building (see Fig. 8.10), tested on the NEES Large High-Performance Outdoor Shake Table at UCSD's Englekirk Structural Engineering under dynamic conditions (Panagiotou et al.,

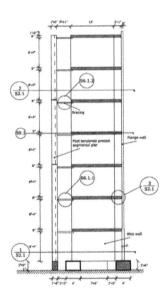

Figure 8.10. RC shear wall building tested at UCSD (from Martinelli and Filippou, 2009).

2006) by applying four subsequent uniaxial ground motions. The structure was designed with the displacement-based capacity approach for a site in Los Angeles, and the design lateral forces were smaller than those at the time specified in U.S. building codes for regions of high seismic risk.

The comparison between experimental and analytical results is shown in Fig. 8.11 for the top displacement time-histories. Again, the analysis is shown to provide a good estimate of the structural response for the entire record, even for large drifts and very high levels of inelasticity.

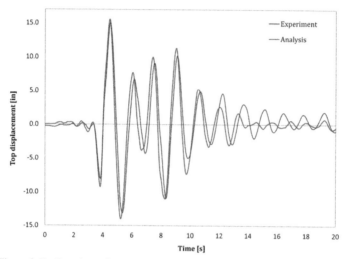

Figure 8.11. Experimental vs. Analytical results – top displacement vs. time (EQ4).

8.4.2.3 Three-storey 3D RC Frame

This is a full-scale, three-storey, three-dimensional RC building, which was designed for gravity loads only, according to the 1954–1995 Greek Code. The prototype building was built with the construction practice and materials used in Greece in the early 70's (non-earthquake resistant construction). It is regular in height but highly irregular in plan (see Fig. 8.12). Details on the structural member dimensions and reinforcing bars can be found in Fardis and Negro (2006). This prototype building was tested at the European Laboratory for Structural Assessment (ELSA) of the Joint Research Centre of Ispra (Italy) under pseudo-dynamic conditions using the Herceg-Novi bi-directional accelerogram registered during the Montenegro 1979 earthquake.

Figure 8.13 presents a comparison between experimental and analytical results, in terms of total displacements, which are again in good agreement between them.

Figure 8.12. Full-scale, three-storey prototype building (Fardis and Negro, 2006).

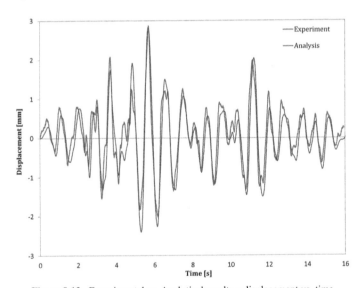

Figure 8.13. Experimental vs. Analytical results – displacement vs. time.

8.4.2.4 Four-storey 3D Steel Frame

This is a full-scale, 3D steel moment resisting frame (see Fig. 8.14) tested under dynamic conditions on the three-dimensional E-Defense shake-table located at Miki City, Hyogo Prefecture (Japan), by applying a scaled version of the near-fault motion recorded in Takatori during the 1995 Kobe earthquake. Details on

Figure 8.14. Four-storey 3D steel moment resisting frame (NRIESDP, 2007).

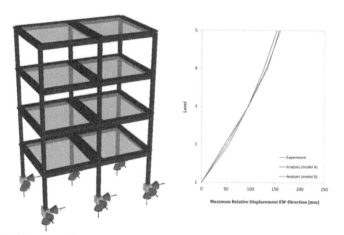

Figure 8.15. Four-storey 3D steel frame model in SeismoStruct, and experimental vs. analytical results comparison.

the prototype building including geometrical and material characteristics are summarized in Pavan (2008).

The analytical model is shown in Fig. 8.15, where the triaxial nature of the input motion is highlighted (by the green arrows applied at the foundations), together with a comparison between experimental and numerical results, in the form of maximum floor displacements recorded in one of the two translation directions. The results from two different models are shown, featuring force-based (model A) and displacement-based (model B) distributed plasticity elements.

8.4.2.5 Four-storey 3D Infilled RC Frame

This is a full-scale, four-storey building, designed according to initial versions of Eurocode 8 (CEN, 1995) and Eurocode 2 (CEN, 1991) and tested at the ELSA laboratory (Joint Research Centre, Ispra) under pseudo-dynamic loading using an artificial accelerogram inspired by the recordings from the 1976 Friuli earthquake. The building, and corresponding SeismoStruct model, is shown in Fig. 8.16, where the presence of infill panels is also conspicuous.

In Fig. 8.17 the base shear time-histories, obtained during the test and computed from the analysis, are plotted against time, where the agreement between the measured values from the test and the analytical predictions can be observed.

Figure 8.16. Four-storey 3D infilled frame tested by Negro et al. (1996).

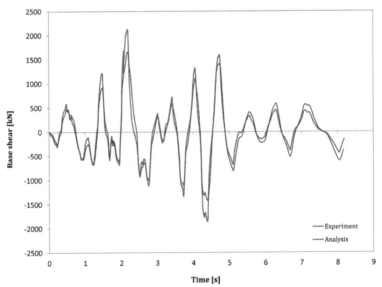

Figure 8.17. Experimental vs. Analytical results—base shear vs. time.

8.5 Recommendations for Advanced Engineering Practice

8.5.1 Structural Members Modelling Choices

As described in the previous sections, there is a large number of models available in the literature and implemented in existing finite element packages for the representation of the inelastic behaviour of beam-column elements. Nonlinear structural analysis models can vary significantly and lead to significantly different predictions with different modelling assumptions. Although it is highly desirable to directly simulate response at a mechanics-based conceptually sound analytical level, ultimately the predicted response must be validated against empirical test data at the component and the global levels. The choice between the phenomenological (more empirical) or fundamental (more analytical) models is not always straightforward and it depends on several factors that need to balance practical design requirements with available modelling capabilities and computational resources.

Some of the points that should be considered, when trying to determine the most appropriate model for each case, are:

i) The balance between the desire for accuracy in the analytical model and other unknowns, since the overall uncertainty is not necessarily reduced when a more sophisticated model is used (e.g., due to the lack of reliable data for the geometric or material characteristics of the building under consideration).

ii) The ability of the engineer to use and calibrate the hysteretic curves and the plastic hinge length of concentrated plasticity models.

iii) The availability of resources (time and effort) and computational tools (analysis software and computing capabilities).

iv) The need for the exact or not simulation of all modes of behaviour, based on the expected structural response and the capacity design methods employed (or not).

v) The need or not to reliably simulate the response of buildings over their full range of response, up to collapse.

vi) The analysis objectives and the required demand parameters, and the need for increased accuracy or not, and at which level of deformation (elastic or inelastic range).

8.5.2 Demand Performance Levels and Acceptance Criteria

As discussed in NIST (2010, 2013), since nonlinear analysis is an advanced method, which involves significantly more effort to perform than linear analysis, it should always be approached by the engineer with a clear view of the analysis objectives. For example, nonlinear analyses to establish equivalence with minimum code requirements for new building designs may differ from those for risk assessment of existing buildings. Depending on the intent of using a nonlinear analysis, the design basis should thus be clearly defined and agreed upon, outlining in specific terms all significant performance levels.

Once the goals of the nonlinear analysis and design basis are defined, the next step is to identify specific demand parameters and appropriate acceptance criteria to quantitatively evaluate the performance levels. In Performance-based engineering, the recommendations and guidelines for modelling the structure and carrying out the analyses must always be accompanied by recommendations and guidelines for employing the results of the analyses in determining if a structure meets the specified acceptance criteria. The demand parameters typically include peak forces and deformations in structural and non-structural components, peak or residual storey drifts, and floor accelerations, and are described in all modern Codes for assessment and retrofit. Other demand parameters, such as cumulative deformations or dissipated energy, may be checked to help confirm the accuracy of the analysis and/or to assess cumulative damage effects. There may also be other performance limits (e.g., the onset of structural damage), which might have major implications on lifecycle cost and functionality of the building, and should be checked (NIST, 2013).

For a given building and set of demand parameters, the structure must be analysed so that the values of the demand parameters are calculated with sufficient accuracy. The performance is checked by comparing the calculated values of demand parameters (in short the 'demand') to the acceptance criteria ('capacity') for the desired performance level. The acceptance criteria and the corresponding safety factors for seismic performance may vary depending on whether static or dynamic nonlinear analysis is used and how uncertainties associated with the demands and acceptance criteria are handled.

Further, acceptance criteria for structural components can generally be distinguished between 'deformation-controlled' actions (for members that can tolerate inelastic deformations), and 'force-controlled' actions (for non-ductile members whose capacity is governed by strength). Typical force-controlled actions are shear acting on any member, or axial force acting on columns. Typical deformation-controlled action is the bending moment or end chord-rotations developed on beams. In nonlinear analysis, the demands of force-controlled actions are typically compared to the strength of the component in question, which is determined using the nominal strength of the component. On the contrary, the demands of deformation-controlled actions are compared against deformation limits that correspond to the selected Performance Level.

8.6 Closing Remarks

The use of nonlinear analysis to assess the seismic response of both existing as well as newly-designed buildings (or other types of framed structures) is becoming more and more recurrent, in recognition of the importance of capturing as faithfully as possible the actual dynamic response of these structures to earthquakes.

Indeed, modern earthquake-resistance design codes now explicitly stipulate the need for seismic assessment of existing buildings to be undertaken through the employment of nonlinear analysis, a requirement that is emphasized even further in guidelines and regulations prescribed by the nuclear power industry and other energy sectors, where it is becoming mandatory for the safety of production plants and facilities (which include numerous types of framed structures) against earthquakes and other dynamic actions to be assessed through the employment of advanced nonlinear structural modelling.

Whilst such type of modelling prerequisites could in the past constitute a challenging task for practitioners, given that it was a field very much confined to the academic research community, structural engineers do now find themselves much better equipped to tackle such regulatory requirements, thanks both to an ever-augmenting computational power at our disposal, as well as to the advancements in the development of practical software tools that are nowadays capable of reproducing the response of framed structures well into the nonlinear inelastic range, as demonstrated by verifications against experimental test results.

Acknowledgements

The preparation of this chapter benefited significantly from the contributions of Zoi Gronti, Fanis Moschas and Dimitra Gerostamoulou, whose precious assistance is therefore gratefully recognized. In addition, the authors would also like to acknowledge the significant inspiration taken from the two reports prepared by the NEHRP Consultants Joint Venture (a partnership of the Applied Technology Council and the Consortium of Universities for Research in Earthquake Engineering) for the National Institute of Standards and Technology (NIST 2010, 2013) that are heavily cited in the main text and also listed below.

References

Antoniou, S. and Pinho, R. 2004. Development and verification of a displacement-based adaptive pushover procedure. Journal of Earthquake Engineering 8(5): 643–661.

[ASCE] American Society of Civil Engineers. 2014. Seismic Evaluation and Retrofit of Existing Buildings (ASCE/SEI 41-13), 2014, Reston, USA.

[ATC] Applied Technology Council. 1996. Seismic Evaluation and Retrofit of Concrete Buildings, ATC-40 Report, Applied Technology Council, Redwood City, USA.

[ATC] Applied Technology Council. 2009. Guidelines for seismic performance assessment of buildings, ATC 58, 50% Draft Report, Applied Technology Council, Redwood City, USA.

Bathe, K.J. 1996. Finite Element Procedures in Engineering Analysis, 2nd Edition, Prentice Hall.

Bianchi, F., Sousa, R. and Pinho, R. 2011. Blind prediction of a full-scale RC bridge column tested under dynamic conditions. Proceedings of 3rd International Conference on Computational Methods in Structural Dynamics and Earthquake Engineering (COMPDYN 2011). Paper no. 294, Corfu, Greece.

Calabrese, A., Almeida, J.P. and Pinho, R. 2010. Numerical issues in distributed inelasticity modelling of RC frame elements for seismic analysis. Journal of Earthquake Engineering 14(1): 38–68.

Carvalho, E.C., Coelho, E. and Campos-Costa, A. 1999. Preparation of the Full-Scale Tests on Reinforced Concrete Frames. Characteristics of the Test Specimens, Materials and Testing Conditions. ICONS

Report, Innovative Seismic Design Concepts for New and Existing Structures. European TMR Network, LNEC.

Clough, R.W. and Johnston, S.B. 1966. Effect of stiffness degradation on earthquake ductility requirements. Proceedings, Second Japan National Conference on Earthquake Engineering 1966: 227–232.

Cook, R.D., Malkus, D.S. and Plesha, M.E. 1989. Concepts and Applications of Finite Elements Analysis. John Wiley & Sons.

Correia, A.A. and Virtuoso, F.B.E. 2006. Nonlinear analysis of space frames. Mota Soares et al. (eds.). Proceedings of the Third European Conference on Computational Mechanics: Solids, Structures and Coupled Problems in Engineering. Lisbon, Portugal.

Crisfield, M.A. 1991. Non-linear Finite Element Analysis of Solids and Structures, John Wiley & Sons.

EN 1998-3. 2004. Eurocode 8: Design of structures for earthquake resistance. Part 3: Assessment and retrofitting of buildings.

Fardis, M.N. and Negro, P. 2006. SPEAR—Seismic performance assessment and rehabilitation of existing buildings. Proceedings of the International Workshop on the SPEAR Project, Ispra, Italy.

Felippa, C.A. 2001. Nonlinear Finite Element Methods, Lecture Notes, Centre for Aerospace Structure, College of Engineering, University of Colorado, USA. Available from URL: http://www.colorado.edu/engineering/CAS/courses.d/NFEM.d/Home.html.

[FEMA] Federal Emergency Management Agency. 1997. NEHRP Guidelines for the Seismic Rehabilitation of Buildings, FEMA 273 Report prepared by the Applied Technology Council and the Building Seismic Safety Council for the Federal Emergency Management Agency, Washington, D.C.

[FEMA] Federal Emergency Management Agency. 2005. Improvement of Nonlinear Static Seismic Analysis Procedures, FEMA 440. Report prepared by the Applied Technology Council for the Federal Emergency Management Agency, Washington, D.C.

[FEMA] Federal Emergency Management Agency. 2009a. Effects of Strength and Stiffness Degradation on Seismic Response, FEMA P-440A Report, prepared by the Applied Technology Council for the Federal Emergency Management Agency, Washington, D.C.

Giberson, M.F. 1967. The Response of Nonlinear Multi-Story Structures subjected to Earthquake Excitation, Doctoral Dissertation, California Institute of Technology, Pasadena, CA., May 1967: 232.

Hellesland, J. and Scordelis, A. 1981. Analysis of RC bridge columns under imposed deformations. IABSE Colloquium, Delft 545–559.

Hilber, H.M., Hughes, T.J.R. and Taylor, R.L. 1977. Improved numerical dissipation for time integration algorithms in structural dynamics. Earthquake Engineering and Structural Dynamics 5(3): 283–292.

Kircher, C.A., Nassar, A.A., Kustu, O. and Holmes, W.T. 1997a. Development of building damage functions for earthquake loss estimation. Earthquake Spectra 13(4): 663–682.

Kircher, C.A., Reitherman, R.K., Whitman, R.V. and Arnold, C. 1997b. Estimation of earthquake losses to buildings. Earthquake Spectra 13(4): 703–720.

Lanese, I., Nascimbene, R., Pavese, A. and Pinho, R. 2008. Numerical simulations of an infilled 3D frame in support of a shaking-table testing campaign. Proceedings of the RELUIS Conference on Assessment and Mitigation of Seismic Vulnerability of Existing Reinforced Concrete Structures, Rome, Italy.

Mari, A. and Scordelis, A. 1984. Nonlinear geometric material and time dependent analysis of three dimensional reinforced and prestressed concrete frames, SESM Report 82-12. Department of Civil Engineering, University of California, Berkeley, USA.

Martinelli, P. and Filippou, F.C. 2009. Simulation of the shaking table test of a seven-storey shear wall building, Earthquake Engineering and Structural Dynamics 38(5): 587–607.

Mpampatsikos, V., Nascimbene, R. and Petrini, L. 2008. A critical review of the R.C. frame existing building assessment procedure according to Eurocode 8 and Italian Seismic Code. Journal of Earthquake Engineering 12(SP1): 52–58.

Negro, P., Pinto, A.V., Verzeletti, G. and Magonette, G.E. 1996. PsD test on a four-storey R/C building designed according to eurocodes. Journal of Structural Engineering—ASCE 122(11): 1409–1417.

Neuenhofer, A. and Filippou, F.C. 1997. Evaluation of nonlinear frame finite-element models. Journal of Structural Engineering 123(7): 958–966.

Newmark, N.M. 1959. A method of computation for structural dynamics. Journal of the Engineering Mechanics Division, ASCE 85(EM3): 67–94.

[NIST] National Institute of Standards and Technology. 2010. Nonlinear Structural Analysis for Seismic Design, A Guide for Practicing Engineers, GCR 10-917-5, Gaithersburg, USA.

[NIST] National Institute of Standards and Technology. 2013. Nonlinear Analysis Research and Development Program for Performance-Based Seismic Engineering, GCR 14-917-27, Gaithersburg, Maryland.

Panagiotou, M., Restrepo, J.I. and Englekirk, R.E. 2006. Experimental seismic response of a full scale reinforced concrete wall building. Proceedings of the First European Conference on Earthquake Engineering and Seismology, Geneva, Switzerland, Paper no. 201.

Pavan, A. 2008. Blind Prediction of a Full-Scale 3D Steel Frame Tested under Dynamic Conditions, MSc Dissertation, ROSE School, Pavia, Italy.

[PEER] Pacific Earthquake Engineering Research Centre. 2010. Tall Buildings Initiative: Guidelines for Performance-Based Seismic Design of Tall Buildings. PEER Report 2010/05, Pacific Earthquake Engineering Research Center, Berkeley, USA.

PEER/ATC. 2010. Modeling and acceptance criteria for seismic design and analysis of tall buildings, PEER/ATC 72-1 Report. Applied Technology Council, Redwood City, USA.

Pinho, R. and Elnashai, A.S. 2000. Dynamic collapse testing of a full-scale four storey RC frame. ISET Journal of Earthquake Technology 37(4): 143–164.

Pinto, A., Verzeletti, G., Molina, F.J., Varum, H., Pinho, R. and Coelho, E. 1999. Pseudo-Dynamic Tests on Non-Seismic Resisting RC Frames (Bare and Selective Retrofit Frames). EUR Report, Joint Research Centre, Ispra, Italy.

Scott, M.H. and Fenves, G.L. 2006. Plastic hinge integration methods for force-based beam–column elements. ASCE Journal of Structural Engineering 132(2): 244–252.

Seismosoft. 2016. SeismoStruct—A computer program for static and dynamic nonlinear analysis of framed structures. Available from URL: www.seismosoft.com.

Silva, S., Crowley, H., Pagani, M., Monelli, D. and Pinho, R. 2014. Development of the OpenQuake engine, the Global Earthquake Model's open-source software for seismic risk assessment. Natural Hazards 72(3): 1409–1427.

Spacone, E., Ciampi, V. and Filippou, F.C. 1996. Mixed formulation of nonlinear beam finite element. Computers & Structures 58(1): 71–83.

Varum, H. 2003. Seismic Assessment, Strengthening and Repair of Existing Buildings, PhD Thesis, Department of Civil Engineering, University of Aveiro.

Willford, M., Whittaker, A. and Klemencic, R. 2008. Recommendations for the seismic design of high-rise buildings. Council on Tall Buildings and Urban Habitat, Illinois Institute of Technology, Chicago, IL.

Zienkiewicz, O.C. and Taylor, R.L. 1991. The Finite Element Method, 4th Edition, McGraw Hill.

9

Seismic Mitigation of Single Pylon Cable-Stayed Bridge

Qiang Han, Jianian Wen* and Xiuli Du

9.1 Introduction

In the last few decades, the single pylon cable-stayed bridge has become one of the most popular types of the bridges in engineering practice due to their aesthetic appeal, excellent spanning capacity, being able to rapidly and easily constructed, and efficient utilization of structural materials (Evangelista et al., 2003; Casciati et al., 2008; Abdel-Ghaffar et al., 1995). Since cable-stayed bridges play an important role in the modern transport networks, the main structural components in the bridge should remain elastic under the design earthquake motions (JTG/T D65-01–2007, 2007). The current practice in the design of single pylon cable-stayed bridges with short to medium span is to integrally construct the pylon, the deck, and the pier, in order to reduce the deformation of the bridge superstructure under regular loads and wind actions (Li, 2006). However, during the ChiChi earthquake, severe earthquake damage was reported in the Chi-Lu Bridge, one of the typical single pylon cable-stayed bridges with the aforementioned configurations. It was reported that flexural plastic hinges were developed at the pylon above the bridge deck in the transverse direction and vertical cracks in the concrete pylon extended upward to the height of the lowest cables. In addition, one of the stay-cables was pulled out from its top and bottom anchorages (Chang et al., 2004). Based on the

Key Laboratory of Urban Security and Disaster Engineering of Ministry of Education, Beijing University of Technology, Beijing 100124, China.
* Corresponding author: wjn@emails.bjut.edu.cn

post-earthquake damage investigation and numerical studies, it was found that although rigid connections at the junction of the deck, the pier and the pylon could reduce the deck displacement, it also significantly increased the shear force and the bending moment in the pylon under seismic loading, which could lead to bridge damage under strong earthquake shakings. One possible solution to this problem in the seismic design of bridge structures is to introduce energy dissipation devices and base isolation systems to the bridge to mitigate the dynamic response of the bridge. The seismic isolation strategy, widely used in seismic active regions to protect structures from earthquake damage for nearly three decades, has been proven as an effective and economical method for seismic control of structures. For the cable-stayed bridge, the seismic isolation devices could decouple the dynamic response of the deck-pylon system from that of the substructure to reduce the seismic inertia forces transmitted to the substructure, and thus effectively reduce the inelastic deformation in the superstructure. Therefore, seismic isolation could be a promising strategy to improve the seismic performance of the single pylon cable-stayed bridges. However, limited research and engineering work has been performed on the use of base isolation techniques on cable-stayed bridges.

In the last two decades, numerical studies have been performed to investigate the effects of different types of isolators with different installation locations, such as between the deck and the top of the bridge pier or between the pylon base and the foundation connection. Chadwell (2003) conducted a series of numerical studies to examine the effectiveness of the isolation system for protecting the Chi-Lu single pylon cable-stayed bridge. Nonlinear time history analyses were performed by Atmaca et al. (2014) to study the seismic performance of the Manavgat cable-stayed bridge isolated by friction pendulum bearings (FPB), which were applied between the pylon base and foundation. Javanmardi et al. (2017) investigated the bi-directional seismic responses of an existing steel cable-stayed bridge isolated by lead rubber bearing (LRB) at the base of the pylon. The elastic 3D beam elements were used to model the box girder and pylon, and the nonlinearity of the cable elements were taken into account in their numerical analyses. Soneji et al. (2008; 2010) studied the isolation effectiveness by placing different isolation devices between the pylon and the deck to protect the superstructure from the seismic damages during earthquakes. The soil-structure interaction was also considered in the numerical analyses. Wesolowsky et al. (2010) examined three cable-stayed bridges equipped with LRBs under near field ground motions. These studies have demonstrated that isolations in the single pylon cable-stayed bridge could substantially reduce the bending moment and shear force in the pylon, but slightly increased the displacements of the superstructure.

The previous numerical studies are mainly concentrated on the seismic performance of the cable-stayed bridge isolated by elastomeric bearings and FPB bearings. The behaviors of the cable-stayed bridge isolated by the double concave friction pendulum (DCFP) and the triple friction pendulum bearing (TFPB) have not been well investigated. DCFP and TFPB usually have larger displacement capacity and better adaptable ability and therefore can be more effective for the seismic isolation of the cable-stayed bridge. Moreover, most of the previous finite element (FE) models used lumped springs or linear elastic soil model to account for the soil-pile interaction (SPI), which do not necessarily lead to accurate estimate of the effects of SPI on the dynamic response of the bridges. Soneji et al. (2008) pointed out that, in the dynamic analyses of the bridge, it is essential to properly consider the nonlinear SPI, which significantly affects the seismic response of bridge structures. As addressed by Priestley et al. (1994), special cares are also needed to be taken during the numerical modeling of the bridges with short piers suffering from typical shear failure under seismic loading, because the traditional fiber beam element is unable to capture the shear or flexure-shear mechanisms of these short piers (Lu et al., 2015). Although nonlinear time-history analyses are most widely used techniques for seismic response analyses of structures, simplified seismic design procedure is needed for engineering practice. In addition, the investigation of simplified design procedure was relatively limitedin the studies aforementioned. Wesolowsky et al. (2010) proposed the design procedure for isolated cable-stayed bridge in the longitudinal direction. The approach is first to specify a design displacement and an estimated design period, and then determine the longitudinal stiffness of the isolated bearings. Chadwell (2003) developed the required stiffness relationship between the ends of the deck and the base of the pylon to optimize the effectiveness of the isolation system.

This chapter is the summarization of the work of Han et al. (2018a, 2018b). The following sections investigate the seismic performance of a single pylon cable-stayed bridge before and after the implementation of different seismic mitigation devices. Section 9.2 describes detailed physical and numerical models of cable-stayed bridge. The most currently used seismic mitigation devices are presented in Section 9.3. The effectiveness of friction sliding bearings and viscous dampers for the seismic performance of a typical single pylon cable-stayed bridge will be discussed in Section 9.4. Section 9.5 provides a seismic design procedure for isolated cable-stayed bridges.

9.2 Modelling of Cable-Stayed Bridge

Clearly establishing the modelling assumptions and simplifications conducted prior to the analysis is a key point. This section describes the physical and numerical models for cable-stayed bridge, including deck, pylon, cable, etc.

9.2.2 Deck

The deck in cable-stayed bridge with lateral cable layout is always modeled by separating girders (longitudinal and transverse girders) and slabs. Both longitudinal and transverse girders are discretized by 3D elastic beams. The slabs are modeled by means of shell elements. The distance between the level of beams and the slab is equal to the separation of the corresponding gravity centers, in order to capture properly the flexural and torsional behavior of the deck (Cámara, 2011), as shown

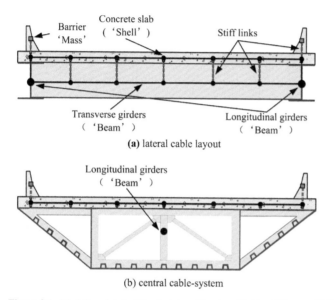

(a) lateral cable layout

(b) central cable-system

Figure 9.1. Modeling detail of the deck in cable-stayed bridge (Cámara, 2011).

in Fig. 9.1a. Generally, for the deck in cable-stayed bridge with central cable layout, the torsional response in bridges is clear, and a single elastic beam is used to model the response of the deck (Cámara, 2011), as shown in Fig. 9.1b.

9.2.3 Pylon and Pier

The pylon is always modeled by means of beam elements through the gravity center of each section (Cámara, 2011). The sections of pylon are discretized into confined and unconfined concrete fibers and reinforcing bars, as shown in Fig. 9.2. The fiber-discretization renders a realistic modelling of the different uniaxial materials. The bilinear and the Menegotto-Pintomodels (Filippou, 1983) are among the most used models for steel, whilst concrete may be characterized by tri-linear, nonlinear with constant orvariable confinement constitutive laws (e.g., Kent-Park Model (Kent, 1971)). In fiber modelling, the sectional stress–strain state of the elements is obtained through the integration ofthe nonlinear uniaxial stress–strain response of the individual fibers in which the section is subdivided, distinguishing steel, confined and unconfined concrete.

The optimum element is related to the length of the plastic hinge (l_p) developed in the concrete member, which can be calculated by

$$l_p = 0.08L_c + 0.022 f_{s,ye}\phi_{bl} \geq 0.044 f_{s,ye}\phi_{bl} \tag{9.1}$$

where, l_p is the plastic hinge length, L_c is the distance between the critical section of the plastic hinge and the point of contra-flexure, ϕ_{bl} is the diameter of longitudinal reinforcing bars. $f_{s,ye}$ is the design yield stress of reinforcing steel.

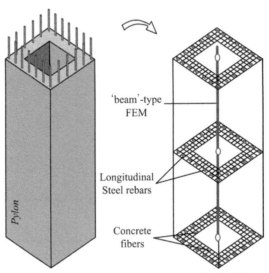

Figure 9.2. Modeling detail of the pylon in cable-stayed bridge(Cámara, 2011).

The fiber beam element is convenient to describe the flexural behavior of the slender structural components, but it is incapable of providing full insight into the short components with the shear or flexure-shear failure mechanisms. When the shear span ratio of RC pier, μ, is small (μ < 2), the shear failure is the dominated failure mode of the pier in the form of diagonal crushing of concrete and opening of inclined cracks (Lu et al., 2015). The smeared multi-layer shell model, proposed by Lu et al. (Lu et al., 2015), is capable of simulating the complex mechanical behavior of shear dominated pier. The pier in the bridge was modeled by the multi-layer shell element to depict the flexure-shear behavior. The shell was discretized into multiple fully-bonded layers consisting of concrete and reinforcing bars. The reinforcing bars were simulated as a smeared orthotropic steel layer of equivalent thickness in the multi-layered shell. The stress over a layer is assumed to be uniform and consistent with the mid-surface (Lu et al., 2015), as shown in Fig. 9.3. Different material properties and thicknesses were assigned to various layers. A two-dimensional material constitutive model of concrete based on the damage mechanics and smeared crack model was adopted. The two-dimensional material constitutive equation of the concrete is defined as:

$$\sigma'_c = \begin{bmatrix} 1-d_1 & \\ & 1-d_2 \end{bmatrix} D_e \varepsilon'_c \tag{9.2}$$

where σ_{c0} and ε_{c0} represent the stress and strain tensors, respectively; D_e is the elastic constitutive matrix; d_1 and d_2 are the damage parameters. Cracks are assumed to form in the layer as the principal tensile stress reaches the specified concrete tensile

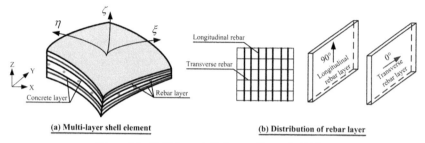

Figure 9.3. Multi-layer shell element (Lu et al., 2015).

strength. After cracking occurs, concrete is no longer isotropic but an orthotropic material, and the relationship between the shear stress τ and shear strain γ in the crack coordinates is defined as:

$$\tau = \beta G \gamma \tag{9.3}$$

where G is the elastic shear modulus and β is the shear retention factor.

9.2.4 Cable-System

The energy transfer between the cables, the tower, and the deck, is a dynamic characteristic of cable-stayed bridge, which can be taken into account by representing each cable with one element. The pre-stressed cables are simulated by the truss elements with initial stress. The axial stiffness of the inclined cables exhibits geometrical nonlinearities caused by cable sag. The geometrical nonlinearities can be idealized as the equivalent modulus of elasticity (E_{eq}) for different cables, and E_{eq} is defined as (Ernst, 1965):

$$E_{eq} = \frac{E}{\left[\dfrac{(WL)^2 AE}{12T^3} \right] + 1} \tag{9.4}$$

where E is the material modulus of elasticity for the straight cable; L is the horizontal projected length of the cable; W is the cable weight per unit length; A is the cross-section area of the cable; T is the pretension force.

9.2.5 Soil-Pile Interaction

The soil-pile interaction(SPI)in the cable-stayed bridge can be simulated by the dynamic p-y method (Boulanger, 1999). As shown in Fig. 9.4a, the nonlinear p-y behavior is conceptualized as consisting of elastic (p-y^e), plastic (p-y^p), and gap (p-y^g) components in series (Mazzoni, 2005). The gap component consists of a nonlinear closure spring (p^c-y^g) in parallel with a nonlinear drag spring (p^d-y^g). Radiation damping is modeled by a dashpot in parallel with the elastic (p-y^e) component. The plastic spring is described by (Boulanger, 1999):

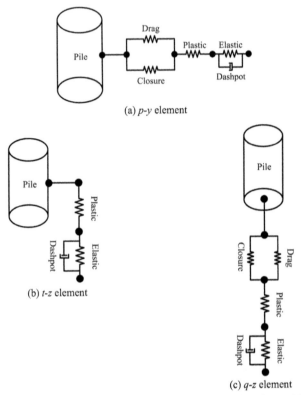

(a) *p-y* element

(b) *t-z* element

(c) *q-z* element

Figure 9.4. Components of nonlinear *p-y*, *t-z*, *q-z* element (Boulanger, 1999).

$$p = p_{ult} - (p_{ult} - p_0)\left[\frac{cy_{50}}{cy_{50} + |y^p - y_0^p|}\right]^n \tag{9.5}$$

where p_{ult} is the ultimate resistance of *p-y* element in the current loading direction; $p_0 = p$ at the start of the current plastic loading cycle; c is the constant to control the tangent modulus; n is the exponent to control sharpness of the *p-y^p* curve; y_{50} is the displacement at the half of the ultimate soil resistance.

The closure spring (p^c-y^g) is described by (Boulanger, 1999):

$$p^c = 1.8 p_{ult}\left[\frac{y_{50}}{y_{50} + 50(y_0^+ - y^g)} - \frac{y_{50}}{y_{50} - 50(y_0^- - y^g)}\right] \tag{9.6}$$

where y_0^+ is the memory term for the positive side of the gap; and y_0^- is the memory term for the negative side of the gap.

The nonlinear drag spring can bedescribed by (Boulanger, 1999):

$$p^d = C_d P_{ult} - (C_d P_{ult} - p_0^d) \left[\frac{y_{50}}{y_{50} + 2|y^g - y_0^g|} \right] \tag{9.7}$$

where C_d is the ratio of the maximum drag force to the ultimate resistance of the *p-y* element; $p_0^d = p^d$, $y_0^d = y^d$ at the start of the current cycle. A typical response of a *p-y* element is shown in Fig. 9.5a.

The nonlinear *t-z* and *q-z* behavior are similar as *p-y* element. The *t-z* element consists of elastic (t^e-z^e) and plastic (t^p-z^p) components in series (Boulanger, 1999), as shown in Fig. 9.4b. The nonlinear *q-z* behavior is conceptualized as consisting of elastic (q-z^e), plastic (q-z^p), and gap (q-z^g) components in series (Boulanger, 1999), as shown in Fig. 9.4c. The typical responses of *t-z* and *q-z* elements are shown in Fig. 9.5b and c.

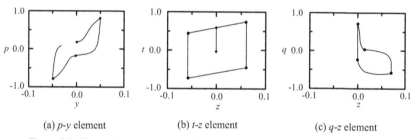

(a) *p-y* element   (b) *t-z* element   (c) *q-z* element

Figure 9.5. Hysteretic behaviour of nonlinear *p-y*, *t-z*, *q-z* element (Boulanger, 1999).

9.3 Seismic Mitigation Devices

9.3.1 General

Several types of isolation bearings and energy dissipation devices have been used in cable-stayed bridge and many new devices are being proposed and investigated. Most currently used seismic mitigation devices include: (i) viscous damper; (ii) metallic dampers; (iii) friction sliding bearings; (iv) rubber bearings; (v) shape memory alloy re-centering devices. This section is focused on the mechanical properties of viscous damper and friction sliding bearings.

9.3.2 Viscous Damper

Viscous dampers dissipate earthquake energy by pushing fluid through orifices, transforming the external work into heat and consequently elevating the temperature of the viscous fluid and its mechanical parts (Cámara, 2011), as shown in Fig. 9.6.

Passive supplement damping can control the level of displacement and dissipate the seismic energy. The force output of a viscous damper is generally considered as a function of the relative velocity (Constantinou et al., 2010):

$$F = C|\dot{\Delta}|^d \, \text{sgn}(\dot{\Delta}) \tag{9.8}$$

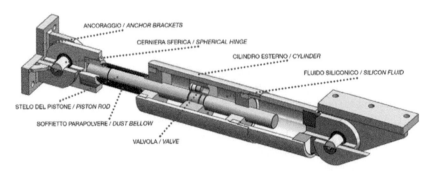

Figure 9.6. Mechanical parts of a viscous damper (Cámara, 2011).

where C is the damping constant, which is dependent on the area of the piston; $\dot{\Delta}$ is the relative velocity between the ends of the damper; α is a velocity exponent ranging between 0.1 and 2.0, however, the range of 0.2 to 1.0 are often used to limit the force output and maximize the energy dissipation (Constantinou et al., 2010).

9.3.3 Friction Sliding Bearings

Friction sliding bearing is a kind of completely passive device which makes use of curved sliding surfaces to generate re-centering force and provide frictional force to exhibit displacement-dependent stiffness and damping. Friction sliding bearing is polished with stainless steel, so that it can satisfy the large vertical bearing capacity, large deformation and excellent durability requirements. Friction sliding bearing can mainly be divided into three different types: (i) Friction Pendulum System (FPS) (Zayas et al., 1990); (ii) Double Concave Friction Pendulum (DCFP) (Fenz et al., 2010a); (iii) Triple Friction Pendulum Bearing (TFPB) (Fenz et al., 2008; Fenz et al., 2010b).

9.3.3.1 Mechanical Behavior of FPB

The classical force-displacement behavior of FPB is described in the following equation:

$$F = \frac{Wd}{R_1} + \mu_1 W \, \mathrm{sgn}\left(\dot{\theta}\right) \tag{9.9}$$

where F is the restoring force; μ_1 is the coefficient of friction; W is the vertical load; R_1 is the radius of the upper spherical surface; d is the relative displacement of the articulated slider and the upper spherical surface; $\mathrm{sgn}\,(\cdot)$ is the signum function, and the value returns -1 or $+1$ which is depended on the negative or positive of the velocity $\dot{\theta}$. The corresponding skeleton curves are shown in Fig. 9.7a.

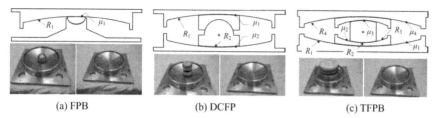

| (a) FPB | (b) DCFP | (c) TFPB |

Figure 9.7. General view of friction sliding bearings.

9.3.3.2 Mechanical Behavior of DCFP

The DCFP used in this numerical study consists of two frictional concaved surfaces with the same configurations and articulated slider, as shown in Fig. 9.7b. The force-displacement relationship of the DCFP is presented in the following equation:

$$F = \frac{W}{(R_1 + R_2)}d + \frac{f_1 R_1 + f_2 R_2}{R_1 + R_2} \tag{9.10}$$

where R_1 and R_2 are the radius of the upper and lower spherical surface; d is the relative displacement of the upper and lower spherical surface; f_1 and f_2 are the friction force of the upper and lower spherical surface. The DCFP can be modeled as two FPB elements with different parameters.

9.3.3.3 Mechanical Behavior of TFPB

The cross section and its details of TFPB are shown in Fig. 9.7c. The center of the TFPB is the rigid slider, which contacts with the upper and lower nested sliders, and the outer of the bearing are upper and lower plates. The TFPB consists of four concave stainless steel surfaces which form three independent sliding systems, where the inner and outer radius of the sliding surfaces are different ($R_{e2}=R_{e3}<<R_{e1}=R_{e4}$). When the coefficients of friction on sliding surfaces are different, the sliding stages can be divided into five different stages, as shown in Table 9.1.

9.4 Seismic Response Analyses of Single Pylon Cable-Stayed Bridge (Han et al. 2018a)

9.4.1 Numerical Model of the Single Pylon Cable-Stayed Bridge

9.4.1.1 Prototype of the Bridge

The prototype of the bridge model used here is the Longwan Bridge crossing the Jian River in Guizhou Province, China. The bridge is a semi-harp type cable-stayed bridge with two unequal spans of 120 m and 114 m, as shown in Fig. 9.8. The single RC pylon is approximately 62 m tall above the deck and sustains the

Table 9.1. Force-displacement relationship of the TFP bearing (Fenz, 2010b).

Sliding mechanism	Description	Force-displacement relationship
	Sliding on surfaces 2 and 3	$F = \dfrac{W}{R_2 + R_3}\,d + \dfrac{f_2 R_2 + f_3 R_3}{R_2 + R_3}$ Valid until: $d = d^* = (\mu_1 - \mu_2)\,R_2 + (\mu_1 - \mu_3)R_3$
	Sliding on surface −1 and 3	$F = \dfrac{W}{R_1 + R_3}\,d + \dfrac{f_1(R_1 - R_2) + f_2 R_2 + f_3 R_3}{R_1 + R_3}$ Valid until: $d = d^{**} = d^* + (\mu_4 - \mu_1)\,(R_1 + R_3)$
	Sliding on surface −1 and 4	$F = \dfrac{W}{R_1 + R_4}\,d + \dfrac{f_1(R_1 - R_2) + f_2 R_2 + f_3 R_3 + f_4(R_4 - R_3)}{R_1 + R_4}$ Valid until: $d = d_{dr1} = d^{**} + d_1\left(1 + \dfrac{R_4}{R_1}\right) - (\mu_4 - \mu_1)(R_1 + R_4)$
	Sliding on surfaces 2 and 4	Valid until: $F = \dfrac{W}{R_2 + R_4}(d - d_{dr1}) + \dfrac{W}{R_1}D_1 + f_1$
	Sliding on surfaces 2 and 3	$F = \dfrac{W}{R_2 + R_3}(d - d_{dr4}) + \dfrac{W}{R_4}D_4 + f_4$ $d = d_{dr4} = d_{dr1} + \left[\left(\dfrac{d_4}{R_4} + \mu_4\right) - \left(\dfrac{d_1}{R_1} + \mu_1\right)\right](R_2 + R_4)$

Note: R_i, f_i, and μ_i are the effective radii, friction force, and coefficient of friction on the i sliding surface, respectively.

superstructure with 16 pairs of steel stay cables. The deck is rigidly connected to the pylon to limit the displacement of the deck under regular loads. Each end of the Longwan Bridge is simply supported on an abutment.

9.4.1.2 Numerical Model of the Bridge

Three dimensional (3D) FE models of the Longwan Bridge with and without isolation system, as illustrated in Fig. 9.9, were developed in OpenSees platform (Mazzoni, 2005), an object oriented, open source FE analysis framework. In the FE model of the example bridge, the fiber elements were used for the pylon and the multi-layer shell elements were used for the thin-wall RC rectangular pier, in order to reproduce the nonlinear characteristics of the critical bridge components under strong ground motions. As a seismic control scheme, friction sliding bearings and viscous dampers were implemented between the pier and the pile cap of the cable-stayed bridge, as indicated in Fig. 9.9. The SPI was simulated by the dynamic *p-y* method. The main longitudinal RC box girder of the cable-stayed bridge was simplified by 3D linear elastic beam elements.

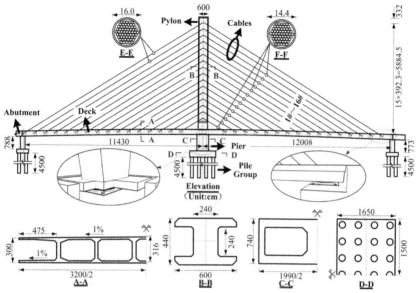

Figure 9.8. General view of the example cable-stayed bridge.

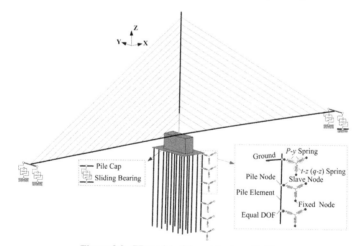

Figure 9.9. FE model of the cable-stayed bridge.

9.4.2 Dynamic Characteristics of Different Bridge Models

The dynamic characteristics of the cable-stayed bridge with four different boundary conditions: the fixed-base original bridge model, the original bridge model considering SPI and the base isolated bridge model with and without SPI are presented in the following section. The mode shapes and the corresponding vibration periods of the first 4 modes are summarized in Table 9.2. Compared to the

Table 9.2. Modal analysis of the different bridge models.

Mode	Original bridge with Fixed base	Original bridge with SPI	Isolated bridge with fixed base	Isolated bridge with SPI
1	1.805s	1.810s	4.247s	4.254s
2	1.158s	1.160s	4.213	4.217s
3	0.938s	0.939s	1.918	1.929s
4	0.461s	0.539s	1.195	1.197s

original bridge model with the fixed base, the primary vibration mode of the original bridge with SPI is the bending mode along the transverse direction, similar to the fixed base model. However, the higher modes (after the 3rd mode) are significantly affected by the SPI. It can be also observed that the dynamic characteristics of the base isolated bridge are not much affected by the SPI consideration. A comparison between the original bridge model and the isolated bridge model show that isolation system significantly increases the fundamental period of the bridge to a value larger than the predominate period of earthquake motions.

9.4.3 Nonlinear Seismic Response

9.4.3.1 Selection of Ground Motion

The seismic ground motion intensity for the Longwan Bridge site is VII in Chinese seismic map, the site classification is III and the corresponding peak ground acceleration (PGA) is 0.15 g. The design response spectrum was constructed with 2% damping following the guidelines in Chinese Specification of Earthquake Resistant Design for Highway Engineering (GBT 7714, 2013), as shown by the black solid curve in Fig. 9.10. Each record consists of two horizontal components. The average PGA of the three selected ground motions is scaled to the design ground motion intensity defined in the seismic code for the site. Figure 9.10 compares the response spectrum of the three PGA scaled ground motion time histories and the averaged response spectrum of the three ground motions with the design response spectrum.

9.4.3.2 Bending Moment at the Deck-Pylon Junction

Figure 9.11a compares the bending moments at the deck-pylon junction in the transverse direction of the bridge under the Artificial-1 input motion, which produced the largest transverse response in the bridge among all the cases. The beading moment (M_y) is normalized to the moment capacity (M_u) of the pylon along its transverse direction. The bending capacity of a cross section is determined by taking into account the axial force and biaxial bending. As shown in Fig. 9.11a, the FPS bearings slightly reduce the bending moment, and the bridge pier is adequate to meet the capacity demand in the transverse direction. Influence of SPI on the bending moments at the deck-pylon junction in the transverse direction under Artificial-1 input motion is shown in Fig. 9.11d. The results indicate that the increase in peak value of the bending moment is 9%.

The demand and capacity of biaxial bending of the cross section at the pylon base along both the longitudinal and transverse directions of the bridge is compared in Fig. 9.12. The close-to elliptical solid curves stands for the bending capacity of the cross section. It is found that when the fixed-base bridge model subjected to Artificial-1 ground motion, the bending moment in the transverse direction of the pylon base is close to its bending capacity. After the implementation of friction sliding bearings, the moment along the longitudinal direction of the bridge decreases significantly, however, the moment in the transverse direction only decreases slightly. These observations explain the damages observed in the Chi-Lu Bridge during the post-earthquake investigation: cracks at the base of the pylon were observed on the transverse faces. However, damage to the opposite faces (longitudinal faces) of the pylon was not as severe.

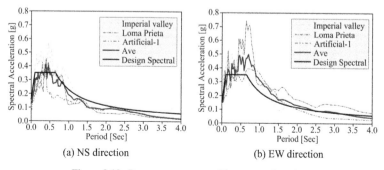

Figure 9.10. Response spectrum of three ground motions.

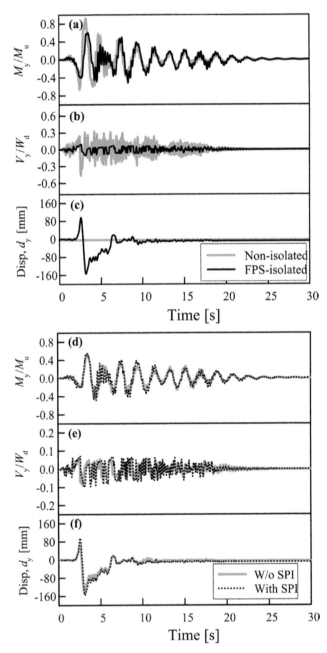

Figure 9.11. Time histories of: (a) and (d) bending moment, (b) and (e) shear force, and (c) and (f) relative displacement in the transverse direction under Artificial-1 ground motion.

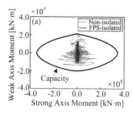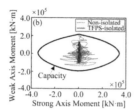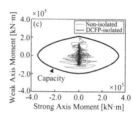

Figure 9.12. Comparison of demand and capacity of biaxial bending of the cross section: (a) FPS; (b) DCFP; (c) TFP.

The bending moments at the deck-pylon junction for all cases are also compared in Fig. 9.13a. The longitudinal moment show a significant decrease and the maximum reductions are 80.9%, 86.4%, and 85.8% after implementation of the FPS, TFPS, and DCFP under El Centro record. However, the transverse moment decreases only slightly with the reduction rates of 10.3%, 10.8%, and 9.9% after implementation of the FPS, TFPS, and DCFP under the Artificial-1 motion. It is because that the bridge pylon, restrained by the cables in its longitudinal direction, is much more rigid than its transverse direction. The friction sliding bearing can effectively reduce the longitudinal bending moment at the pylon base by lengthening the fundamental period of the bridge. However, implementing the friction sliding bearings has only insignificant effect on the vibration modes in the transverse direction. It can be observed in Table 9.2 that the time periods of the bending deformation of the bridge pylon in the transverse direction for both the non-isolated and isolated bridge are around 1.8s. Since the base isolation device does not significantly affect the vibration mode of the bridge pylon in the transverse direction, the response in the transverse direction is therefore not significantly changed. Consequently, the reduction rates of the transverse bending moment at the pylon base is much less than those in the longitudinal direction.

9.4.3.3 Base Shear Force

Figure11b shows the shear force generated at the base of the pier in the transverse direction under Artificial-1 ground motion. The shear force (V_y) is normalized by the weight of deck (W_d). From the figure, it can be observed that the reduction in base shear is substantial after implementation of friction sliding bearings. Considering SPI results in a slightly larger, about 7%, base shear response of the pier along the transverse direction than that predicted without considering SPI, as shown in Fig. 9.11e.

The maximum shear force obtained from analysis for all cases under earthquake excitations are given in Fig. 9.13b. The peak base shear force in the longitudinal direction is 2.24×10^5 kN under LGPC ground motion and the reductions are 82.5%, 89.8%, and 90.4% when the FPS, TFPS, and DCFP are utilized at the base of the pylon. Similarly, the peak shear force in the transverse direction is 1.10×10^5 kN under El-Centro ground motion, and the reductions of the peak value are 87.9%,

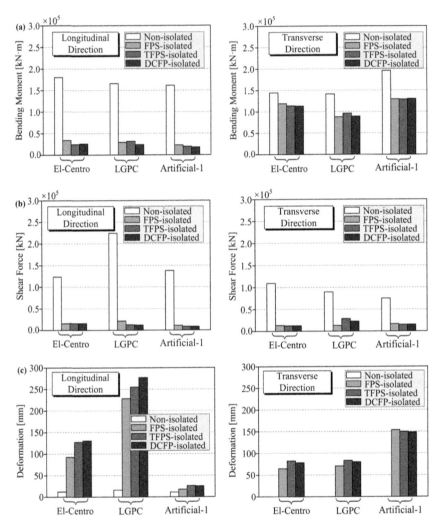

Figure 9.13. Summary of isolation results: (a) bending moment at the deck-pylon junction; (b) base shear force; (c) displacement at the abutment.

88.6%, and 88.8% after implementation of the FPS, TFPS, and DCFP. Obviously, the base isolated cable-stayed bridge can significantly reduce the seismic inertia forces transmitted to the substructure and eliminate inelastic deformation in the RC pier.

9.4.3.4 Deck Displacement and Hysteretic Curves of Isolation Bearing

Figure 9.11c presents the relative displacement between the pier and foundation with or without isolation bearings in the transverse direction under Artificial-1 ground motion. It can be observed that the displacement of the isolated bridge increases significantly when compared with that of the fixed-base bridge. Considering SPI also increases the relative displacement between the pier and foundation and isolation bearings by 4%, as shown in Fig. 9.11f. The peak values of the deck displacement with and without sliding bearings in all cases are shown in Fig. 9.13c. The maximum longitudinal deck displacement increases from 16 mm to 277 mm after implementation of the DCFP subjected to the LGPC records from Loma Prieta earthquake. In the transverse direction, the maximum deck displacement increases from 0 mm to 154 mm after implementation of the TFPS under the Artificial-1 ground motion. Obviously, the relative displacement of isolated cable-stayed bridge increases significantly in the both directions, because the boundary conditions at the pylon-foundation connection and abutments are changed.

The force-displacement relationships of the FPS, TFPS, and DCFP are presented in Fig. 9.14 under El Centro ground motion. The hysteretic curves are plotted with or without considering the bi-directional coupled behavior. The uncoupling behavior of the bearing is simulated by MultiLinear uniaxial material. The area of the hysteretic loops has a slight reduction when the bi-directional coupled behavior is considered. Thus, it will lead to less dissipation of seismic energy and relatively large peak displacement.

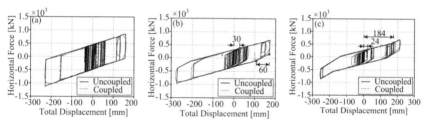

Figure 9.14. Hysteretic curves for different isolators: (a) FPS; (b) DCFP; (c) TFP.

9.4.3.5 Cable Response

Because the cable-stayed bridge is approximately symmetric in longitudinal direction, only the responses of the cables on the right side numbered from #1 to #16 are presented here, as shown in Fig. 9.15.

From the nonlinear numerical analyses, as indicated in Fig. 9.15a, c, and e, it is obvious that the longest cable (#16) experiences the highest tensile stress compared to the other cables. Moreover, it is shown in Fig. 9.15b, d, and f that, subject to ground motion excitations, the increase in the stress from the initial stress of the longest cable (#16) is most significant among all the cables, because the longest cable has lager deformation. From the figure, it is also observed that the base isolations can reduce the peak stress in the cables by up to 8%, 15%, and 8% under El-Centro, LGPC, and Artificial-1 ground motions, and all the peak stresses of the cables are within the allowable stress level.

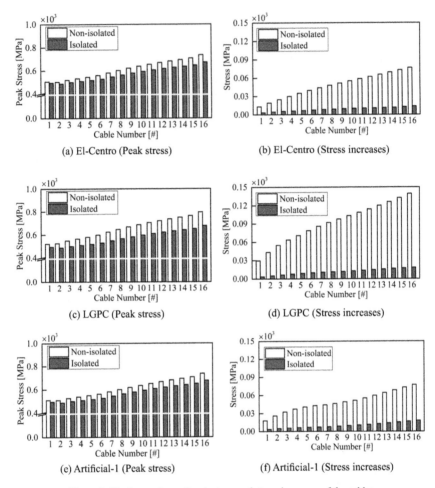

Figure 9.15. Comparison of peak stress and stress increases of the cables.

9.4.3.6 Pile Foundation Response

In this section, the bending moment of the piles and the lateral displacement of the soil are studied as the main parameters to investigate the effect of base isolation on the SPI. Figure 9.16a to f present the bending moments along the depth of the pile for different rows. The reduction in the bending moment of the piles is significant after the implementation of the FPS. The maximum bending moment in the longitudinal direction is 1.71×10^5 kN·m subjected to LGPC ground motion and the reduction is 90.2%. Similarly, the peak value in transverse direction is 1.35×10^5 kN·m under El Centro ground motion, and the reduction of the peak value is 85.5%. Hysteretic loops for soil reactions at the top of soil layers are shown in Fig. 9.16g to l. It is indicated that the soil plasticity has been developed for the non-isolated bridge,

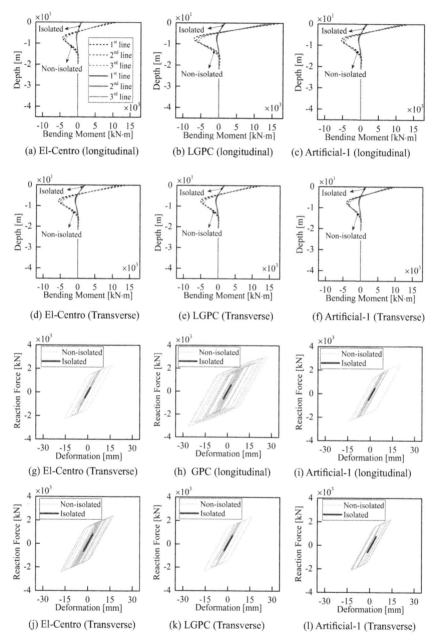

Figure 9.16. Responses of substructure: (a) to (f) bending moment distribution along the lengths of the piles; (g) to (l) hysteretic response of top soil layer.

while the base isolation is found to be very effective in reducing soil deformation and the top soil layer remains elastic for the three considered ground motions.

9.4.3.7 Pier Response

Figure 9.17 presents that the principal stress contours on the core concrete under El-Centro, LGPC, and Artificial-1 ground motions, which is visualized by GiD postprocessor. The FE results show large diagonal stress for the non-isolated cable-stayed bridge, indicating possible shear failure of the pier. During nonlinear time history analysis, peak values were observed to initiate at the bottom of pier and radiate out from this location at approximately 45° angles with respect to the diagonal orientation. However, the distributions of the principal stress on the core concrete are different after implementation of the base isolation system. Figure 9.16 illustrates that the peak values were observed at the top and bottom of the pier, and reduction in the amplitudes of the principal stress is prominent.

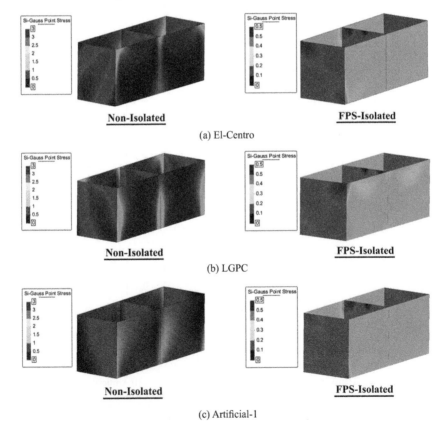

Figure 9.17. Comparison of principal stress distribution of the pylon.

9.4.4 Seismic Response of the Bridge with Viscous Dampers

In the particular base isolated cable-stayed bridge, connecting the deck to the abutments and the pylon to the foundation by means of viscous dampers is proposed. A series of velocity exponent (0.3, 0.6 and 1.0) and the damping constants (1000, 2000, 3000, 4000 and 5000 kN/(m·s)^a) are selected to evaluate the effectiveness of the viscous dampers on controlling the bridge deck displacement under El Centro ground motion. An index J is adopted to describe the effectiveness of viscous damper in controlling the seismic responses of the bridge:

$$J = \frac{D-E}{D} \times 100\% \tag{9.11}$$

where D and E are seismic responses of the base isolated bridge without and with viscous dampers, respectively.

As shown in Fig. 9.18a, it is observed that the displacement of the superstructure decreases with increasing values of the damping constants. The nonlinear viscous damper with a smaller velocity exponent can reduce the maximum displacement. The displacement of the deck can decrease by 3% to 35%, depending on the damping

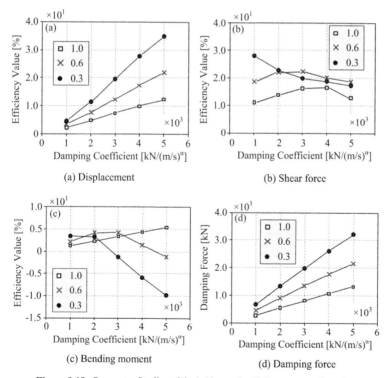

(a) Displacement

(b) Shear force

(c) Bending moment

(d) Damping force

Figure 9.18. Parameter Studies of the bridge under El Centro ground motion.

constant and velocity exponent. Figure 9.18b and c show the comparisons of the base shear for the bridge implemented with viscous dampers of different parameters. As shown the base shear forces and bending moments are not always reduced, such as in the case with C = 3000, 4000, 5000 kN/(m·s)$^\alpha$ and α = 0.3, 0.6. This is because larger damping constant and smaller velocity exponent result in larger damping force (Fig. 9.18d), this, in turn, would increase the base shear by exerting counteracting forces. The maximum damping force in the devices is 3150 kN (Fig. 9.18c), which is lower than the capacity of the commercial fluid viscous damper.

9.5 Seismic Design of Base Isolated Cable-Stayed Bridge (Han et al. 2018b)

In this section, a direct displacement based seismic design (DDBD) (Priestley et al., 2007) procedure for base isolated cable-stayed bridge under transverse seismic excitation is presented.

9.5.1 Substitute Structure for Base Isolated Cable-Stayed Bridge

The first step of the DDBD procedure is to identify the target displacement profile of the bridge deck, which is commonly determined by the first inelastic deformation shape (Adhikari et al., 2010), before the transformation of the bridge model to an equivalent single-degree of freedom (S-DOF). In a base isolated cable-stayed bridge, isolation bearings are installed at the pylon base and the abutments of the bridges, as shown in Fig. 9.19a. The shape of the transverse vibration modes of the bridge deck is closely dependent on the relative flexibility between the isolation system placed at the base of the pylon and at the ends of the deck. For the case when the transverse stiffness of the isolation system placed at the ends of the deck is significantly larger than that of the isolation system placed at the base of the pylon, the fundamental transverse vibration mode is illustrated in Fig. 9.19b. Figure 9.19c presents the fundamental transverse vibration mode of the bridge for the opposite case, in which the transverse stiffness of the isolation system placed at the ends of the deck is negligible compared to that of the isolation system placed at the base of the pylon. The other vibration modes are bounded between these two modes. To create an effective isolation system in the transverse direction, the stiffness of the isolation system placed at the base of the pylon should be tuned to the stiffness of the isolation system placed at the ends of the deck of the bridge so as to achieve uniform translation of the bridge as shown in Fig. 9.19d. Otherwise, the bending of the deck along the strong axis (bending around the z-axis, as shown in Fig. 9.19a) may result in severe damage at the connection of the deck and the pylon.

To determine the target stiffness ratio between the isolation system placed at the base of the pylon and the ends of the deck of the bridges that minimize transverse bending of the deck, a simplified two-degree of freedom (2-DOF) model was built in the generalized coordinates, which can be described by two shape functions, as

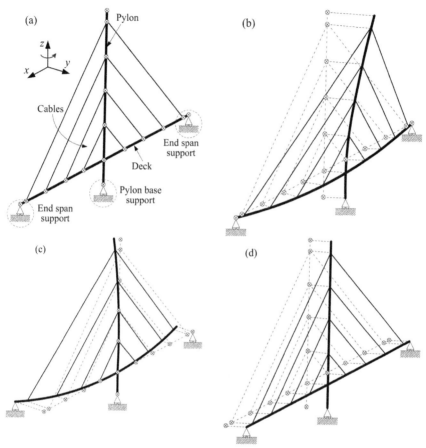

Figure 9.19. Transverse modes of vibration: (a) bridge model; (b) end span support restrained; (c) pylon base support restrained; (d) target displacement profile.

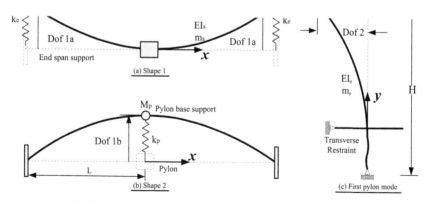

Figure 9.20. Generalized function and interpolation of superstructure (Chadwell, 2003).

shown in Fig. 9.20a and b. In the first shape function, φ_{1a}, assuming that the stiffness of the bearings at the abutments is significantly smaller than that of the bearings under the pylon, the corresponding shape function is the bending of deck along the strong axis. On the contrary, in the second shape function, φ_{1b}, considering the rigid bearings at the abutments, the pylon and the deck move along the transverse direction of the bridge. It is important to note that, in the shape function, the pylon is assumed to be rigid.

The characteristic equation of the 2-DOF model is given by:

$$\boldsymbol{K\Phi}_n = \omega_n^2 \boldsymbol{M\Phi}_n \tag{9.12a}$$

$$\begin{pmatrix} k_{11} & k_{12} \\ k_{21} & k_{22} \end{pmatrix} \begin{pmatrix} \varphi_{1a} \\ \varphi_{1b} \end{pmatrix} = \omega_n^2 \begin{pmatrix} m_{11} & m_{12} \\ m_{21} & m_{22} \end{pmatrix} \begin{pmatrix} \varphi_{1a} \\ \varphi_{1b} \end{pmatrix} \tag{9.12b}$$

where $\boldsymbol{K}$ and $\boldsymbol{M}$ are the stiffness and mass matrices; ω_n and φ_n are the natural frequencies and modes. The stiffness and mass matrix elements are defined as follows:

$$k_{ij}^e = \int_0^l EI(x)\varphi_{1i}''(x)\varphi_{1j}''(x)dx \tag{9.13}$$

$$m_{ij}^e = \int_0^l m(x)\varphi_{1i}(x)\varphi_{1j}(x)dx \tag{9.14}$$

The shape functions are given as (Chadwell, 2003):

$$\varphi_{1a}(x) = \frac{3}{2}\left(\frac{x}{L}\right)^2 - \frac{1}{2}\left(\frac{x}{L}\right)^3 \tag{9.15}$$

$$\varphi_{1b}(x) = 1 - \left[\frac{3}{2}\left(\frac{x}{L}\right)^2 - \frac{1}{2}\left(\frac{x}{L}\right)^3\right] \tag{9.16}$$

where, L is the length of one span.

Therefore, the elements in the stiffness and mass matrices in Eq. (9.12b) are derived as:

$$k_{11} = \frac{6EI_s}{L^3} + 2k_e, \ k_{12} = k_{21} = -\frac{6EI_s}{L^3}, \ k_{22} = \frac{6EI_s}{L^3} + k_p \tag{9.17}$$

$$m_{11} = \frac{33}{70}m_sL, \ m_{12} = \frac{39}{140}m_sL, \ m_{22} = \frac{34}{35}m_sL + M_p$$

where, M_p is the total mass of the pylon; m_s is the mass of the superstructure (per length); k_p and k_e are the stiffness under the pylon and the stiffness at the end of the deck.

Given that the optimized deformation shape is the uniform translation of the cable-stayed bridge, the modal coordinates of the generalized shape functions are $\varphi_{1a} = \varphi_{1b} = 1$. The Eq. (9.12a) can be expanded to the following equations:

$$\begin{cases} k_{11} + k_{12} = \omega_n^2 \left(m_{11} + m_{12} \right) \\ k_{21} + k_{22} = \omega_n^2 \left(m_{21} + m_{22} \right) \end{cases} \qquad (9.18)$$

Substitution of Eq. (9.17) into Eq. (9.18), the relationship of the stiffness of the bearings under the pylon and the bearings at the end of the deck can be calculated as follows:

$$\frac{k_p}{k_e} = \frac{10}{3} + \frac{16}{3} \left(\frac{M_p}{M_s} \right) \qquad (9.19)$$

where, M_s is the total mass of the superstructure.

In the derivation of Eq. (9.19), the pylon is assumed to be rigid. The flexibility of the pylon will be considered in the following steps (Fig. 9.20c). In order to take the stiffness of the pylon into consideration, another simplified 2-DOF system is built, as shown in Fig. 9.21. DOF 1 represents the uniform translation of the cable-stayed bridge and DOF 2 indicates the bending deformation of the pylon. The effective mass and stiffness of the pylon can be computed by:

$$\begin{aligned} m_{eff} &= \frac{1}{U_m^2} \sum_j P_j U_j^2 \\ k_{eff} &= \frac{g}{U_m^2} \sum_j P_j U_j \end{aligned} \qquad (9.20)$$

where, P_j is the mass of the jth node in the discretized model; U_j is the displacement of the jth node in the discretized model; g is the gravitational acceleration; U_m is the transverse displacement of the node at the top of the pylon.

The stiffness and mass elements in the characteristic equation of the 2-DOF system, Eq. (9.1a), for the new 2-DOF system are derived as:

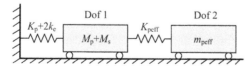

Figure 9.21. A simplified 2-DOF model of pylon.

$$\begin{pmatrix} k_p + 2k_e + k_{eff} & -k_{eff} \\ -k_{eff} & k_{eff} \end{pmatrix} \begin{pmatrix} \varphi_{1a} \\ \varphi_{1b} \end{pmatrix} = \omega_n^2 \begin{pmatrix} M_p + M_s & 0 \\ 0 & m_{eff} \end{pmatrix} \begin{pmatrix} \varphi_{1a} \\ \varphi_{1b} \end{pmatrix} \tag{9.21}$$

Combining Eq. (9.21) and Eq. (9.19), the following equation can be obtained:

$$\left(k_{eff} - \omega_n^2 m_{eff}\right) \left[k_p \left(\frac{8M_p + 8M_s}{5M_p + 8M_s} \right) + k_{eff} - \omega_n^2 \left(M_p + M_s\right) \right] = k_{eff}^2 \tag{9.22}$$

The relation between the stiffness of the pylon base and the target period can be derived from Eq. (9.22):

$$k_p = \frac{1}{2} \frac{\pi^2}{T^2 - T_1^2} \left[\left(1 + \frac{m_{eff}}{M_p + M_s} \right) - \left(\frac{T_1}{T} \right)^2 \right] \left(5M_s + 8M_p\right) \tag{9.23}$$

where, T is the first period of vibration; T_1 can be calculated using the following equation:

$$T_1 = 2\pi \sqrt{\frac{m_{eff}}{k_{eff}}} \tag{9.24}$$

In the DDBD procedure, it is required to convert the nonlinear M-DOF model of the structure into an equivalent linear S-DOF system, which involves the assumption of a target displacement profile. As shown in Fig. 9.21, the transverse displacement profile of the single pylon cable-stayed bridge includes theuniform translation of the whole bridge (DoF 1) and transverse bending of the pylon (DoF 2). It is found that the effective mass of the DoF 2 is significantly smaller than the DoF 1. Therefore, the effects of the pylon in transverse bending are ignored in the proposed DDBD procedure, which greatly simplifies the design procedure. In general, the base isolated cable stayed-bridge can be transferred into an S-DOF equivalent structural model with characteristic equation given by Eq. (9.23).

9.5.2 Design Algorithm

The primary objective of the DDBD methodology is to determinethe critical characteristics of the isolators, which deforms at the target displacements under given intensity of ground motions. Figure 9.22 presents the flowchart of the procedure for DDBD of the base isolated cable-stayed bridge. The basic input data are determined in Step1, including the geometry, the mass, the type and the design displacement of the isolators. Step 2 includes the selection of the design displacement of the deck, which is calculated as follows:

$$\Delta_d = \frac{\sum_{i=1}^{n} \left(m_i \cdot \Delta_i^2\right)}{\sum_{i=1}^{n} \left(m_i \cdot \Delta_i\right)} \tag{9.25}$$

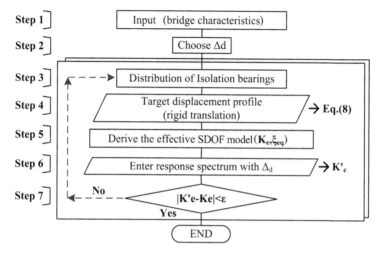

Figure 9.22. Design algorithm of DDBD.

where m_i and Δ_i are the masses and displacements of the *i*th significant mass location respectively.

As discussed in the previous section, the contribution of the pylon to the deformation is insignificant, Eq. (9.25) can then be reduced to $\Delta_d = \Delta_i$. Thus, the target displacement is determined by the design displacement of the isolation devices.

Steps 3 and 4 include the preliminary selections of the design parameters of the isolation devices. It should be noted that the stiffness of the isolators must satisfy Eq. (9.19) to ensure the target displacement profile, namely, uniform translation of the deck.

In Step 5, the equivalent stiffness,K_e, and the equivalent damping, ξ_e, of the isolation system, are obtained using Eqs. (9.26) and (9.27):

$$K_e = \sum K_i \tag{9.26}$$

$$\xi_e = \frac{\sum K_i \xi_i}{K_e} \tag{9.27}$$

where K_i and ξ_i are the equivalent stiffness and damping for the *i*th isolator, respectively.

In Step 6, the design displacement spectrum is determined for the equivalent S-DOF structureat a specific site with the given equivalent damping (ξ_e) of and the soil type. In the DDBD procedure, the Damping Reduction Factor (CEN, 1998) has been utilized to derive high damping (larger than 5%) response spectra using the following expression:

$$DRF = \sqrt{\frac{10}{5+\xi}} \qquad (9.28)$$

Then, the target displacement of the S-DOF system (Δ_d) is read directly from the spectrum at a given effective period T_e. A new equivalent stiffness (K'_e) of the system corresponding to the target displacement can be calculated by:

$$K'_e = 4\pi^2 \cdot m_e / T_e^2 \qquad (9.29)$$

Where m_e is the effective mass of the equivalent system.

If the absolute value between the new equivalent stiffness (K'_e) and the previous one in Step 5 is more than the specified tolerance, iterations are needed to update the parameters of the isolators (Step 3) as shown in Fig. 9.22, until the convergence is reached. The design base shear is then calculated by multiplying the effective stiffness by the target displacement:

$$V_b = K_e \Delta_d \qquad (9.30)$$

Finally, as suggested by Cardone et al. (2009; 2010) and seismic design codes, in the procedure of DDBD for base isolated structures, the limit values of isolation ratio are still needed, which is defined as the ratio between the fundamental period of the isolated and non-isolated structure as estimated by Eq. (9.23). The effective periods of base isolated structures are defined between $2T_{fb}$ (T_{fb} is the fundamental periods of non-isolated structure) and 4 s (Cardone, 2010).

9.5.3 Verification Study

The prototype of the bridge model used here is the Longwan Bridge, as shown in Fig. 9.8. The FE model has been presented in Section 9.4. A suite of six ground motion records are artificially generated to match the design response spectrum, as shown in Fig. 9.23.

Figure 9.24 compares the target displacement profiles of the deck obtained from DDBD procedure with the peak displacements recorded in the NLTH analyses along the transverse direction of the bridge. For the base isolated cable-stayed bridge, a range of possible target displacements (120 mm to 200 mm) is considered. Therefore, the range for target displacements can be approximately estimated through aforementioned method. It should be noted that because of the symmetry of the cable-stayed bridge about the pylon in the longitudinal direction, the results are listed for only the left half of the bridge. It can be seen from Fig. 9.24 that the displacement profiles of the deck obtained from the DDBD are in close agreement with the mean values of NLTH analyses. It indicates that the isolation systems of the bridge designed following the proposed procedure is able to achieve the desired uniform rigid translation of the deck in the transverse direction of the bridge. The results in Fig. 9.24 also indicate that the proposed DDBD procedure is capable of capturing the peak deck displacements.

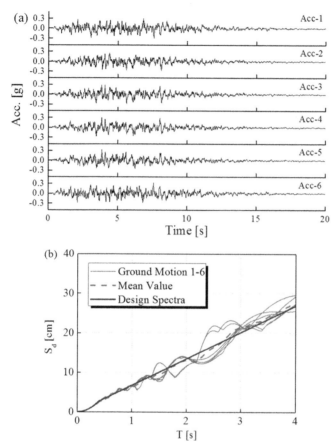

Figure 9.23. (a) Acceleration time histories employed in the verification study; (b) Comparison between the design spectrum and the average response spectrum.

As shown the displacement profiles estimated by DDBD method are in general smaller than the average values of the 6 nonlinear time history analyses, and the differences between the DDBD estimated displacement profile and NLTH analyses decrease with the increase of the target displacement response. The differences between the displacement profiles of the deck of NLTH and the target profile are listed in Table 9.3, labeled as Profile Error. The profile errors are defined as the absolute difference of the peak displacement between the two ends and the center of the deck normalized by the peak displacement at the end of the deck. Well agreement of displacement profiles can be seen from the comparison of the results obtained using NLTH and DDBD with all errors below 5.00%.

The displacement error is defined as Eq. (9.31) labeled with Displacement Error in Table 9.3:

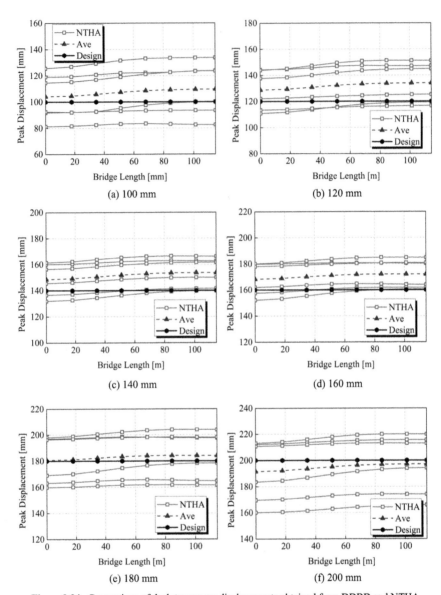

Figure 9.24. Comparison of deck transverse displacements obtained from DDBD and NTHA.

$$\hat{u}^{DDBD}_{\text{max,NLTH}} = \frac{\left|u_{\text{max,NLTH}} - u_{DDBD}\right|}{\left|u_{DDBD}\right|} \tag{9.31}$$

where $\hat{u}^{DDBD}_{\text{max,NLTH}}$ is the absolute difference between the maximum displacement of NLTH and DDBD results normalized by NLTH results. It can also be observed

Table 9.3. Comparison between NLTH and design objectives for the base isolated cable-stayed bridge.

Case study [#]	Δ_d [mm]	T_e [s]	ξ_{eq} [%]	Displacement error, $\hat{\Delta}^{DDBD}_{max,NLTH}$ [%]	Shear force error, $\hat{R}^{DDBD}_{max,NLTH}$ [%]	Profile error [%]
1	100	2.560	24.49	9.60	7.00	5.00
2	120	2.767	18.94	11.73	8.60	4.00
3	140	2.922	14.61	9.75	7.99	3.27
4	160	3.101	11.12	7.52	7.31	2.21
5	180	3.160	7.49	2.25	0.29	1.90
6	200	3.195	6.49	1.43	0.35	2.85

from Table 9.3 that the maximum deck displacement is underestimated by DDBD procedure. Obviously, the errors increase as the damping ratios of the isolators increase andthe maximum displacement error is 11.73%.

The maximum base shear force is also reported in Table 9.3 labeled with Shear Force Error. Similarly, the shear force error is defined by Eq. (9.32):

$$\hat{R}^{DDBD}_{ave,NLTH} = \frac{|R_{ave,NLTH} - R_{DDBD}|}{|R_{DDBD}|} \tag{9.32}$$

where $\hat{R}^{DDBD}_{max,NLTH}$ is the absolute value of the difference between the average base shear force of NLTH and DDBD results normalized by NLTH results. The shear force is underestimated by DDBD procedure compared to NLTH with a maximum error of 8.60%.

9.6 Conclusions and Further Studies

The seismic behavior of single pylon cable-stayed bridge with seismic mitigation devices has been analyzed in this chapter. It can be concluded that using base isolation is effective in mitigating the seismic responses of the bridge pier and pile foundation, but increased the deck displacement response. To control the deck displacement responses, viscous dampers were added to the base isolated bridges. Adding viscous dampers to the bridge model effectively reduced the deck displacement responses, but led to larger bending moment and shear forces in the pier, thus making the base isolation less effective. Based on the previous nonlinear time history analyses, a seismic design procedure is presented in this chapter for the base-isolated single pylon cable-stayed bridge. Generally, the DDBD procedure for the single pylon base isolated cable-stayed bridge in transverse direction is simple and straightforward to be implemented in engineering practice.

The seismic behavior of cable-stayed bridge is a broad topic involving many specific aspects of dynamics of structures, which need to be thoroughly addressed. The present study has been focused on the seismic behavior of single pylon cable-stayed bridge, and obviously further research is still needed. The following items are suggested for future works: (1) Extending the research on the passive protection systems for cable-stayed bridge, studies about the effectiveness of different seismic mitigation devices; (2) A seismic design procedure is developed for the base-isolated single pylon cable-stayed bridge in its transverse direction. Further studies are needed to fully account for the effects of bidirectional excitation in DDBD procedure.

References

Abdel-Ghaffar, A.M. and Ali, H.E.M. 1995. Modeling of rubber and lead passive-control bearings for seismic analysis. Journal of Structural Engineering 121(7): 1134–1144.

Adhikari, G., Petrini, L. and Calvi, G.M. 2010. Application of direct displacement based design to long span bridges. Bulletin of Earthquake Engineering 8(4): 897–919.

Atmaca, B., Yurdakul, M. and ŞevketAteş 2014. Nonlinear dynamic analysis of base seismic soil-pile-structure interaction experiments and analyses. Journal of Geotechnical and Geoenvironmental Engineering 125(9): 750–759.

Cardone, D., Dolce, M. and Palermo, G. 2009. Direct displacement-based design of seismically isolated bridges. Bulletin of Earthquake Engineering 7(2): 391–410.

Cardone, D., Palermo, G. and Dolce, M. 2010. Direct displacement-based design of buildings with different seismic isolation systems. Journal of Earthquake Engineering 14(2): 163–191.

Casciati, F., Cimellaro, G.P. and Domaneschi, M. 2008. Seismic reliability of a cable-stayed bridge retrofitted with hysteretic devices.Computers & Structures 8(17): 1769–1781.

Chadwell, C.B. 2003. Seismic response of a single tower cable-stayed bridge. PhD dissertation, University of California, USA.

Chang, K.C., Mo, Y. L., Chen, C. C., Lai, L.C. and Chou, C.C. 2004. Lessons learned from the damaged chi-lu cable-stayed bridge. Journal of Bridge Engineering 9(4): 343–352.

CEN ENV-1-1 European Committee for Standardisation. 1998. Eurocode 8: design provisions for earthquake resistance of structures, Part 1.1: General rules, seismic actions and rules for buildings.

Constantinou, M.C. and Symans, M.D. 2010. Experimental study of seismic response of buildings with supplemental fluid dampers, Structural Design of Tall Buildings 2(2): 93–132.

Cámara Casado, A. 2011. Seismic behaviour of cable-stayed bridges: design, analysis and seismic devices. PhD dissertation University Politecnica de Madrid.

Ernst, J H. 1965. Der E-Modul von Seilen unter berucksichtigung des Durchhanges. Der Bauingenieur 40(2): 52–55 (in German).

Evangelista, L., Petrangeli, M.P. and Traini, G. 2003. IABSE symposium on structures for high-speed railway transportation. Antwerp, 2003. Antwerp: IABSE.

Fenz, D.M. and Constantinou, M.C. 2008. Modeling triple friction pendulum bearings for response-history analysis. Earthquake Spectra 24(4): 1011–1028.

Fenz, D.M. and Constantinou, M.C. 2010a. Behaviour of the double concave friction pendulum bearing. Earthquake Engineering & Structural Dynamics 35(11): 1403–1424.

Fenz, D.M. and Constantinou, M.C. 2010b. Spherical sliding isolation bearings with adaptive behavior: theory. Earthquake Engineering & Structural Dynamics 37(2): 163–183.

Filippou, F., Popov, E. and Bertero, V. 1983. Effects of bond deterioration on hysteretic behavior of reinforced concrete joints. Report No. UCB/EERC-83/19. Berkeley, USA.

GBT 7714. 2013. Ministry of Communications of the People's Republic of China. Specification of Earthquake Resistant Design for Highway Engineering.

Javanmardi, A., Ibrahim, Z., Ghaedi, K., Jameel, M., Khatibi, H. and Suhatril, M. 2017. Seismic response characteristics of a base isolated cable-stayed bridge under moderate and strong ground motions. Archives of Civil & Mechanical Engineering 17(2): 419–432.

JTG/T D65-01-2007. 2007. Ministry of Communications of the People's Republic of China. Guidelines for Design of Highway Cable-stayed Bridge.

Kent, D. 1971. Flexural members with confined concrete. Journal of Structural Division Asce 97: 1969–1990.

Li, X.L. 2006 Study for design theories of single pylon cable-stayed bridges. PhD dissertation, Tongji University. (in Chinese)

Lu, X., Xie, L., Guan, H., Huang, Y. and Lu, X. 2015. A shear wall element for nonlinear seismic analysis of super-tall buildings using OpenSees, Finite Elements in Analysis & Design 98(C): 14–25.

Mazzoni, S., McKenna, F. and Fenves, G.L., 2005. Open Sees command language manual. http:// opensees. berkeley. edu/. Pacific earthquake engineering research.

Priestley, M.J.N., Ravindra, Verma and Yan, Xiao. 1994. Seismic shear strength of reinforced concrete columns, Journal of Structural Engineering 120(8): 2310–2329.

Priestley, M.J.N., Calvi, G.M. and Kowalsky, M.J. 2007. Displacement-based seismic design of structures. IUSS Press, Pavia, Italy, 720 pp.

Qiang Han, Jianian Wen, Xiuli Du, Zilan Zhong and Hong Hao. 2018a. Nonlinear seismic response of a base isolated single pylon cable-stayed bridge. Engineering Structures 175: 806–821.

Qiang Han, Jianian Wen, Xiuli Du, Zilan Zhong and Hong Hao. 2018b. Simplified seismic resistant design of base isolated single pylon cable-stayed bridge. Bulletin of Earthquake Engineering (Doi. org/10.1007/s10518-018-0382-0).

Soneji, B.B. and Jangid, R.S. 2008. Influence of soil-structure interaction on the response of seismically isolated cable-stayed bridge. Soil Dynamics & Earthquake Engineering 28(4): 245–257.

Soneji, B. and Jangid, R.S. 2010. Response of an isolated cable-stayed bridge under bi-directional seismic actions. Structure & Infrastructure Engineering 6(3): 347–363.

Wesolowsky, M.J. and Wilson, J.C. 2010, Seismic isolation of cable-stayed bridges for near-field ground motions. Earthquake Engineering & Structural Dynamics 32(13): 2107–2126.

Zayas, V.A., Low, S.S. and Mahin, S.A. 1990. A simple pendulum technique for achieving seismic isolation. Earthquake Spectra 6(2): 317–333.

10

Principles of Noise Control

James K. Thompson

10.1 Introduction

Noise control is an applied field. It is much more than the application of formulae or rules of thumb. In complex acoustic environments with multiple sources, multiple paths for sound propagation, and multiple receiver points; the development of an understanding of the sound field can be quite challenging. Most often this understanding requires a series of measurements that may include sound pressure measurements, vibration or acceleration measurements, the collection of operational information for the machines or operations generating noise, and possibly dosimetry data (measurements of the noise exposure of employees). In the simplest terms noise control is most often a combination of experimental and analytical work.

As noted above, often noise control requires the development of a complete understanding of complex problems. The problem may be complex for a number of reasons. It may be the complexity of an acoustic environment. For instance, a factory or refinery with closely spaced sources of noise (electric motors, pumps, fans, control valves, vents, compressors, turbines, etc.), and a wide range of randomly spaced reflective surfaces (concrete pads, metal grating, pipes, machine surfaces, etc.), present a daunting challenge to understand. Several detailed measurements and a comprehensive computer model may be required to fully understand the sound field. There certainly is the occasion when there is one dominant source and the means of noise control is straight forward. However, as equipment manufacturers and design engineers learn more about noise control engineering, there instances are becoming rare.

JKT Enterprises, 3962 Polly Court, Williamsburg, VA 23188.
Email: JKT.JKTEnterprises@outlook.com

After building the required understanding, the ideal solution may not be feasible or affordable. Replacing an expensive machine with a new, much quieter model may be the ideal solution to reduce employee noise exposure. However, the cost of the new machine and the down time involved with its installation may not be economically feasible. In such circumstances, the noise control engineer must develop alternative solutions that still protect workers' hearing or reduce noise levels in the nearby community and that are feasible to be implemented.

The noise control engineer must be adept at capturing critical data, understanding the physics of sound reduction, the physiology of hearing and hearing damage, and the psychology of those exposed to environmental noise. To those new to the field this may seem like an impossible challenge, but there has been much work done to support the noise control engineer in meeting these challenges. There is good information on the effects of noise exposure and on the potential reaction of people exposed to environmental noise. There is a significant amount of data on the noise emissions of machines or how to predict noise emissions based on operational conditions. Finally, there is a great deal of data and practical information on the application and effectiveness of noise controls. The key is to understand how to use this information properly and in an effective manner.

A fundamentally important point to remember in noise control is that hearing protection devices such as ear plugs and ear muffs are not noise control devices. They are useful in helping to protect individuals, but they do not control the noise, only the exposure of the individual (receiver as will be defined later). Around the world laws and regulations treat the use of hearing protectors differently. In some cases, they are an accepted step in the protection of employees. In other cases, they are considered a temporary expedient until noise controls can be implemented to sufficiently reduce the noise levels to which the employee is exposed. It is also important to note that hearing protectors are much less effective than their rated noise reduction. Scientific studies have shown that employees are much less protected than the rated performance of the hearing protection used (Berger et al., 1996; Brueck, 2009). Many employees do not properly fit their hearing protection and drastically reduce the effectiveness. Others takeoff or remove hearing protection to hear instructions, at breaks, or for other reasons and fail to reinsert or replace them. Due to perspiration or physical motion, protectors may become dislodged and lose their effectiveness. A good rule of thumb is that hearing protectors seldom provide more than 10 dB of protection in long term usage in real world situations.

Before closing this introduction, a look at a couple classic noise control problems may be helpful. As noted above, the first step in any noise control solution is to understand the problem. To build this understanding may be quite easy or it may require a lot of measurements, interviews, etc. A simple example is the company that installs a new fan on the roof of their facility and suddenly there are noise complaints from the nearby residential neighborhood. It is clear the fan has changed the background noise in the community. A few measurements should document whether the noise from the fan violates local ordinances or causes environmental noise levels to exceed acceptable limits. If either of the above are true, one can then begin investigating the best way to reduce the noise to acceptable levels.

A more complex example is the need to reduce employee noise exposure in a complex factory space. There may be dozens of sources contributing to the noise in the space. Due to the nature of their work, some employees may be in the problem space eight hours a day and others may only be there for a few hours but work near one or more of the noise sources. To provide an effective solution in such a case, the noise control engineer will have to perform dosimetry studies in conjunction with time-motion studies to understand whether employees are overexposed to noise and what machines or portions of the work function results in the overexposure. It will also be necessary to rank each of the sources in the area in terms of their contribution to employee overexposure. Once the dominant sources are determined, the noise control engineer can begin to work on how to reduce the noise. This may require understanding operational schedules, specific aspects of machine operation, machine mechanics, transmission paths, etc. It is easy to see that such a project can be very time consuming.

This chapter will provide a fundamental understanding of the noise control methodology for the basic types of noise control issues. It will not attempt to explain the detailed mechanics of noise sources or noise control methods. There are other texts that provide this sort of information. The goal of this chapter is to provide the foundation in terms of method and concepts that can be used to resolve noise issues. Both structural and airborne noise control will be discussed since ultimately all structural noise must become airborne to be heard. In addition, noise control on only the structural transmission components would be incomplete and unsuccessful in resolving most noise issues. Noise control is a very diverse topic and a single chapter cannot attempt to provide a comprehensive treatment. Instead this chapter will provide a good outline with specific treatment of the best techniques and methods used in a wide array of noise control solutions.

10.2 Noise Control Methodology

One of the first steps in any noise control project is to understand the nature of the problem. Noise may be a problem for many reasons. There may workers who are overexposed to noise and are having their hearing damaged. There may be an environmental noise problem where neighbors and surrounding communities are annoyed by noise from a factory or other industrial area. Most industrialized countries have regulations concerning the overexposure of workers that define limits and impose penalties for failing to protect workers (Lie et al., 2015; Concha-Barrientos et al., 2004). There are a wide range of national and local regulations with regard to noise in communities. In many cases these are generic limits related to sleep disturbance or the disturbance of typical activities in the home (International Institute of Noise Control Engineering, 2009; Ontario Ministry of the Environment, 2013). In other cases, these regulations are specific to the source. For instance, many countries regulate and require monitoring aircraft noise near an airport

(Transportation Research Board of the National Academies, 2008; Koopmann and Hwang, 2014; UK Environmental Research and Consultancy Department of the Civil Aviation Authority, 2014). In other countries noise from traffic and trains are also specifically regulated (International Institute of Noise Control Engineering, 2009; International Institute of Noise Control Engineering Working Party on Noise Emissions Of Road Vehicles (WP–NERV), 2001; Japan Automobile Manufacturers Association, Inc., 2013).

There are also other potential issues related to noise. Since high levels of noise can cause vibration in structures, excessive noise may represent structural fatigue and damage issues. An excellent example is the payload in a rocket nose cone. Often these satellites or other spacecraft contain fragile components that can be damaged by the high sound pressures generated by rocket motors. Noise control engineering is often required to develop sound insulation or other controls to limit the noise impinging on these fragile structures. Due to the need to keep weight to an absolute minimum, these can be challenging noise control projects.

Perhaps the most common need for noise control is in the consumer product industry. The development of products as diverse as automobiles, dishwashers, vacuum cleaners, cell phones, and many other consumer products require a very high level of noise control engineering. For automobiles the sound of a slammed door, the engine exhaust note, interior squeaks and rattles, and many more items are part of the noise control engineer's concerns. The sounds an automobile makes are part of the customer's evaluation of the quality and safety of the vehicle. In addition, the quality of the sound can be very important to the customer's perception of performance.

For a dishwasher the consumer wants the lowest level of noise possible. Years ago, it was unthinkable to stand in the kitchen next to a running dishwasher and hold a conversation. Now, that is considered a reasonable expectation. One of the challenges for the noise control engineer in making the dishwasher ever quieter is that noises previously too subtle to be perceived now become apparent. Noise from the operation of valves and switches must now be controlled because they have become apparent in the low sound levels from modern appliances.

It is important to note that noise control is not simply reducing noise levels. For employee exposure and environmental noise this is generally true. However, changing the spectrum of the noise moving energy from low to high or high to low frequencies can be helpful. High frequency sound does not propagate as effectively over distance as does low frequency sounds. Thus, if one can move some of the sound energy to a higher frequency it may be less objectionable to the nearby community. On the other hand, human hearing is more quickly damaged by high frequency sounds. Moving sound energy to lower frequencies can help to prevent hearing damage.

When it comes to consumer products both the level and the quality of sound are important. Automobile consumers have expectations with regard to normal and abnormal sounds. For instance, electric seat motors that are quiet but irregular sounding are more objectionable than motors that may be slightly louder but do not vary in sound during operation. Recently a major refrigerator manufacturer was surprised to find consumers complaining about knocking and swooshing sounds. A careful study found that these sounds were not new. The various sounds included solenoid operations, pressure reducing valves, and compressor transients. They had always been present. However, the refrigerator and the other appliances in the kitchen have become so quiet that these sounds were not apparent to consumers. The level of sound from the refrigerator was not the issue. It was the quality of the sound. This type of work involves a different aspect of noise control, the perception of sounds and their annoyance.

From this discussion, it should be apparent that it is important to understand the noise issue that is to be addressed and clearly define the objectives for the noise control to be performed. For employee exposure issues in may be necessary to do sound surveys in the workspace to identify high noise locations. Dosimeters may have to be used to accurately track the exposure of workers in the area. If the tasks being done are complex, time-motion studies may be required to what task or operation has been primarily responsible for the overexposure.

For community noise it is most likely necessary to make measurements at the facilities property line and in the community to determine the sound levels and to see if they are associated with the operation of particular equipment or specific operations. This may be only a few hours of measurement or several days depending on the nature of the issue. Airport noise is often monitored 24/7 due to the irregular operations and the need to have records of performance.

10.2.1 Source – Path – Receiver Principle

Perhaps the most important concept in this chapter is the use of source, path, and receiver analysis in noise control. It works for every noise issue and is the fundamental method used in all noise control problems. Sound or noise is generated by a source. It then must propagate through some medium to a receiver. The simplest example is a noise source in an open field and sound propagating through the air to a person standing at the edge of the field. However, in the practice the source and path are difficult to define as shown in Fig. 10.1. The source seems apparent. It is the compressor shown. If the compressor were sitting in an open field, the sound pressure level would decay as shown in the upper right of the figures. A short distance from the source there would be spherical spreading and the sound pressure level would decay by 6 dB each time the distance from the source is doubled. In the case of the compressor in a field, the path is direct transmission through the air with to a lesser extent some sound energy reflected from the ground. The receiver for all cases discussed in this chapter will be a person. In this case, a person is near this compressor.

It is very unlikely that one will find a compressor operating in an open field except a pipeline pumping station or for a controlled acoustic test. The more likely scenario is a compressor in an industrial plant or a building with multiple reflecting surfaces and concern about occupants in the room and in the immediately adjoining rooms. There may be significant noise energy radiating from the compressor to the

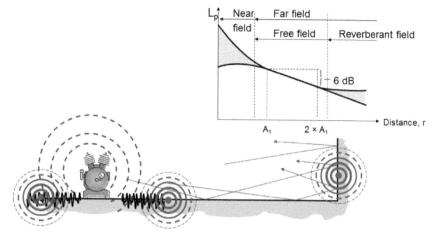

Figure 10.1. Fundamental illustration of source, path, and receiver.

air. However, there is also energy being transmitted into the structure that can be radiated back into the compressor room or adjoining rooms from panel vibration. In addition, there is the airborne sound energy that is reflected from the walls. The primary source is the compressor. However, it is a source of both airborne sound and structural vibration that may contribute to the noise at the receiver's ears. The paths include direct transmission in the air, reflections from the walls, and transmission through the structure and radiation from vibrating wall surfaces.

As this figure shows, defining the source and path can be quite complex. The receiver is usually well defined for this discussion. It may be one or more employees or residences or locations in a community. Considering an industrial plant, the sources may be numerous in multiple locations. In this scenario the paths may include propagation in the air, in the structure, in pipes and ducts, and with multiple reflections for both hard and soft surfaces. Simply tracing all the possible paths can be quite challenging.

10.2.2 Sources

There are a wide variety of mechanisms for generating noise and therefore a variety of sources. A few of these will be described in greater detail later in the chapter. In most industrial situations there are several sources present. In such cases it is critical that the primary sources be identified. Reducing the sound radiated from a secondary source may have little to no impact on the overall noise level.

Consider the situation shown in Fig. 10.2. There are multiple sources closely spaced together. The noise control engineer needs to be effective in treating the sources. Which sources should be treated? Since decibels are added and subtracted logarithmically, it is not immediately obvious what strategy is most effective. Starting from the bottom of the list of possible solutions, one could eliminate both

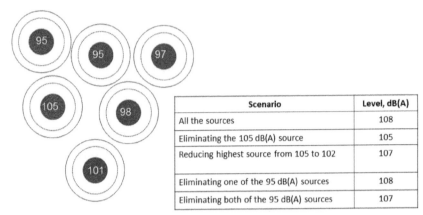

Scenario	Level, dB(A)
All the sources	108
Eliminating the 105 dB(A) source	105
Reducing highest source from 105 to 102	107
Eliminating one of the 95 dB(A) sources	108
Eliminating both of the 95 dB(A) sources	107

Figure 10.2. Complex source scenario.

of the 95 dB(A) sources. This only reduces the overall sound level to 107 dB(A) from 108 dB(A). This is not much improvement. What about reducing the highest source sound level from 105 to 102 dB(A). This only reduces the overall sound level to 107 dB(A)—again not a major reduction. What if the 105 dB(A) source could be eliminated? This would lower the overall sound level to 105 dB(A). Note that this reduction of 3 dB represents a 50% reduction in the total sound energy. Clearly identifying and controlling the primary noise source or sources is extremely important for effective noise control.

In this example the sound levels from each source were already identified. In most noise control problems this is not the case. Quantity and ranking the noise sources can be a difficult problem. If the sources are close together, it may be difficult to distinguish one source from the other. In some cases, it may be able to obtain data for the sound power produced by each source from the manufacturer. An alternative is to test the source in a controlled environment like a reverberant or anechoic room to quantify the sound power radiated by the source.

It is important to make the distinction between sound power and sound pressure. Sound pressure or sound pressure level is the magnitude of the oscillating pressure created by the source. This is what is heard. Its magnitude is dependent on the distance from the source, the acoustic environment (are there reflective surfaces or is it open to acoustic propagation), and atmospheric conditions. Sound power is the magnitude of power radiated as sound from the source. This is a property of the source and does not change with distance, environment, or for the most part atmospheric conditions. Therefore, sound power or sound power level is a much better way to characterize sources.

The process of noise source identification and source ranking is a major sub-discipline in noise control engineering. Where sources are not close to each other or they can be easily separated from other sources for testing, noise source identification is straightforward. One only needs to compute the source sound power using standard means. However, where sources are closely spaced together, more advanced methods must be used. One approach is to operate each source selectively so that its individual contribution can be quantified. A variation on this approach is to cover all the other sources with high attenuation noise control treatments in succession to quantify the contribution and thus rank each source.

In complex machines it may be difficult to determine what the sources are precisely. As an example, consider an automobile engine. One could lump the entire engine as a source and consider it in comparison to other sources in the vehicle. For basic analysis this is a reasonable approach. However, for refined analysis to optimize the interior compartment noise, the engine component sources must be separated and considered separately. Acoustic intensity measurement provides the sound intensity magnitude and vector from a source. This vector is calculated by multiplying the acoustic pressure, a scalar, and the acoustic particle velocity (a vector). Using signal analysis techniques this complex quantity can be estimated accurately using simple instruments. By making numerous measurements, one can identify where most of the sound energy is coming from the engine and locate the 'hot spots' or sources. Such measures must be done near the engineer surfaces and hundreds of measurements may be required to completely map the engine.

Near-field acoustic holography and beam-forming employ large arrays of microphones (30–120) located some distance from the engine surface to again find the 'hot spots' of sound energy on the engine. Using the information from the array of microphones accounting for the positional differences and very small differences in phase (the arrival times of waves), an accurate reconstruction of the sound field at a surface can be made. The large number of microphones required for this type of measurement can be a significant cost and the size of the arrays may be a problem for some measurement environments.

One of the most difficult cases is the presence of multiple sources that are linked structurally. Again, using the engine example, there are combustion pressures acting inside each cylinder, there are pistons slapping the cylinder walls, there may be high pressure injection systems, there is a valve train with many moving parts and impacts, there is the motion and force transmission of the driveshaft, and many other sources. One can measure the hot spots on the engine surface with the techniques noted above. However, it is not clear in many cases whether a hot spot is due to piston slap, combustion pressures, or some other source. This is because the engine block and other metal components are efficient conductors of mechanical energy. As combustion pressures and pistons impact the engine block, the energy is transmitted through the metal block. Measuring on the outside, one can see the noise radiated from the block due to its surface vibration. With advanced techniques, one may be able to say a certain percentage of the radiated noise is due to piston slap or combustion pressure. However, since these two sources are at least partially correlated, it is difficult to make this distinction for the entire frequency spectrum of noise.

The final step in noise source identification in such cases is modeling. By modeling the mechanisms of the sources, the flow of energy through the engine block, and the radiation of noise from the complex block geometry it is possible to separate the contributions from each source. To perform such analysis accurately requires highly detailed models and a great deal of experimental validation. Going back to the example of the engine developing a finite element model to even the mid-frequency range 500–1000 Hz may require millions of degrees of freedom. At high modal densities it may be possible to use statistical energy analysis (SEA), but it may be not be possible to distinguish between the different possible sources using this technique.

Ultimately, one would like to understand the fundamental source of the noise, for example combustion pressure versus piston slap in the previous example. This knowledge will permit the most efficient and effective noise control.

10.2.3 Types of Sources

It is important to understand that there are a wide variety of noise source mechanisms. There are several mechanical sources resulting from impacts, larger forces, and other mechanical operations. Industrial machinery is a good example of these types of sources. There are also many sources that involve changes to fluid flow. These include pumps, compressors, exhausts, fans, control valves, and many more. Another way to characterize sources is the way in which the sound energy propagates. In many cases the sound energy directly causes sound waves in the air that we hear as noise. An exhaust vent, a fan, and the impact of a hammer are just a few examples. In some instances, the noise is radiated from vibrating surfaces and is often called a structural source. For instance, a pump casing can radiate significant amounts of noise and is often treated to reduce noise radiation. Many industrial machines are contained in metal enclosures that serve as excellent radiators of sound. It is important to understand how noise is created and propagated to effectively treat it.

10.2.4 Noise Control at the Source

It is nearly always best to control noise at the source if possible. Typically, this will result in the most effective controls with the greatest noise reduction. Developing a strategy for noise control at the source is important. As noted above, the dominant sources must be identified. It is also necessary to determine a strategy for the control of each dominant source. Different methods may be required for each source. It is critical to evaluate all the viable methods in terms of effectiveness, practicality, and cost. Often the best solution is a compromise between these factors. The ideal solution may be too expensive or present difficult issues with respect to operation or maintenance.

One method of source control is to modify the source. It may be possible to reduce speed, add a silencer, or an enclosure to reduce the radiated noise. In some cases, it may be possible to relocate the source to an area that will mean less exposure for workers or away from the nearby community. More and more there are low noise alternatives available commercially. The manufacturer may have a quieter model of the machine or a quiet package that can be added to reduce noise. A special case of quieter alternatives is the control valves commonly used in industrial systems. Where there are large pressure drops, such valves can generate high noise levels radiated to the air at the valve and down the piping. Control valve manufacturers provide a range of solutions to both predict and reduce this noise effectively at the valve. This is an excellent example of a case where modifying the source is highly effective in terms of noise reduction, cost, and maintenance. One should always evaluate treating the source first. This is the most effective means of noise control and provides the broadest possible reduction of noise.

10.2.5 Evaluate Noise Control Along the Path

If treatment of the source is not possible or if it does not provide sufficient noise reduction, one should next evaluate noise controls along the path. For interior spaces one of the most common controls is to add sound absorptive materials to the room surfaces. This treatment will reduce the reverberant sound energy in the space and thus reduce the noise levels. Also, barriers may be used to protect specific locations whether in an interior or exterior space. The effectiveness of barriers is limited, and a later segment will provide some guidance to the use of such controls.

Where the transmission path is structural, other means of path controls should be considered. The first line of defense is vibration isolation at the connection point of the source to the structure. It is relatively common practice to mount engines, pumps, compressors, fans, and various machines on vibration isolators. It is also common for these isolators to be short circuited by flanking paths (other non-isolated attachments to the structure) or to be inadequate to provide sufficient isolation. In industrial applications isolators are often selected for performance at low frequencies to prevent fatigue or gross vibration issues. However, they may not be as effective at higher frequencies where noise issues occur. In highly engineered products like automobiles, sophisticated isolators incorporating multiple chambers and even active elements are used to suppress noise transmission. Structural isolation will be discussed in greater detail in a later segment.

As noted above, flanking paths can make well designed isolators ineffective. One should always carefully examine all connection to structure to be sure that flanking paths are not being created that short circuit the designed isolation. Piping and other miscellaneous connections are commonly problems. A pump may be well isolated but piping to or from the pump may be rigidly connected to the structure at some point by oversight or poor maintenance.

For vents, engine exhaust, and blowers, silencers can provide effective noise reduction. There are silencers designed for a wide range of applications. When used properly, they can be highly effective in providing major noise reductions. A special case for the use of silencers is in automobiles where quality of the sound is just as important as the level. Automotive mufflers are carefully designed to provide the desired engine sound characteristics as well as reduce or eliminate unwanted noise.

It must be remembered that noise control along the path has limitations. Near the source, it may not be highly effective because of direct radiation. Also, in many industrial environments it may be a challenge to maintain absorptive treatments, isolators, and even silencers.

10.2.6 Evaluate Noise Control at the Receiver

As noted previously, for this discussion the receiver will always be a person. Therefore, control at the receiver is limited. The use of hearing protection has already been discussed. While useful as a temporary control, hearing protectors should never be thought of as a permanent noise control. In some instances, it may be possible to provide a small enclosure for a worker to reduce his/her exposure while in a noisy area.

In other situations, controls at the receiver might include masking of background noise. In offices where speech privacy and sound quality are an issue, random background noise can be used to mask nearby speakers and protect privacy. There has been a lot of work on how to do this effectively and not result in high background noise levels that simply force everyone to speak louder.

In non-reverberant enclosed spaces or in open spaces, increasing the distance from the source for the receiver can effectively reduce the noise exposure. Maybe a worker's workstation can be moved. For environmental noise it has been effective in some cases to purchase nearby properties to eliminate high noise exposures. Airport zoning laws have sometimes been used to prevent residential areas too close to the flight paths of aircraft.

The final approach to treat the receiver, especially in work exposure issues, are administrative controls. One can limit the time a worker spends in a high noise area to protect him/her from overexposure to noise. While this can be effective, it is often difficult to implement due to changing work assignments and schedules and the presence of multiple high noise areas.

While controls at the receiver can be effective, they should be the choice of last resort. They often only protect a few workers and leave a high noise area that may cause safety and stress issues for other workers. Controls at the receiver tend to be expensive solutions that have to be modified when plant operations change, or equipment is modified.

10.2.7 Buy Quiet

A special class of solutions are non-engineering solutions to noise control issues. One of the most effective preventative measures that can be taken for noise control is to 'buy quiet'. Purchase machines and components with low noise operation in mind. Many manufacturers and suppliers have quiet options available, but often this is not part of the purchasing specification. Companies showing leadership in protecting worker hearing have very specific purchasing specifications and will not purchase equipment that will lead to worker exposures over 80–85 dB(A) when regulations only specify maximum levels of 85 or 90 dB(A) for an eight-hour day. Preventing the noise problem in the beginning is the most effective noise control practice. Clearly not all equipment and component suppliers provide quiet options. However, working with those knowledgeable in noise control, a machine can be modified before or shortly after delivery to prevent it from being a noise problem. This approach is generally much more cost effective than coming in after complete installation and attempting to retrofit the machine to reduce noise.

10.2.8 Selection of the Best Noise Control Methods—An Example

Consider a simple example noise problem. In an industrial plant there are two motor driven pumps as shown in Fig. 10.3. An operator spends approximately 4 hours per day near these pumps. If this is his only high noise exposure, the highest permissible level for this individual is 88 dB(A). Clearly significant noise reduction is required.

Fortunately, one is able to operate the pumps independently and this allows the quantification of the contribution from each pump. Measurements find that Pump 1 operating alone provides a sound pressure level of 93 dB(A) at the operator's location. For Pump 2 operating alone, the level is 105 dB(A). Combining the two sound levels the worker is currently exposed to 105 dB(A) for the four hours which is confirmed by measurements during normal operations.

Using the methods noted above possible solutions are investigated. The most promising noise controls are summarized in Table 10.1. Considering these potential solutions, possible combinations should be considered. If only Pump 2 is enclosed the cost is $4000 and the noise level at the operator is reduced to 93.6 dB(A). If the silencer is used on Pump 1 and the enclosure used on Pump 2, the cost is $4600, and the resulting sound level will be 89.8 dB(A). If an enclosure is used for both pumps the cost is $7000 and the sound level at the operator is reduced to 85.8 dB(A). Therefore, the only viable solution is the third combination and unfortunately the most expensive.

Before leaving this example, it is useful to note that the installation employs good practices with the piping being heavily lagged for both thermal and noise

Figure 10.3. Noise control example—Motor driven pumps.

Table 10.1. Example noise controls.

Pump	Solution	Noise Reduction, dB	Cost
1	Silencer	5	$600
1	Enclosure	15	$3000
2	Silencer	10	$750
2	Enclosure	20	$4000

issues. However, the motors appear to be rigidly connected to a base with no isolation. It is possible that there is isolation between the frame supporting the motors and the rest of the structure that is not apparent in this photo.

10.2.9 Summary

The steps to a good noise control process have been defined in this segment. One must begin by determining the nature of the noise problem. Is it a worker exposure issue, is there an environmental noise problem, or is the goal to provide sound quality that is not annoying? Once the nature of the problem is understood, the next step is to identify the primary noise sources. This can be a complicated process and one should be careful to do this properly. The effectiveness of any subsequent noise control is dictated by the accuracy in identifying and quantifying the primary sources. The next step is to step through the process of evaluating noise control at the sources, along the path, and at the receiver in this order of preference. The final step is to select the best methods of control considering cost, effectiveness, and other operational considerations.

10.3 Noise Control Methods

The key to any noise control project is developing and implementing controls that are effective in reducing noise. In this segment some of the basic approaches to noise control will be discussed and some tools provided for designing and evaluating noise control solutions.

10.3.1 Radiation from Plates

In structural acoustics the primary noise source is the motion of plates and shells that have sufficient surface area and velocity to create significant sound fields. The analysis of the radiation from such panels can be quite complex (Lamancusa Eschenauer, 1994; Cremer and Heckl, Translated by Ungar, 1988; Ver, 2006; and Bies and Hansen, 2009). This chapter will not attempt to provide a thorough examination of sound radiation from structural elements. Instead a few examples will be used to illustrate the basic principles.

In its simplest form radiation from a panel can be defined in terms of the panel normal velocity and a radiation efficiency. The radiation efficient can be described as the following.

$$\sigma_{rad} = \frac{W_{rad}}{v_n^2 \rho_0 c_0 S}$$ (10.1)

where:

σ_{rad} = the radiation efficiency

W_{rad} = the radiated sound power,

v_n = the normal component of the space-time average mean-square vibration velocity on the surface,

ρ_0 = the ambient air density,

c_0 = the speed of sound in air, and

S = the surface area of the panel.

This simple expression would seem to indicate that one only needs to know an average normal surface velocity and the area of the panel to estimate the sound power radiated. This is true. However, determining the space and time average velocity for complex modes of panels with various end conditions is not a trivial matter. Some basic expressions can be derived for classic cases as shown below.

$\sigma_{rad} = 1$ for a rigid piston

0 for $\lambda_B < \lambda_0$ for an infinite plate in bending

$[1-(\lambda_0/\lambda_B)]^{-1/2}$ for $\lambda_B > \lambda_0$ for an infinite plate in bending

$(k_0 * a)^2/(1 + (k_0 * a)^2)$ for pulsating sphere

where:

λ_B = the bending wavelength in the plate, all terms with the 0 subscript reference to ambient air, and

a = diameter of the sphere.

While these values may be useful to bracket actual panel configurations, they cannot accurately predict sound power radiation.

To obtain more useful expression, a different approach is required. A fundamental concept in a sound radiation cases is the critical frequency. As shown in Fig. 10.4, the sound wave transmitted from a panel will be at an angle to the propagating bending wave in the panel. The transmitted sound wave length is defined as λ. The wavelength in the plate is $\lambda/\sin\theta$. When these two wavelengths match sound radiation efficiency of the plate is maximized. The frequency at which this match, occurs is called the critical frequency. Using this wavelength relationship, it can be shown that the lowest coincident frequency for grazing incidence ($\Theta = 90°$) is

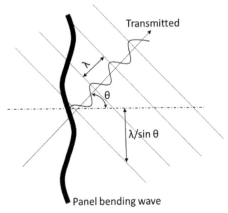

Figure 10.4. Panel bending and sound transmission.

$$f_c = \frac{c^2}{1.8\ tc_l} \tag{10.2}$$

where:

f_c = the critical or lowest coincident frequency, Hz

c = the speed of sound in air

t = the thickness of the panel

c_l = the longitudinal wave speed in the panel

Below the critical frequency, little of the surface motion is converted to generating sound. At and above the critical frequency the radiation efficiency increases.

For typical panels and surfaces due to wavelength effects, radiation efficiency is low at low frequencies and rises to one at some higher frequency threshold above the critical frequency.

Considering a plate suspended between rigid supports as shown in Fig. 10.5, the bending motion at low order modes causes significant spatial variations in normal velocity. In fact, at the lowest frequencies, most of the sound energy is radiated from the margin—near the clamped edges. The large amplitudes toward the center of the plate merely push and pull on the air molecules cancelling out any net sound energy emission producing little sound radiation.

With some work it can be shown that the radiation efficiency for a point-excited lightly damped rectangular plate is given by the following expressions (Ver, 2006):

Figure 10.5. Simple bending in plate between rigid supports (Cremer and Heckl, Translated by Ungar, 1988).

$$\sigma = \frac{P\lambda_c}{\pi^2 S}\sqrt{\frac{f}{f_c}} \qquad \text{for } f \ll f_c$$

$$= 0.45\sqrt{\frac{P}{\lambda_c}} \qquad \text{for } f = f_c$$

$$= 1 \qquad \text{for } f \gg f_c$$

where:

σ = radiation efficiency
P = the perimeter dimension of the plate
S = the plate surface area
λ_c = the wavelength in the plate at the critical frequency
f = the frequency of interest
f_c = the critical frequency for the plate.

10.3.2 Walls and Room Partitions

Obstructions to the transmission of sound are the most common controls found in practice. Examples include noise barriers on highways, room walls, partial enclosures to protect workers, and shields on machinery or equipment. By considering barriers and walls we can then move on to whole enclosures and their performance.

The transmission loss level in decibels of a wall or barrier is defined by the simple equation:

$$TL = 10Log\left(\frac{W_i}{W_t}\right) \tag{10.3}$$

where:

W_i is the incident sound power
W_t is the transmitted sound power

Another was to write this equation is in terms of the transmission coefficient:

$$\tau = \frac{W_t}{W_i} = \frac{p_t^2}{p_i^2} \tag{10.4}$$

If this expression is substituted into Eq. (10.3)

$$TL = 10Log\left(\frac{1}{\tau}\right) \tag{10.5}$$

To compute the transmission loss of a partition or barrier can be very complicated considering all the possible modes of vibration and the details of materials and their attachments. For simple structures a good approximation was developed by Beranek (Bies and Hansen, 2009) which will be described here. This approach assumes that the transmission loss is solely dependent on the materials used with a characteristic plateau and valley in the transmission loss curve. Figure 10.6 provides a fundamental guide for this computation with information for common materials. This chart assumes a reverberant sound field on the source side and approximates the behavior around the critical frequency with a horizontal line or plateau.

To compute the transmission loss, a four-step process is defined.

1. Compute the first bending resonance frequency for the panel.

$$f_r = \frac{\pi}{n} \sqrt{\frac{B}{\rho_s}} \left[\left(\frac{1}{L_x}\right)^2 + \left(\frac{1}{L_y}\right)^2 \right] \tag{10.6}$$

where:

f_r = first bending resonant frequency, Hz

n = 2 for simply supported edges

    1 for fixed edges

B = bending stiffness39

    = $E*I/L_y$

E = modulus of elasticity

I = moment of inertia

ρ_s = surface mass density (density multiplied by thickness)

L_x = panel length

L_y = panel width

    Coincidence occurs when the wavelength of the sound in air and in the panel become equal. At this point there is a significant decrease in transmission loss. This is illustrated in Fig. 10.7. The angle, θ, is between the normal to the panel and the normal to the wave front. The length of the forced wave in the panel, λ_p, to the trace wavelength of the sound wave $\lambda/\sin\theta$. Coincidence occurs when $\lambda_p = \lambda/\sin\theta$.

    Using this wavelength relationship, it can be shown that the lowest coincident frequency for grazing incidence ($\theta = 90°$) is as shown in Eq. (10.2).

2. Plot the field incidence transmission loss $2f_r$ to $f_c/2$ employing:

$$TL_{FI} = 20Log(fps) - 47 \text{ (for MKS units)} \tag{10.7}$$

where:

TL_{FI} = the field incidence transmission loss as shown in Fig. 10.6.

f = frequency, Hz.

3. At the coincidence frequency,

$$TL = TL_{FI}(@f_c) - C \tag{10.8}$$

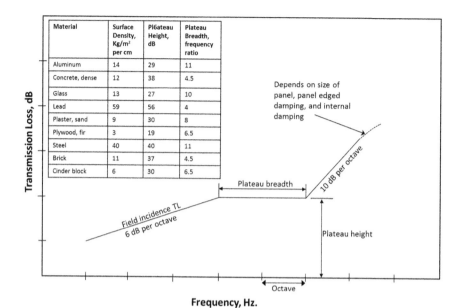

Material	Surface Density, Kg/m² per cm	Pl6ateau Height, dB	Plateau Breadth, frequency ratio
Aluminum	14	29	11
Concrete, dense	12	38	4.5
Glass	13	27	10
Lead	59	56	4
Plaster, sand	9	30	8
Plywood, fir	3	19	6.5
Steel	40	40	11
Brick	11	37	4.5
Cinder block	6	30	6.5

Figure 10.6. Approximate design chart for estimating the sound transmission loss of single panels (Ver, 2006).

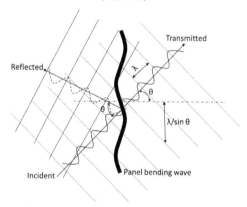

Figure 10.7. Geometry of sound waves impinging on a panel and relative wave fronts (Ver, 2006).

where

$C = 9$ for high panel damping

$C = 12$ for medium panel damping

$C = 24$ for low panel damping

4. t $2f_c$ from Eq. (10.8):

$$TL = TL(@f_c) + 9$$

.(10.9)

5. Connect points at $f_c/2$, f_c, and $2* f_c$ with straight lines.
6. Below $f_r/2$ no predictions are possible.

To clarify this process, an example calculation is made for an aluminum panel that is 3.2 mm thick and 2 m by 1.5 m (Ver, 2006). The first step is to calculate the lowest resonant frequency using Equation 4. This gives f_r of approximately 4 Hz. Next using Eq. (10.2), $f_c = 4000$ Hz.

The computed transmission loss is compared to the calculated in Table 10.2.

The method presented is a good first estimate for transmission loss of panels. However, it is no substitute for the detailed information provided by manufacturers and trade associations. Most noise control material suppliers can provide detailed information accounting for methods of attachment and other issues that may affect performance.

10.3.3 Multiple Walls and Complex Constructions

The calculation of transmission loss for multiple partitions and built-up structures can be quite complex. Two different types of transmission systems need to be considered: parallel and series. For common occurrences like doors and windows in a panel, parallel transmission occurs. Sound energy is transmission through the wall, door, and windows simultaneously. For such a case the composite transmissibility can be calculated from the following expression.

$$\tau = \frac{\tau_1 S_1 + \tau_2 S_2 + \cdots \tau_n S_n}{S_1 + S_2 + \cdots + S_n} \tag{10.10}$$

where

τ_n = transmission coefficient for each of the different panels
S_n = the surface area of each of the different panels

From this expression, the transmission loss for such a panel will always be less that the highest transmission loss sub-panel. Another important concept demonstrated by this formulation is the effects of leakage. One of the critical

Table 10.2. Comparison of measured and calculated transmission loss for an aluminum panel.

Frequency, Hz.	Calculated TL by Beranek (Bies and Hansen, 2009) Method, dB	Calculated TL by Method Above, dB	Measured TL, dB
125	14	12	19
250	20	18	24
500	26	22	27
1000	32	31	32
2000	38	38	34
4000	26	32	24
8000	35	40	32

considerations in construction enclosures is that each of the panels must be well sealed. Consider a case where a 2 m by 2 m panel has a transmission loss of 30 dB or a transmission coefficient of 0.001 per Eq. 4. If there is an opening equivalent to 1% of the panel area, there will be a significant reduction in the effectiveness of the panel. This opening will have a transmission loss of 0 dB and a transmission coefficient of 1. From Eq. (9), $\tau = 0.011$ and the transmission loss will be 19.5 dB or more than 10 dB reduction in the performance of the panel. Sealing is a critical part of noise control enclosures.

The other type of complex structure is the series case when there is more than one panel or a wall of multiple elements. A good example is a wall made of two outer surfaces with an inner layer of fiberglass. There is not a simple formula to predict the performance of such constructions. Clearly multiple layers should be better than one, but the prediction of the actual performance is difficult. Generally, if the distance between panels is greater than one-quarter of the wavelength of the sound frequency in air, the panel effects are additive. If the cavity between the panels is filled with a light absorbent material like fiberglass or even air at high frequencies (> 2000 Hz.) the combined performance can be greater than simply logarithmic addition of the performance of each panel.

In the architectural and building industries a standard has been developed that greatly simplifies the prediction of sound transmission for walls. This is the sound transmission class (STC) rating. This is a single number rating of the transmission loss performance for panels and partitions. The STC is computed using the 16 one-third octaves from 125 to 4000 Hz. The assumed contour for the transmission loss curve is illustrated in Table 10.3 (Bies and Hansen, 2009). The STC value is the transmission loss in the 500 Hz one-third octave. Given an actual transmission loss curve the correct STC value can be determined using this contour noting that there can be no more than 8 dB difference between the transmission loss and the STC value for any one-third octave. Also, the total difference from the STC contour summed by one-third octave cannot be more than 32 dB.

Using STC, manufacturers, architects, and builders have an easy to use method to select and rate panels and partitions. Some examples of STC values for common construction materials are shown in Table 10.4.

There is one last point regarding panels and partitions that also applies to all noise controls. Looking back to the original discussion of transmission loss, it was formulated in terms of sound power loss. This is a convenient manner to deal with noise controls since environmental effects and distances to the receiver are not considered. An alternative approach is to use noise reduction which is defined below.

$$NR = L_{p1} - L_{p2} \tag{10.11}$$

where:

NR = noise reduction, dB and

L_{p1} and L_{p2} = the sound pressure levels on either side of the partition and panel.

Table 10.3. STC contour (Bies and Hansen, 2009).

STC reference contour								
1/3-Octave Band [Hz]	125	160	200	250	315	400	500	630
Reference Contour [dB]	−16	−13	−10	−7	−4	−1	+0	+1
1/3-Octave Band [Hz]	800	1000	1250	1600	2000	2500	3150	4000
Reference Contour [dB]	+2	+3	+4	+4	+4	+4	+4	+4

Table 10.4. Example STC values for common construction materials.

Construction	Mass per unit area kgm^{-2}	STC
0.5 inch gypsum wallboard	10	28
Two 0.5 inch gypsum wallboards bonded together	22	31
2 inch by 4 inch studs on 16 inch centers, 0.5 in gypsum wallboards on both sides	21	33
4 inch hollow block, 0.5 in plaster on both sides	115	40
4 inch brick, 0.5 in plaster on both sides	210	40
9 inch brick, 0.5 in plaster on both sides	490	52
24 inch stone, 0.5 in plaster on both sides	1370	56

To illustrate the difference between transmission loss and noise reduction, consider a panel with a reverberant room on one side. L_{p1} is the pressure acting on the panel and L_{p2} is the sound pressure inside the reverberant. Note that it is assumed that the two sound pressures are diffuse and thus uniform across the surface on both sides of the panel. For this case the governing equation would be as shown below.

$$L_{p2} = L_{p1} - TL - 10Log\left(\frac{S_w}{R}\right) \tag{10.12}$$

where

S_w = the wall surface area of the reverberant room and
R = is the room constant (typically $S_w^* \alpha$) for the reverberant room.

Thus,

$$NR = TL + 10\,Log\left(\frac{S_w}{R}\right) \tag{10.13}$$

from Eq. 12. In this particular case the noise reduction is greater than the transmission loss due to the effect of the room wall absorption.

10.3.4 Enclosures

When panels are erected to fully enclose a source, the same principles regarding attenuation apply. The easiest type of enclosure to analyze are large enclosures where the distances from the source to the walls are much greater than the wavelengths for the sound frequencies of interest. Another way of looking at this criterion is that the large enclosure assumption works at higher frequencies with shorter wavelengths. An often-used rule of thumb is that the enclosure walls should be at least one meter from the source. This means that the large enclosure assumption is good for frequencies above 1000 Hz.

With this assumption the enclosure can be treated like a room and the equation predicting sound pressures outside the enclosure becomes the following.

$$L_{p2} = L_{p1} - TL - 10Log\left(\frac{S_e}{R_e}\right) \tag{10.14}$$

where:

L_{p2} = the sound pressure level at the exterior of the enclosure, dB
L_{p1} = the sound pressure level in the interior of the enclosure
S_e = the surface area of the enclosure interior
R_e = the room constant for the enclosure.

Another measure of the performance of the enclosure is the insertion loss.

$$IL = L_{po} - L_{p2} \tag{10.15}$$

where:

IL = insertion loss, dB
L_{po} = the sound pressure level at a position before the enclosure was installed, dB.

It is important to note that the noise reduction and the insertion loss are different and will be different values. Too often these are confused in noise control programs.

Considering an enclosure large enough to have a diffuse sound field, the sound pressure at a given point without the enclosure is defined by this expression.

$$L_{po} = L_{wo} + 10Log\left(\frac{Q_o}{4\pi r^2} + \frac{4}{R}\right) \tag{10.16}$$

The equivalent expression for the sound pressure level at the same point after the installation of the enclosure is shown below.

$$L_{p2} = L_{w2} + 10Log\left(\frac{Q_o}{4\pi r^2} + \frac{4}{R}\right) \tag{10.17}$$

where:

L_{Wo} and L_{W2} = the sound power levels for the source without the enclosure and with the enclosure, respectively

Q_o and Q_2 = the source directionality for before and after the enclosure

r = the distance to the point of interest

R = the room constant for the space in which the source is housed

If the point of interest is in the reverberant field (far enough from source), the differences between these two quantities are insignificant. In this case, substituting in Eq. 15, the following expression is obtained.

$$IL = L_{Wo} - L_{W2} \tag{10.18}$$

Using Eqs. (10.16) and (10.17), the expressions for the sound pressure level with the enclosure can be written as follows.

$$L_{W2} = L_{p1} + 10Log(S_e) \tag{10.19}$$

$$= L_{Wo} + 10Log\left(\frac{Q_o}{4\pi r^2} + \frac{4}{R_e}\right) - 10Log(S_e) - TL \tag{10.20}$$

If the sound field is the enclosure is diffuse and then is averaged, the sound power with the enclosure becomes the following.

$$L_{W2} = L_{W0} + 10Log\left(\frac{1}{\alpha_e}\right) - TL \tag{10.21}$$

$$= L_{W0} - 10Log\left(\frac{\alpha_e}{\tau}\right) \tag{10.22}$$

Employing Eq. (18), the expression for insertion loss can be written as the following.

$$IL = 10Log\left(\frac{\alpha_e}{\tau}\right) \tag{10.23}$$

Although this is a significant simplification, it can provide a good first estimate of the performance of a large enclosure. As an example, consider an enclosure lined with fiberglass with an absorption coefficient of 0.5 at the frequency of interest and walls providing 20 dB of transmission loss. The insertion loss is estimated by Eq. (10.23) to be 17 dB. Is this a surprise? Because of the reverberant build-up of sound energy in the enclosure one does not see the full transmission loss performance of the walls. To get an insertion loss of 20 dB, the walls of the enclosure would have to be 100% absorptive, having an absorption coefficient of 1.0.

To help in setting expectations for large enclosure performance, the following rules of thumb are helpful.

- For an enclosure with no absorption
  IL = TL – 20

- For an enclosure with partial absorption
  IL = TL – 15

- For an enclosure lined with absorption material
  IL = TL – 10

It is important to note the above discussion applies to large enclosures where the distance from the source to the enclosure walls is much more than the wavelength of the frequencies of interest. Predicting the performance of smaller or tight-fitting enclosures is much more difficult. This difficulty stems from the potential presence of standing waves or acoustic resonances between the source and the walls of the enclosure. These resonances can lead to high noise levels and the potential to degrade enclosure performance well below the transmission loss. This is especially true for unlined enclosures. For enclosures with significant absorptive material, this is less of a problem. In many cases an accurate estimate of the performance of a small enclosure can be made using Eq. (10.23) if there are deep layers of absorption materials on all the inner surfaces of the enclosure.

10.3.4.1 Partial Enclosure

Another unique enclosure found in industrial noise control is the partial enclosure. Often partial enclosures are used to protect an operator who must stand in one location for a period but needs to move frequently to other locations. They are also used when one is only concerned about noise radiation in a particular direction or one set of locations near a source. However, a partial enclosure has many limitations. As was noted previously, sealing is a critical factor in enclosure design. A partial enclosure is by its nature unsealed. It is good to think of a particle enclosure as a partial barrier that is only effective for a very limited space. In many industrials spaces there is too much reverberation to effectively use a partial enclosure. The partial enclosure only is effective in acting as a barrier for sound coming from a particular direction. In a highly reverberant field, sound is coming from every direction and a partial enclosure may provide little benefit unless it is located very close to the problem source. In the worst-case scenario, a partial enclosure may actual increase the sound level in a location.

It is important to look at the more common case where a partial enclosure is being used in an enclosed space, which is not highly reverberant. Many industrial spaces with a mixture of acoustically hard and soft areas and perhaps even open segments would meet this criterion. The basic equation for this situation is similar to what was used for a large enclosure.

$$L_p = L_W + 10Log\left(\frac{Q_B}{4\pi r^2} + \frac{4}{R}\right) \tag{10.24}$$

In this case Q_B is the directivity with the barrier. To a good approximation, it can be defined as the following.

$$Q_B = Q\sum_{i=1}^{3}\left(\frac{\lambda}{3\lambda + 20\delta_i}\right) \tag{10.25}$$

This is the Fresnel diffraction approximation about three exposed edges of the barrier or partial enclosure.

where: $\delta_i = A_i + B_i - d$. These dimensions are shown in Fig. 10.8. Note that *i* refers to each of the three edges of the barrier.

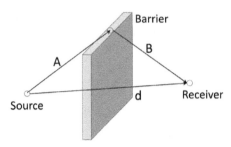

Figure 10.8. Critical Dimensions for Partial Enclosure Analysis.

10.3.4.2 Pipe Lagging

Another special case of an enclosure is pipe lagging. This is commonly used in process plants, refineries, compressor stations, and in other locations where noise radiation from piping is a concern. This is simply a close-fitting enclosure for a pipe. As with any enclosure, there is a need for sound absorption material and a high transmission loss layer for pipe lagging to work effectively.

Figures 10.9 and 10.10 show examples of pipe lagging in common industrial practice. As discussed with enclosures, the first layer of material is acoustic absorption material. Commonly mineral wool and fiberglass industrial applications. Next, if required, is thermal insulation. On the outside of this built-up treatment is a thin aluminum sheet or a fiber reinforced mastic. Note that this outer layer is important for protection the insulation and adds surface mass to improve transmission loss performance at lower frequencies.

An example of the insertion loss for pipe lagging is shown in Table 10.5. This is for a 30.5 mm (12 in) steel pipe with an aluminum outer jacket or 0.25 lb/ft^2

Figure 10.9. Example of pipe lagging in an industrial plant.

Figure 10.10. Example of pipe lagging composition.

Table 10.5. Typical Performance of Pipe Lagging.

| Material | Transmission Loss, dB | | | | |
| | Octave-Band Center Frequency, Hz | | | | |
	250	500	1000	2000	4000
2.5 cm thick	1	6	14	19	26
5.0 cm thick	1	6	15	21	28
7.6 cm thick	2	8	18	23	30

(Lord and Evensen, 1980). These data were obtained using glass fiber insulation of the thicknesses shown. As would be expected the lagging performs best at high frequencies. Since pipe lagging is being used to attenuate noise that it is being propagated down the pipe, it is critical that the lagging cover a significant length of piping. Covering a short section only provides limited attenuation in the specific area.

10.3.5 Added Absorption

The addition of absorption in rooms, enclosures, and ducts can be an effective means of noise control along the path. Absorption works by dissipating acoustic energy in one of three ways:

- Viscous loses during acoustic propagation in the material. This often is due to tight and tortuous passages.
- Reactive losses occur when reflections are used to induce phase shifts that partially cancel incoming acoustic energy.
- Mechanical losses are the result of the energy used in bending or mechanically working materials.

Some of the critical parameters related to the performance of absorption materials include mounting, flow resistance, depth, face area, density, and stiffness.

The absorption coefficient is defined by the expression below.

$$\alpha = \frac{W_A}{W_I} \tag{10.26}$$

where:

W_A = wound power absorbed

W_I = Incident sound power

For the purposes of this discussion the absorption coefficient used throughout will be the value measured by the Sabine method in a reverberant room. One can also measure sound absorption in impedance tubes to obtain a normal incidence value that can be different than the Sabine value used in this discussion.

The absorption coefficient varies greatly with frequency and therefore Eq. (10. 26) must be evaluated at multiple frequencies or over frequency ranges like a one-third octave. To simplify the comparison of materials or treatments, the noise reduction coefficient is sometimes used as shown below. It averages the absorption coefficients for the 250, 500, 1000, and 2000 Hz octaves.

$$NRC = \frac{\alpha_{250} + \alpha_{500} + \alpha_{1000} + \alpha_{2000}}{4} \tag{10.27}$$

where: NRC = the noise reduction coefficient

$\alpha_{250}, \alpha_{500}, \alpha_{1000},$ and α_{2000} = the absorption coefficients for the respective octave bands

Fortunately, most providers of acoustic absorption material provide data on performance and there is a great deal of data on the Internet for various materials. Absorption values for common materials are shown in Table 10.6.

All absorption data is highly dependent on how the material is mounted. In most cases specific data such as that in Table 10.6 will describe the mounting conditions. If the same material is to be used with different mounting conditions the performance will be different than what has been tabulated (Acoustical and Insulating Materials Association Bulletin, 1941–1974). Some standard mounting practices are shown in Fig. 10.11. In some cases, absorption coefficients may be provided with a notation of only mounting 1 or 4 from this or similar standards.

An air space between the absorption material and a hard backing surface can improve low frequency performance. In some sense it makes the material appear to be deeper at low frequencies where there is a simple reflection from the hard surface. The opposite concern is a material that is hung as a space absorber or freely hung in a space, not on a hard surface. The performance of such applications is generally reduced compared to that tabulated for the material mounted on a hard surface.

To determine the effectiveness of additional absorption, consider the noise reduction equation shown below.

$$NR = L_{po} - L_{pa} \tag{10.28}$$

$$= 10Log\left(\frac{Q}{4\pi r^2} + \frac{4}{R_o}\right) - 10Log\left(\frac{Q}{4\pi r^2} + \frac{4}{R_a}\right) \tag{10.29}$$

where:

L_{pa} = the sound pressure level after absorption added, dB
L_{po} = the sound pressure level for the original configuration, dB
Q = the source directivity
r = the distance to the point of interest

Table 10.6. Sound absorption coefficients for common materials.

Material	125Hz	250Hz	500Hz	1000Hz	2000Hz	4000Hz	NRC Number*
Walls (1–3, 9, 12) Sound-Reflecting:							
1. Brick, unglazed	0.02	0.02	0.03	0.04	0.05	0.07	0.05
2. Brick, unglazed and painted	0.01	0.01	0.02	0.02	0.02	0.03	0
3. Concrete, tough	0.01	0.02	0.04	0.06	0.08	0.10	0.05
4. Concrete block, painted	0.10	0.05	0.06	0.07	0.09	0.08	0.05
5. Glass, heavy (large panes)	0.18	0.06	0.04	0.03	0.02	0.02	0.05
6. Glass, ordinary window	0.35	0.25	0.18	0.12	0.07	0.04	0.15
7. Gypsum board, 1/2 in thick (nailed to 2 x 4s, 16 in oc)	0.29	0.1	0.05	0.04	0.07	0.09	0.05
8. Plywood, 3/8-in paneling	0.28	0.22	0.17	0.09	0.10	0.11	0.15
9. Steel	0.05	0.10	0.10	0.10	0.07	0.02	0.10
10. Venetian blinds, metal	0.06	0.05	0.07	0.15	0.13	0.17	0.10
11. Wood, 1/4-in paneling, with airspace behind	0.42	0.21	0.10	0.08	0.06	0.06	0.10
12. Wood, 1-in paneling with airspace behind	0.19	0.14	0.09	0.06	0.06	0.05	0.10
Sound-Absorbing:							
13. Concrete block, coarse	0.36	0.44	0.31	0.29	0.39	0.25	0.35
14. Shredded-wood fiberboard, 2 in thick on concrete (mtg. A)	0.15	0.26	0.62	0.94	0.64	0.92	0.6
15. Thick, fibrous material behind open facing	0.6	0.75	0.82	0.8	0.6	0.38	0.75
16. Wood, 1/2-in paneing, perforated 3/16-in-diameter holes, 11% open area, with 2 1/2-in glass fiber in airspace behind	0.4	0.9	0.8	0.5	0.4	0.3	0.65
Ceilings (6, 8–10) Sound-Reflecting:							
17. Concrete	0.01	0.01	0.02	0.02	0.02	0.02	0
18. Gypsum board, 1/2 in thick	0.29	0.10	0.05	0.04	0.07	0.09	0.05
19. Gypsum board, 1/2 in thick, in suspension system	0.15	0.10	0.05	0.04	0.07	0.09	0.05
20. Plaster on lath	0.14	0.10	0.06	0.05	0.04	0.03	0.05
21. Plywood, 3/8 in thick	0.28	0.22	0.17	0.09	0.10	0.11	0.15

R_o = the original room constant

R_a = the room constant with added absorption

Since only the reverberant field is changed, Eq. (29) can be reduced to the following.

$$NR = 10Log\left(\frac{R_a}{R_o}\right) \tag{10.30}$$

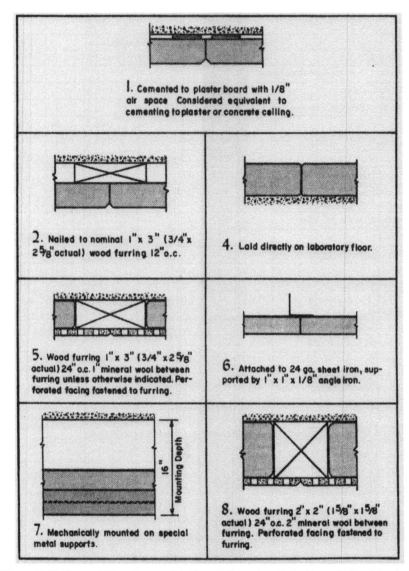

Figure 10.11. Standard Mountings used in Sound Absorptions Tests (Acoustical and Insulating Materials Association Bulletin, 1941–1974).

With very few exceptions in such circumstances R_a/R_o is generally less than 10 and thus, NR is generally less than 10 dB. One should not expect extremely high levels of noise reduction due to the addition of absorption in a room. As shown in Fig. 10.12, a large change in the absorption ratio is required to obtain a significant noise reduction. In simple terms, adding absorption where there is practically no absorption can be effective. Adding more absorption where there is already significant absorption, is not generally effective.

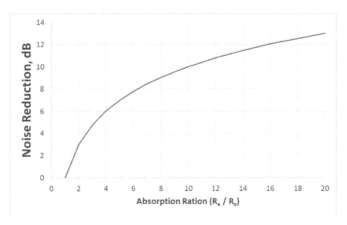

Figure 10.12. Effectiveness of Adding Absorption.

10.3.6 Vibration Isolation

Reducing the transmission of vibrational energy into a structure from a source can be a highly effective means of noise control. As noted previously, the noise problems associated with vibrating machines are not just the machine itself. If the machine conveys vibrational energy into a structure, the energy may be transmitted to some other point and reradiated from panels and surfaces as noise. The purpose of vibration isolation is to break this transmission path and reduce potential noise issues.

It should be noted that vibration isolation is also used to reduce the energy transmitted that leads to vibration problems in other parts of the structure. When large compressors are next to office spaces or sensitive instruments are housed in a building with large mechanical equipment, vibration isolation becomes critical.

The first point of view for vibration isolation is to reduce the motion of the machine in question. There may be concern about the noise radiated from the machine, delicate instruments on the machine that may be damaged, or connected piping that may fatigue. If a machine is considered as a lumped mass with one or several isolators treated as a simple spring and damper, a simple one degree of freedom expression can be used to describe the effect of the isolator. From the equations for a forced single degree of freedom system one can derive the following relationship for the ratio of the static displacement to the vibrational displacement.

$$MF = \frac{X}{X_0} = \frac{X}{F}k = \frac{1}{\sqrt{(1 - r^2)^2 + (2\zeta r)^2}} \tag{10.31}$$

where:

MF = the magnification factor

X = the displacement magnitude

X_0 = the static displacement due to the static application of the force magnitude
    ($X_0 = F/k$)

$r = \omega/\omega_n$

ω = forcing frequency $(2 * \pi * f)$

ω_n = the natural frequency of the system

F = the force magnitude

k = the spring constant

ζ = the critical damping ratio = c/c_c

c = damping coefficient

c_c = critical damping coefficient

Figure 10.13 was generated using Eq. (10. 31). It clearly shows that with low damping ($\zeta = 0.05$) the amplitude of the displacement at the resonant frequency can be quite large. In this case 10 times the static displacement. With higher levels of damping, this large response can be reduced significantly. Therefore, when one is working to suppress displacement for a machine on isolators, damping in the isolators is important.

From this figure it is also clear that there are regions of response related to characteristics of the system. Where ω is much less ω_n, is a stiffness-controlled region and $X = F/k$. In this region $MF = 1$. Where ω is much greater than ω_n, is a mass-controlled region and $X = F/m* \omega_n^2$ which approaches zero for large values of ω. Finally, the large peak in response, where $\omega = \omega_n$, is the damping-controlled region and $X = F/(c * \omega_n)$. A good rule of thumb is to keep the operating frequency at least two times the natural frequency of the machine on its isolators.

The other issue of concern is the transmission of force to the structure from the vibrating machine. This mechanical energy may lead to noise or vibration in other parts of the structure that are a concern. Automakers use vibration isolators to mount engines to the car structure to ensure that passengers in the interior are not disturbed by noise or vibration from the engine. Using the same assumption about a forced one degree of freedom system, the force ratio can be defined by the following equation.

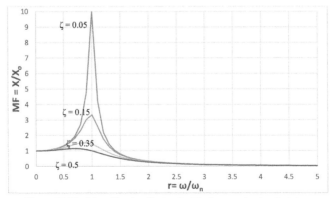

Figure 10.13. Magnification Factor for Displacement using Isolators.

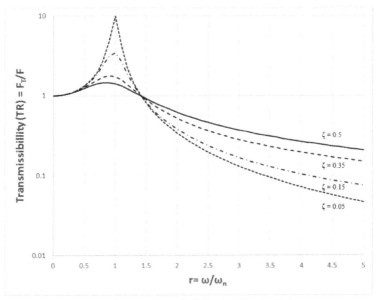

Figure 10.14. Force Transmissibility for Vibration Isolators.

$$TR = \frac{F_T}{F} = \frac{\sqrt{1 + (2\zeta r)^2}}{\sqrt{(1 - r^2)^2 + (2\zeta r)^2}} \qquad (10.32)$$

where

TR = transmissibility

F_T = transmitted force

Figure 10.14 was generated using Eq. (10.32). It shows that the transmission of force is much different than the displacement characteristics. The transmitted force ratio drops quickly at higher frequencies. However, above the resonant frequency high damping ratios lead to greater force transmitted. This means that one must be careful in selecting isolator parameters to minimize force transmission in the region of major concern. As noted above, a good rule of thumb is to design isolators such that the operating frequencies are well above the natural frequency of the machine on the isolator. For force isolation a ratio of three times the natural frequency may be a better rule of thumb than the two used for displacement control.

Often vibration isolator providers support the selection of isolators based on static displacement, the X_o term used above. This simplifies the selection process but it does not address the damping provided by the isolators. Looking at this in terms of the governing equations, the following expression can be derived.

$$X_o = \frac{F}{k} = \frac{mg}{k} = \frac{g}{w_n^2} \qquad (10.33)$$

Rearranging the terms, one obtains.

$$f_n = \frac{1}{2\pi}\sqrt{\frac{g}{X_o}}$$

(10.34)

where:

g = the gravitational constant

f_n = the natural frequency, Hz.

This permits one to select the static displacement based on natural frequency or the natural frequency based on static displacement. As one would expect, to achieve a low natural frequency a large static displacement is required. Minimizing the static displacement (stiffening the isolator) leads to a higher natural frequency and possibly less isolation.

Values from this expression are shown in Table 10.7.

This analysis can be taken further. From Eq. (10.32) if damping is neglected ($\zeta = 0$) and if the operational frequency far exceeds the natural frequency ($\omega \gg \omega_n$) the equation can be reduced to the following.

$$TR = \frac{1}{r^2-1} = \frac{\omega_n^2}{\omega^2-\omega_n^2}$$

(10.35)

This can be rearranged to the following expression for static displacement.

Table 10.7. Relationship between static displacement and natural frequency for isolators.

Static displacement, mm	Natural frequency, Hz.
0.5	22.29
1.5	12.87
2.5	9.97
3.5	8.42
4.5	7.43
5.5	6.72
6.5	6.18
7.5	5.75
8.5	5.41
9.5	5.11
10.5	4.86
11.5	4.65
12.5	4.46
13.5	4.29
14.5	4.14
15.5	4.00

$$X_o = g \frac{(1 + TR)}{(TR \; \omega^2)} \tag{10.36}$$

This expression can be a good way to quickly predict the required static displacement for a desired transmissibility and a known operating frequency.

10.3.6.1 Inertia Mass

In many instances one or more machines are mounted on an inertia mass or structural frame to provide isolation. A basic representation of such a system is shown in Fig. 10.14. The transmissibility for this system can be written as shown below (Bies and Hansen, 2009).

$$TR = \left| \frac{(1 - R^2)}{1 - r^2 * \left(G^2 - \frac{R^2}{M} \right)} \right| \tag{10.37}$$

where:

$R = f/f_s$

f_s = the natural frequency of the support without the machine in place

$G = f_s / f_m$

f_m = natural frequency of the machine without inertia mass, m_s

$M = m / m_s$

m = machine mass

Some sample results using Eq. (10.37) are shown in Fig. 10.15.

The higher order effects shown in this chart would be reduced or eliminated with damping considered in the machine and inertia mass mounts. It should be apparent that it is possible to make significant adjustments in the isolation of a machine using an inertia mass.

Whether using isolators alone or with an inertia mass the selection of the proper isolator is critical. There are many types of vibration isolators in use and commercially available. In many cases the suppliers of isolators provide detailed information to help in the selection of isolators and make recommendations for applications. Commercial isolators include simple springs, elastomeric pads, elastomeric isolators, and sophisticated multiple chamber and active control units. The later can provide isolation over a range of frequencies. Figure 10.16 summarizes the most common application ranges for common types of isolators (Bies and Hansen, 2009).

There is one last unique case of vibration isolation that must be considered for noise control. In some cases, the vibration introduced by a machine is too large to be sufficiently suppressed by the addition of an inertial mass or the installation of isolators. For instance, consider the installation of a large reciprocating

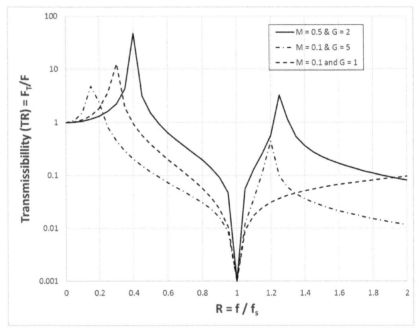

Figure 10.15. Transmissibility for a two mass system.

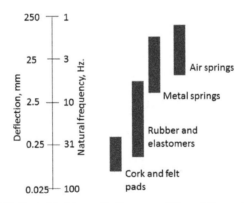

Figure 10.16. Typical isolator application ranges (Bies and Hansen, 2009).

compressor on the same structural slab as a conference room. In such instances the best solution may be to install the compressor on a separate foundation from the rest of the building. Although an expensive solution, such a noise control can be highly effective in reducing the noise and vibration transmitted into the structure. For such an installation care must be taken to prevent ancillary connections for plumbing, HVAC, electrical, or other purposed from short-circuiting the isolation of the separate foundation.

10.3.7 Tuned Vibration Absorber

Similar to the use of an inertia mass, a small additional mass may be used to detune a system to reduce or eliminate the response at a resonance frequency. The principle is that a small mass like that shown in Fig. 10.17 is attached to a large machine or structural component and tuned such that it eliminates resonance response at a problem frequency. They are widely used to solve crucial resonance problems. In tall buildings large masses are suspended in elevator shafts to act as a pendulum to detune a resonance excited by wind or seismic forces. In the past automobiles used bumpers (the heavy steel kind) and cantilever masses off transmissions to suppress problem resonances. Another good example still in wide usage is the torsional vibration absorber on the crankshaft of modern automobile engines.

The concept behind a vibration absorber is to have the small mass vibrate with high amplitudes at the frequency of concern. This will absorb all the energy at this frequency and effectively suppress the transmission of force to the base structure at this frequency. In the simplest case one simply selects m_2, k_2, and c_2 to provide a first resonance at the problem frequency. Unfortunately, the machine speed may

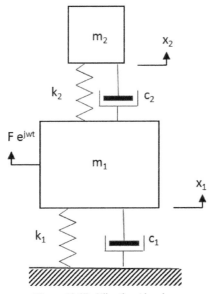

Figure 10.17. Vibration Absorber.

vary, and the frequency of concern may vary. This may result in a need to suppress force transmission over a range of frequencies.

To examine the performance of the vibration absorber the following equation is used (Bies and Hansen, 2009).

$$\frac{X k_1}{F} = \sqrt{\frac{\left(\frac{2\zeta_2\,\Omega m_1}{m_2}\right)^2 + \left(\Omega^2 - \frac{k_2 m_1}{k_1 m_2}\right)^2}{\left(\frac{2\zeta_2\,\Omega m_1}{m_2}\right)^2 \left(\Omega^2 - 1 + \frac{m_2\Omega^2}{m_1}\right)^2 + \left(\frac{k_2}{k_1}\Omega^2 - (\Omega^2 - 1)\left(\Omega^2 - \frac{k_2 m_1}{k_1 m_2}\right)\right)^2}}$$

(10.38)

where:

$\Omega = \omega * (m_1 / k_1)^{0.5} = f/f_0$

f_0 = frequency of concern

$\zeta_2 = c_2/2 * (k_1 * m_1)^{0.5}$

Figure 10.18 shows the performance predicted by this equation for a series of values of the main parameters.

With proper design, the vibration absorber can be highly effective in suppressing vibration and noise at a particular frequency. The simplest approach it to simply design the added mass, spring, and damper to have the first resonance at the problem frequency.

From this discussion it is clear that there are a number of methods to isolate or minimize the mechanical energy from a source being transmitted into the structure.

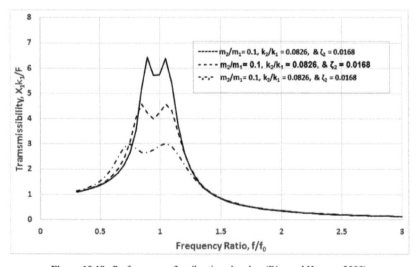

Figure 10.18. Performance of a vibration absorber (Bies and Hansen, 2009).

This does not mean that structural noise transmission is not an issue. It is a major noise control problem even with proper isolation and source controls the amount of energy transmitted through the structure may still cause noise issues. Therefore, it is necessary to consider other means of control along the path.

10.3.8 Vibration Damping

The use of damping can be both a source modification and a path treatment when used in engineering noise control. Noise control damping is most often applied to vibrating panels to reduce the displacement amplitude. By doing so, the radiated noise is decreased. When these panels are part of a noise source this treatment represents control at the source by source modification. When the radiating panel is being excited by energy that has traveled through structure, the treatment may be seen as control along the path. Damping materials simply contribute a loss component that reduced the amount of vibrational energy. Typically, elastomeric materials are used that convert the mechanical energy into heat.

Based on the mechanical characteristics of a panel damping can be applied sparingly and be highly effective. The key is to place the damping material where it will reduce the energy available for modes of vibration that contribute the most to the radiated noise. For damping materials to work well, the material must experience high flexural energy or displacement. Placing damping materials at a node point will be highly ineffective. For this same reason, damping works best on thin panels. A good rule of thumb is that panels should be no more than 7 mm in thickness. It is important to note that a small patch of damping placed in the right place may be as effective as coating a whole panel with damping material.

Sound radiation from panels is highly dependent on the frequency of interest relative to the coincident frequency. Equation 2 can be rewritten as the following.

$$f_c = \frac{c^2}{2\pi} \sqrt{\frac{\rho_s t}{EI}} \tag{10.39}$$

where:

c = speed of sound in air

ρ_s = surface density of panel

t = panel thickness

E = panel modulus of elasticity

I = moment of inertial of panel cross section

For frequencies much less than the coincidence frequency, damping should be applied at the panel edges and attachment points. For a rectangular plate supported on all four sides this would mean damping should be applied along the panel edges. This will be where most of the radiated noise initiates at such low frequencies. While damping can be effective in this case, the amount of noise control possible is limited.

For frequencies greater that he coincidence frequency damping should be applied at the antinodes of the frequencies of concern. Damping can be highly effective in these cases.

To best define the effectiveness of damping, the damping loss factor, η, is commonly used. It is defined by the equation below.

$$\eta = 2 \left(\frac{c}{c_c} \right) r = 2\zeta r \tag{10.40}$$

This equation is useful in defining the loss factor and in relating loss factor to the critical damping ratio used in so many expressions in defining vibration and transmission loss.

Since damping is generally applied at resonance, Eq. (10.40)

can be written as the following:

$$\eta = 2\zeta \tag{10.41}$$

This loss factor is used in defining the complex modulus employed to describe the characteristics of damping materials.

$$E^* = E\,(1+j\eta) \tag{10.42}$$

Fortunately, the performance of most damping materials used in noise control is well documented. However, there are well established means for measuring the loss factor. It is important to note that the loss factor for elastomeric materials is a function of temperature and frequency. The performance of a material can change drastically depending on temperature and frequency.

There are well defined procedures for testing damping materials using beams and plates. In these procedures great care is given to quantify damping accurately. In some case a log decrement procedure is used. As shown in Fig. 10.19, the calculation is based on the decreasing amplitude of the vibration.

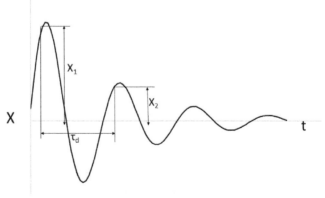

Figure 10.19. Logarithmic decrement analysis.

From Figure 10.19, the logarithmic decrement is simply.

$$\delta = \ln\left(\frac{X_1}{X_2}\right) \tag{10.43}$$

It can be shown that the loss factors can be calculated as shown in the following equation.

$$\eta = \frac{2\delta}{\sqrt{(2\pi)^2+\delta^2}} \tag{10.44}$$

For small δ (usually the case), this expression becomes the following.

$$\eta = \frac{2}{\pi} \tag{10.45}$$

There are also beam tests done to assess loss factor. In this case the damping material is placed on a beam of known properties and the shift in damped resonant frequency measured. For this case the loss factor can be approximated by the expression below.

$$\eta = \frac{\Delta f}{f} \tag{10.46}$$

where:

Δf = the change in natural frequency from that for the beam without the damping material

f = the natural frequency for the beam without the damping material.

There are two ways to apply damping to a vibrating plate. The first is homogeneous damping. This is a free viscoelastic layer applied on top of the plate. The damping occurs because of extensional deformation of the material. Often plastic or asphaltic materials are used for this sort of application. The damping material thickness must be equal to or greater than that of the plate to be highly effective. The loss factor is proportional to the mass of the damping materials and the damping material surface mass should be at least 20% of that of the panel. The minimum dimensions for such a treatment should be at least 40% of the flexural wavelength.

The more effective and most costly damping treatment is constrained layer damping as shown in Fig. 10.20. In this type of application there is a stiff outer layer on top of the damping treatment. The shear between this surface treatment and the panel is what introduces the damping loss in this application. Since high

Figure 10.20. Constrained layer damping treatment.

shear stresses can be generated in a short space between the panel and the stiff layer, a high level of damping can be provided. This method minimizes the amount of damping material required and the added mass to the panel. One can often use an outer sheet of aluminum or steel that is one-third the thickness of the plate and get effective damping. For this type of application, the damping surface mass should be 20% or less of that of the panel. Also, the maximum dimensions of the treatment should be 60% of the flexural wavelength.

10.3.9 Silencers

For vents to the atmosphere, fan noise, HVAC duct systems, and many other applications involving noise in a fluid stream, silencers can be highly effective noise controls. Silencers are referred to using a variety of names including mufflers, snubbers, resonators, and filters to name a few. This segment will cover all the major types of such devices.

To begin, it is important that the definitions in considering the effectiveness of silencers be reviewed. Equation 1 defines transmission loss as 10 times the log of the ration of the input and output sound power. This is an effective way to characterize a silencer since it is independent of what is upstream and downstream of the silencer. However, it is very difficult to measure directly. The insertion loss is defined in Eq. 15. It is the difference in two sound pressure levels measured with and without a noise control treatment. This quantity is relatively easy to measure, but it is difficult to calculate for complex environments. With silencers, one may encounter dynamic insertion loss which refers to the insertion loss with fluid flowing in the system. In some cases, it is much easier to measure insertion loss without flow. The final term discussed previously was noise reduction as defined in Eq. (11). For silencers noise reduction is the difference in sound pressure levels measured at the inlet and the outlet of the silencer.

For HVAC systems and other similar flow components, the noise control industry often uses the term attenuation to describe the effectiveness of components. This is simply the reduction in sound power between two points in the system. In this sense it is much like transmission loss. One may treat attenuation as a rough approximation to transmission loss. With the assumption of plane waves where the wavelength of the sound is much larger than the dimensions of the duct, the change in sound power level is equivalent to the change in sound pressure level. This makes attenuation a very convenient term for rough calculations. However, one must remember that, at higher frequencies when the wavelength of sound approaches the dimensions of the duct, the equivalence of sound pressure level and sound power level breaks down.

For the purposes of this discussion, silencers will be divided into two types: reactive and dissipative. A reactive silencer attenuates sound by cancellation using out of phase reflections. One can think of such silencers as similar to resistance in electrical circuits. The dissipative silencer employs absorptive materials to attenuate sound. The dissipative silencer is generally more effective at higher frequencies

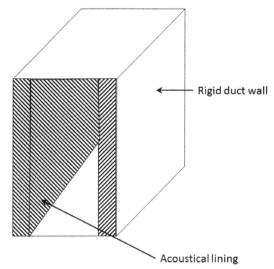

Figure 10.21. Simple duct silencer.

and has broader application. In some cases, both types of silencers are combined in one to provide a broad frequency range of attenuation.

The simplest dissipative silencer is shown in Fig. 10.21. This is the use of fiberglass or other absorptive materials as linings in a duct. The attenuation is provided by the parallel absorptive baffles that absorb sound. The attenuation for this type of silencer can be expressed as the following.

$$A = 1.05\alpha^{1.4}\frac{P}{S}$$
(10.47)

where:

A = attenuation (dB/m)

P = perimeter of flow area, m

S = flow area, m^2

α = random incidence absorption coefficient.

Note that his is a commonly used expression which is a simplification of the room acoustics approach. It is at best a good approximation. There are several restrictions to the use of this expression. Flow velocity should not be too high or self-generated noise will exceed the attenuation achieved. A good rule of thumb is to keep the flow velocity below 1200 m/min. This formulation assumes a plane wave propagating down the duct. This requires the large transverse dimension be kept to less than $\lambda/10$. The length of the insulation or parallel baffles as they would be called in a silencer must be a minimum length to provide the predicted performance. The rules of thumb are not consistent here. However, if the length is at least three times the larger duct transverse dimension, Eq. 10.47 should provide

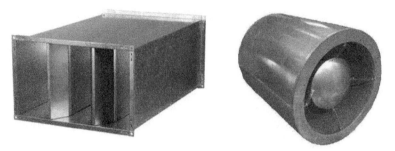

Figure 10.22. Examples of duct silencers.

a good approximation. Also, the performance of this type of silencer has an upper bound. No matter how long the treated section, the maximum attenuation seldom exceeds 20 dB. In real world applications it is quite difficult to get this level of attenuation.

To improve performance, it is necessary to expose a greater surface area of insulation and have insulation in the center of the flow stream. To do this a number of different types of silencer geometries are used. Some examples are shown in Fig. 10.22. There are a wide variety of geometries used in these types of silencers. It must be noted that primary considerations in the design and selection of such silencers beyond noise performance include pressure drop, self-generated noise, plugging with particulates in the flow, and durability. These considerations can have major impact on finding the best silencer for a particular application. There will more discussion on this topic at the end of the segment.

An important tool for noise control in ducts is the right-angle bend. Because of the potential to present a large absorptive surface to the flow stream, lining bends in ductwork can be highly effective in reduction propagated noise. An illustrative example is shown in Fig. 10.23. Note that the lining is shown extending well beyond

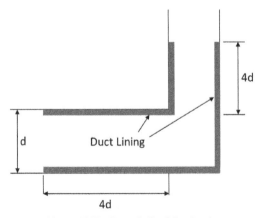

Figure 10.23. Example lined duct bend.

Table 10.8. Attenuation for lined elbow without turning vanes (Kingsbury, 1979).

Duct width, cm	Frequency, Hz							
	63	125	250	500	1000	2000	4000	8000
15	0	0	0	1	7	12	14	16
30	0	0	1	7	12	14	16	18
60	0	1	7	12	14	16	18	18
120	1	7	13	15	15	18	18	18

the actual bend. This is very important to achieve good performance. Because of the turbulence (noise) created in the bend it is critical that the lining extend at least four times the largest transverse dimension of the duct on either side of the bend.

While there are different approaches to calculating the attenuation of lined duct bends, Table 10.8 (Acoustical and Insulating Materials Association Bulletin, 1941–1974) provides values for a lined 90° elbow with no turning vanes. This chart assumes a 2.54 cm thickness of insulation, insulation for a sufficient distance before and after the turn, and no turning vanes. For more complex cases the American Society of Heating and Refrigeration Engineers (ASHRE) and others provide data. Attenuation in duct systems can be considered for branches, outlets, turning vanes, and many other aspects of the system design.

10.3.9.1 Plenum Chambers

Plenum chambers are also used for noise control in duct systems. When lined and properly designed plenum chambers like the one shown in Fig. 10.24 can be effective in reducing the sound power propagating down the duct. For the design

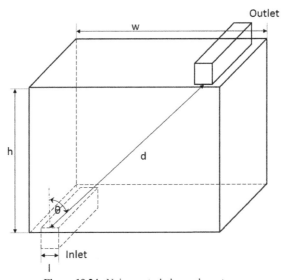

Figure 10.24. Noise control plenum layout.

shown the sound transmission loss for the chamber can be calculated from the equation below (Ver, 2006).

$$TL = 10Log\left(S\left(\frac{\cos\theta}{2\pi d^2} + \frac{1-\alpha}{\alpha\, S_w}\right)\right) \qquad (10.48)$$

where

S = the inlet or exit area

Θ = the angle between the inlet and outlet

d = the distance between the inlet and outlet

α = the average absorption coefficient for the plenum interior surfaces

S_w = the wall area inside the plenum

This expression is accurate where λ is less than the major dimensions of the plenum. When λ is equal to or greater than the dimensions the predictions will tend to underestimate the performance of the plenum by 5–10 dB. Designing a plenum in which there is no line of sight between the inlet and outlet (using baffles or a tortuous path) can give performance far exceeding the predictions of Eq. (48).

10.3.9.2 Reactive Silencers

As noted before, reactive silencers use reflected or out of phase sound to cancel unwanted sound. Their design can be quite different than dissipative silencers. The simplest and most common reactive silencer is an expansion chamber. This is commonly found on compressors, blowers, and small internal combustion engines. This silencer is just what the name implies. It is a sudden expansion in the flow pipe or duct. This sudden expansion leads to a reflection of the acoustic energy that reduces the propagating energy. A typical expansion chamber is shown in Fig. 10.25.

The transmission loss for the expansion chamber is defined in the equation below (Ver, 2006).

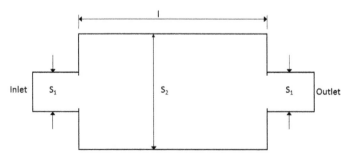

Figure 10.25. Expansion chamber silencer.

$$TL = 10Log\left(\cos^2 kl + 0.25\left(m + \frac{1}{m}\right)^2 \sin^2 kl\right) \qquad (10.49)$$

where: $k = \frac{2\pi f}{c} = \frac{\omega}{c}$ (wave number)

$m = S_2/S_1$

S_1 = cross-sectional area of inlet and outlet (assumed to be the same)

S_2 = cross-sectional area of the expansion chamber

l = length of the chamber

For an expansion chamber to be effective, m must be four or greater. Also, the dimensions of the chamber must be greater that $0.8 * \lambda$ for the lowest frequency of interest. For maximum attenuation there are a series of optimum lengths. These are defined by the equations below.

$$kl = \frac{n\pi}{2}, n = 1, 3, 5, \dots \qquad (10.50)$$

or

$$l = \frac{n\lambda}{4}, n = 1, 3, 5, \dots \qquad (10.51)$$

Equation (10.49) neglects the effects of flow velocity. This is a good assumption for $M < 0.1$.

An example calculation using Eq. (10 49) is shown in Fig. 10.26. The rise and fall of the transmission loss with multiples of $n*\pi/2$ is clearly evident in these data. In actual applications the transmission loss will diminish at higher multiples of n as the wavelength of sound approaches or goes below the cross-sectional dimensions of the chamber. Often expansion chamber silencers act like high pass filters. They are most effective at low frequencies where λ is much greater than the cross-sectional dimensions. The reduced effectiveness at higher frequencies means that most of the energy at high frequencies is "passed" through the silencer with little or no attenuation.

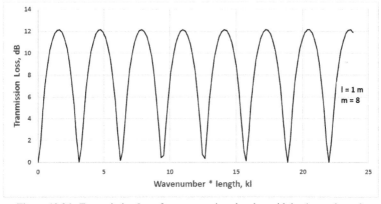

Figure 10.26. Transmission Loss for an expansion chamber with $l = 1$ *m* and *m* = 8.

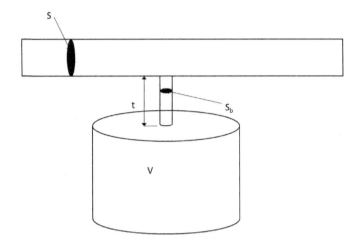

Figure 10.27. Helmholtz side branch resonator.

Another common reactive silencer is the side branch Helmholtz resonator as shown in Fig. 10.27. This silencer can be effective in reducing sound by generating an out of phase wave at the side branch that can be tuned to significantly reduce the noise propagating down the main fluid path. It is sometimes convenient to think of the volume of the resonator as a mass oscillating on a spring represented by the side branch tube. This analogy will be discussed in greater detail later.

The equation which describes the effectiveness of the Helmholtz resonator is shown below (Kinsler and Sanders, 1982).

$$TL = 10Log\left(1 + \frac{c^2}{4S^2\left(\frac{\omega l}{S_b} - \frac{c^2}{\omega V}\right)}\right) \tag{10.52}$$

where:

S = the cross-sectional area of the main duct or pipe

S_b = the cross-sectional area of the side branch tube

l = length of the side branch

$$= t + 0.8\sqrt{\frac{S_b}{n}}$$

N = number of branches

The maximum effectiveness occurs at the resonance frequency, f_R, of the side branch as given by the following.

$$f_R = (c/(2*\pi)) * (S/(l*V))^{0.5} \tag{10.53}$$

$$f_R = \frac{c}{2\pi}\sqrt{\frac{S}{lV}} \tag{10.54}$$

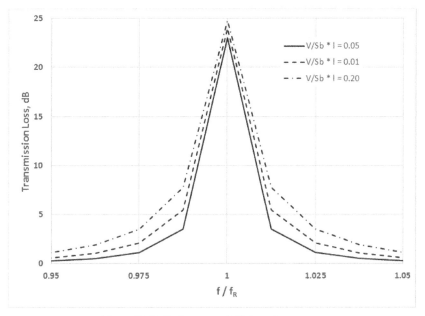

Figure 10.28. Performance of side branch resonator.

Note that the resonant frequency expression is much like that for a mass on spring, the term S/l is equivalent to stiffness and V is equivalent to the mass. Therefore, one can lower the frequency of the resonator by increasing the volume or mass equivalent. Also, one can decrease the resonant frequency by lowering the equivalent stiffness by increasing 1 or reduction S. Generally, one only considers changing 1 since S is normally fixed in a control situation. The effectiveness of a side branch resonator is shown in Fig. 10.28 with dimensionless ratios of the primary factors.

Unlike the expansion chamber filter, the sider branch resonator type silencer acts more like a band pass filter. It can provide quite effective attenuation around the resonant frequency as shown in Fig. 10.28. However, above and below this

Figure 10.29. Examples of commercial reactive silencers.

frequency the amount of attenuation drops off. One can tune the Helmholtz resonator to have broad or sharper response. As shown in Fig. 10.28, as the factor $(V/S_b)*1$ can be treated as an approximate damping factor. As this term gets larger the response is broader. One trades off a larger attenuation at the resonant frequency for a broader frequency range of attenuation. Note that to increase this factor decreasing S_b or increasing l are the common approaches. It can also be seen intuitively that, if there is a column of air moving back and forth in a narrower or longer tube between the volume and the main pipe, there would be greater losses. Typically, increasing volume is not a good solution since the space required becomes a problem.

As with dissipative silencers, there are a wide range of commercially available reactive silencers. As with dissipative silencers, design and selection criteria focus on more than simply noise control. Considerations include durability, pressure drop, clogging, and water liquid collection/drainage to name a few. Some examples of commercial reactive silencers are shown in Fig. 10.29.

While there are many HVAC silencers that are purely dissipative, most commonly reactive silencers contain dissipative elements as well. The purple area in the right silencer in Fig. 10.23 would normally be a glass or mineral wool fiber to provide dissipative attenuation in addition to the reactive elements. In these two silencers one can clearly see the expansion chamber at the entrance to the silencer and the side branch Helmholtz resonators in the many holes in the tubes in the silencers. Note that the length of the side branch tube is on only the depth of the hole or equivalently the thickness of the tube in which it is cut.

This combination of elements is used to get a broad range of attenuation and to tune the silencer to provide the desired noise control or transmission loss characteristics. When greater higher frequency attenuation is needed, a larger or more densely packed dissipative section could be used. By changing the length of the inner tubes, the size of holes and the volume of the expansion chambers, further tuning can be done. In addition, these changes must also be done in consideration of pressure drop, clogging, fluid retention, and other design parameters. Commonly such silencers are used on HVAC systems, compressors, automobiles, aircraft engines, turbine intakes and exhausts, commercial blowers, and many other applications.

One specific application that is worth special consideration is the vent silencer. These silencers are often used when air or steam are vented to the atmosphere and can represent both a worker exposure and environmental noise problem. Such silencers use the same principles at noted previously, however there may be additional considerations in terms of high thrust loads for energetic vents, environmental concerns with long term exposure and corrosion, and other concerns. Often the cost of such silencers with high flow rates can be large and justifying such cost for a silencer that sits unused until an upset or other rare condition occurs can be hard to justify. The author has seen instances where it was necessary to have such a silencer nearly 10 m in diameter to handle an extreme upset condition. Justifying the cost and space requirements was a daunting task in this case.

10.4 Conclusion

Noise control is often a complex process which involves the understanding of both airborne and structural transmission of energy. It may be necessary to separate the airborne and structural paths to understand how to best reduce or modify the noise exposure of workers or individuals in the environment. However, both transmission paths must be considered in every noise control process to insure success.

What has been presented in this chapter is only an introduction to the basic principles. Fortunately, all noise control processes revolve around the use of the source, path, and receiver model. Applying this model is the first step in any noise control process. One should also first seek to apply noise controls at the source. In all cases, source controls should be the most effective and efficient. There are often cases where control at the source are not possible due to cost, production operations, or other issues. The next step is to develop controls along the path. Much of this chapter has discussed the basic principles of such controls. This should provide a good start. In addition, there are numerous materials and noise control devices available commercially for path noise control. Using these basic principles, one can understand their effectiveness and how they can best be applied. Note that the inexpensive "magic" material or device that provides a double-digit reduction in decibels is not often going to live up to the performance claims. However, well designed noise controls addressing the source and path can produce double digit reductions when properly applied.

References

Acoustical and Insulating Materials Association Bulletin, 1941–1974, Performance Data, Architectural Acoustical Materials.

Berger, E.H., Franks, J.R. and Lindgren, F. 1996. International review of field studies of hearing protector attenuation. pp. 361–77. *In*: Axlesson, A. et al. (eds.). Scientific Basis of Noise-Induced Hearing Loss. Thieme, New York.

Brueck, Liz. 2009. Real world use and performance of hearing protection. Prepared by Health and Safety Laboratory for the Health and Safety Executive 2009, RR720, Health and Safety Laboratory Harpur Hill Buxton Derbyshire, UK.

Lie, Arve, Skogstad, Marit, Johannessen, Håkon, A., Tynes, Tore, Mehlum, Ingrid Sivesind, Nordby, Karl-Christian, Engdahl, Bo,and Tambs,Kristian, 2015,Occupational noise exposure and hearing: a systematic review. Int. Arch. Occup. Environ Health: 1–22.

Concha-Barrientos, Marisol, Campbell-Lendrum, Diarmid and Steenland, Kyle. 2004. Occupational noise—Assessing the burden of disease from work-related hearing impairment at national and local levels, World Health Organization Protection of the Human Environment, 33 p.

International Institute of Noise Control Engineering, survey of legislation, regulations, and guidelines for control of community noise,2009,International Institute of Noise Control Engineering Publication 09-1, Final Report of the I-INCE technical study group on noise policies and regulations (TSG 3). 50 p.

Ontario Ministry of the Environment, Environmental Approvals Access and Service Integration Branch andEnvironmental Approvals Branch. 2013. Environmental Noise Guideline Stationary and Transportation Sources – Approval and Planning Publication NPC-300. 65 p.

Transportation Research Board of the National Academies. 2008. Effects of Aircraft Noise: Research Update on Selected Topics – A Synthesis of Airport Practice, 99 p.

Koopmann, Jonathan, Solman, Gina Barberio, Ahearn, Meghan, and Hwang, Sunje, John A. Volpe National Transportation Systems Center (U.S.). 2014. Aviation Environmental Design Tool user guide version 2a. 184 p.

UK Environmental Research and Consultancy Department of the Civil Aviation Authority, 2014, Aircraft noise, sleep disturbance and health effects, CAP 1164. 33 p.

International Institute of Noise Control Engineering Working Party on Noise Emissions of Road Vehicles (WP–NERV), 2001, Noise Emissions of Road Vehicles Effect of Regulations Final Report 01-1, 56 p.

Japan Automobile Manufacturers Association, Inc. 2013. Report on Environmental Protection Efforts Promoting Sustainability in Road Transport in Japan, 38 p.

Lamancusa, John, S. and Eschenauer, Hans, A. 1994. Design optimization methods for rectangular panels with minimal sound radiation. AIAA Journal 32(3): 472–479.

Cremer, L. and Heckl, M., Translated by Ungar, E.E. 1988. Structure-Borne Sound: Structural Vibrations and Sound Radiation at Audio Frequencies, Springer-Verlag, Berlin.

Ver, Istvan, L. 2006. Interaction of sound waves with solid structures. pp. 389–515. *In*: Istvan, L. Ver and Leo, L. Beranek (eds.). Noise and Vibration Control Engineering: Principles and Applications. John Wiley & Sons, Inc., New York.

Bies, David, A. and Hansen, Colin, H. 2009. Engineering Noise Control: Theory and Practice, Fourth Edition, Spoon Press, New York.

Lord, H., Gately, W.A. and Evensen, H.A. 1980. Noise control for engineers. McGraw Hill, New York.

Kingsbury, H.F. 1979. Heating, Ventilating, and Air-Conditioning Systems. pp. 28–5. *In*: Harris, C. M. (ed.). Handbook of Noise Control. McGraw Hill, New York.

Kinsler, L.E., Frey, A.R., Coppens, A.B. and Sanders, J.V. 1982. Fundamentals of Acoustics, 3rd edn. John Wiley & Sons, New York.

Index

Printed and bound by CPI Group (UK) Ltd, Croydon, CR0 4YY

24/10/2024

01778304-0017